PHOTOSYNTHESIS

Genetic, Environmental and Evolutionary Aspects

Research Progress in Botany

PHOTOSYNTHESIS

Genetic, Environmental and Evolutionary Aspects

Philip Stewart, PhD

Head, Multinational Plant Breeding Program; Author; Member, US Rosaceae Genomics, Genetics and Breeding Executive Committee; North Central Regional Association of State Agricultural Experiment Station Directors, U.S.A.

Sabine Globig

Associate Professor of Biology, Hazard Community and Technical College, Kentucky, U.S.A.

Apple Academic Press

Research Progress in Botany Series

Photosynthesis: Genetic, Environmental and Evolutionary Aspects

First Published in the Canada, 2011
Apple Academic Press Inc.
3333 Mistwell Crescent
Oakville, ON L6L 0A2
Tel. : (888) 241-2035
Fax: (866) 222-9549
E-mail: info@appleacademicpress.com
www.appleacademicpress.com

The full-color tables, figures, diagrams, and images in this book may be viewed at www.appleacademicpress.com

ISBN 978-1-926692-63-0

Philip Stewart, PhD
Sabine Globig

Printed at Thomson Press (I) Ltd.
9 8 7 6 5 4 3 2 1

Cover Design: Psqua

Library and Archives Canada Cataloguing in Publication Data
CIP Data on file with the Library and Archives Canada

CONTENTS

INTRODUCTION

Although photosynthesis can happen in different ways in different species, some features are always the same. For example, the process always begins when energy from light is absorbed by proteins called photosynthetic reaction centers that contain chlorophylls. In plants, these proteins are held inside organelles called chloroplasts embedded within the cell membranes, while in bacteria they are embedded in the plasma membrane. This membrane may be tightly folded into cylindrical sheets called thylakoids, or bunched up into round vesicles called intra cytoplasmic membranes. These structures can fill most of the interior of a cell, giving the membrane a very large surface area and therefore increasing the amount of light that the bacteria can absorb. Meanwhile, a typical plant cell contains about 10 to 100 chloroplasts. The chloroplast is enclosed by a membrane composed of a phospholipid inner membrane, a phospholipid outer membrane, and an intermembrane space between them. Within the membrane is an aqueous fluid called the stroma. The stroma contains stacks (grana) of thylakoids, which are the site of photosynthesis. The thylakoids are flattened disks, bounded by a membrane with a lumen or thylakoid space within it. The site of photosynthesis is the thylakoid membrane, which contains integral and peripheral membrane protein complexes, including the pigments that absorb light energy, which form the photosystems.

Although all cells in the green parts of a plant have chloroplasts, most of the energy is captured in the leaves. The cells in the interior tissues of a leaf, called themesophyll, can contain between 450,000 and 800,000 chloroplasts for every square millimeter of leaf. The surface of the leaf is uniformly coated with a water-resistant waxy cuticle that protects the leaf from excessive evaporation of water and decreases the absorption of ultraviolet or blue light to reduce heating. The transparent epidermis layer allows light to pass through to the palisade mesophyll cells where most of the photosynthesis takes place.

At its most basic level, the genetic, environmental and evolutionary aspects of photosynthesis can be summarized as we have done here, but research into this process is ongoing. This volume brings to light the most recent studies of this topic. The detailed information provided here will allow readers to stay current with our ever-developing knowledge of the structures that carry out photosynthesis.

— Philip Stewart, PhD

Chloroplast Two-Component Systems: Evolution of the Link between Photosynthesis and Gene Expression

Sujith Puthiyaveetil and John F. Allen

ABSTRACT

Two-component signal transduction, consisting of sensor kinases and response regulators, is the predominant signalling mechanism in bacteria. This signalling system originated in prokaryotes and has spread throughout the eukaryotic domain of life through endosymbiotic, lateral gene transfer from the bacterial ancestors and early evolutionary precursors of eukaryotic, cytoplasmic, bioenergetic organelles—chloroplasts and mitochondria. Until recently, it was thought that two-component systems inherited from an ancestral cyanobacterial symbiont are no longer present in chloroplasts. Recent research now shows that two-component systems have survived in chloroplasts as products of both chloroplast and nuclear genes. Comparative genomic analysis of photosynthetic eukaryotes shows a lineage-specific distribution of chloroplast two-component systems. The components and the systems they comprise have homologues in extant cyanobacterial lineages, indicating their ancient cyanobacterial origin. Sequence and functional characteristics of chloroplast two-component systems point to their fundamental role in linking photosynthesis with gene expression. We propose that two-component systems provide a coupling between photosynthesis and gene expression that serves to retain genes in chloroplasts, thus providing the basis of cytoplasmic, non-Mendelian inheritance of plastid-associated characters. We discuss the role of this coupling in the chronobiology of cells and in the dialogue between nuclear and cytoplasmic genetic systems.

Keywords: cytoplasmic inheritance, endosymbiosis, redox response regulator, redox sensor kinase, signal transduction, transcription

Two-Component Systems Enter the Eukaryotic Domain of Life

The name 'two-component system' is used to describe members of a class of signal transduction pathways found in eubacteria and made up of two conserved protein components (Stock et al. 1985; Nixon et al. 1986). These two conserved protein components are a sensor kinase and a response regulator (figure 1).

Of these two, the component that is first to detect and respond to an environmental change is the sensor kinase. A sensor kinase is a histidine protein kinase that combines a variable sensor domain with an invariable kinase domain (Figure 1). The sensor domain perceives different specific signals in different histidine sensor kinases. The nature of the signal sensed and the structure of the protein's sensor domain are specific for each histidine sensor kinase. By contrast, the kinase domain is highly conserved in structure and function, being made up of an independent dimerization motif and a catalytic core. The catalytic core of the kinase domain consists of five conserved amino acid motifs: H-box, N, G1, F and G2 (figure 1). The H-box contains the conserved histidine residue that is the site of phosphorylation and is usually located in the dimerization motif. N, G1, F and G2 boxes form the ATP-binding pocket of the catalytic core (Stock et al. 2000).

The second component of any two-component system is its response regulator. The response regulator protein is also made up of two domains (figure 1). The first domain of a response regulator is its invariable receiver domain. The second domain of a response regulator is its variable effector domain, which mediates the specific output response. The chemistry of signal transduction is common to different pathways: two-component systems use a phosphotransfer mechanism from the invariant kinase domain of the sensor to the invariant receiver domain of the response regulator (figure 1). The sequence of events that leads to an output response from a two-component signalling pathway begins when histidine kinases, in their functional dimeric form, and upon sensing the signal, undergo an ATP-dependent trans-autophosphorylation reaction, whereby one histidine kinase monomer phosphorylates a second monomer within the dimer. The phosphate group becomes covalently, though weakly, bound to a conserved histidine residue of the catalytic core. The receiver domain of the response regulator protein then catalyses the transfer of the phosphate group from the histidine residue of the kinase to a conserved aspartate residue within the receiver domain of the response regulator protein. This creates a high-energy acyl phosphate that activates the effector domain of the response regulator (Stock et al. 2000).

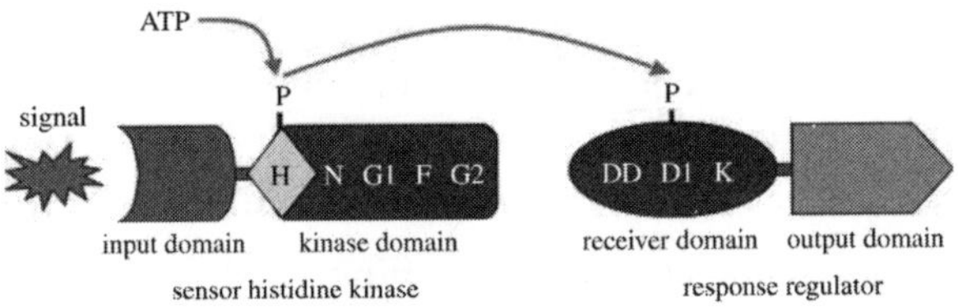

Figure 1. Schematic of a two-component system. The sensor kinase and the response regulator components and their domain architecture are shown. The kinase domain consists of a dimerization domain (diamond) and an ATP-binding domain (rectangle). The H, N, G1, F and G2 motifs of the kinase domain are indicated. Conserved sequence motifs of the response regulator receiver domain, DD, D1 and K, are also indicated. The phosphotransfer signalling mechanism of the two-component systems is depicted as phosphate group transfer from ATP to the conserved aspartate residue of the response regulator via the conserved histidine residue of the sensor kinase.

There are actually two reactions catalysed by the 'sensor kinase' enzyme. The first is transfer of the γ-phosphate of ATP to a histidine side chain of the protein itself, to form a phosphoamide linkage (autophosphorylation; equation (1.1)). The second reaction is transfer of the phosphate moiety from the histidine of the sensor kinase to an aspartate on the corresponding response regulator (equation (1.2)). Thus, the phosphohistidine acts as a covalent chemical intermediate in transfer of the phosphate group between ATP and the response regulator, and so sensor kinases are, in a biochemical sense, really 'response regulator kinases' (equation (1.3)).

$$\text{Sensor-His} + \text{ATP} \rightleftharpoons \text{Sensor-His}{\sim}\text{ADP} \quad (1.1)$$

$$\text{Regulator-Asp} + \text{Sensor-His}{\sim}\text{P} \rightleftharpoons \text{Regulator-Asp} \sim \text{P} + \text{Sensor-His} \quad (1.2)$$

$$\text{Sum:Regulator-Asp} + \text{ATP} \rightleftharpoons \text{Regulator-Asp} \sim \text{P} + \text{ADP} \quad (1.3)$$

The histidine sensor kinase becomes autophosphorylated (equation (1.1)) if, and only if, the specific environmental precondition is met, and on an invariant histidine residue. The phosphate group is then transferred to the aspartate residue of one or more response regulators (equation (1.2)). Phosphorylation of the response regulator then activates the appropriate response to the environmental change that produced the original phosphorylation on histidine.

Although many two-component systems follow the above archetype in modular design (figure 1) and phosphotransfer mechanism, variations exist on this basic theme. These variations may involve additional conserved domains in histidine kinases or separate modules, and more complex phosphorelay pathways.

Two-component systems are known to regulate diverse physiological responses in bacteria. Most of these physiological responses involve transcriptional regulation, mediated by the response regulator protein acting as a transcription factor. These signalling systems must have originated very early in evolution (Stock et al. 1989) and are now found in both eubacteria and archaebacteria ('Archaea'; Koretke et al. 2000). The two-component systems are nevertheless typically found in eubacteria, including alpha-proteobacteria and cyanobacteria, and probably became widespread in the eukaryotic domain of life through symbiotic, lateral gene transfer from the ancestors of chloroplasts and mitochondria (Koretke et al. 2000).

Chloroplasts and mitochondria retain functional genomes and each houses a complete, cytoplasmic apparatus of gene expression that is separate from that of the nucleus and cytosol. The two-component systems that couple regulatory signals to gene expression were predicted to be present in these eukaryotic subcellular organelles (Allen 1993a). The regulatory signals that thus modulate organellar gene expression are now known, notably as changes in the redox state of key components of energy transduction (Pfannschmidt et al. 1999a). If chloroplasts and mitochondria were responsible for the acquisition of two-component systems by eukaryotes, it is interesting to ask whether these organelles themselves retain the two-component systems from their bacterial ancestors. The answer for chloroplasts is now known. It is 'yes'. The two-component systems from cyanobacteria have survived in chloroplasts as products of both chloroplast and nuclear genes (Duplessis et al. 2007; Puthiyaveetil & Allen 2008a; Puthiyaveetil et al. 2008). Here, we describe the properties of two-component systems of chloroplasts, their phylogenetic distribution and the functional implications of these analyses for regulatory coupling between photosynthesis and gene expression.

Component 1. Chloroplast Sensor Kinases Come in Different Hues

Sequencing of chloroplast genomes is a routine exercise in molecular systematics of plants and algae. Chloroplast DNA sequencing has thus resolved many cladistic disputes. It has also revealed some unexpected genes in chloroplasts, such as subunits of NAD(P)H dehydrogenase. Other unexpected genes show sequence similarity to genes for regulatory proteins that were once thought to be confined to bacteria. One of these bacterial-type regulatory proteins is a histidine sensor kinase variously known as Dfr in Gracilaria tenuistipitata, Tsg1 in Heterosigma akashiwo and as ycf26 in Porphyra purpurea (Duplessis et al. 2007). As more plastid genomes were sequenced, it became apparent that this histidine sensor kinase gene was limited in its phylogenetic distribution: ycf26 is found only in red algal, raphidophyte and haptophyte chloroplasts, and is completely unknown in chloroplasts of green algae and land plants (table 1; Duplessis et al. 2007). The phylogenetic distribution of ycf26 in non-green algae is also found to be discontinuous, as it is absent from chloroplasts of the ancient red alga Cyanidioschyzon merolae and from chloroplasts of the diatoms Phaeodactylum tricornutum and Thalassiosira pseudonana (table 1; Duplessis et al. 2007).

Table 1. Distribution of chloroplast sensor kinases in photosynthetic eukaryotes. (A tick (√) indicates the presence and a cross (×) indicates the absence of ycf26 or CSK. A dash (—) indicates that the complete genome sequence for that taxon is not available, so the presence or absence of ycf26 or CSK is unknown. An asterisk (*) denotes that the occurrence is based on an earlier report (Duplessis et al. 2007). The taxonomic group 'Viridiplantae' means 'green plants', and includes green algae, lower plants and higher plants.)

taxonomic group/organism	ycf26	CSK
glaucophytes		
Cyanophora paradoxa	×	—
rhodophytes		
Porphyra purpurea	√*	—
Porphyra yezoensis	√*	—
Porphyridium aerugineum	—	—
Cyanidioschyzon merolae	×	√
Cyanidium caldarium	√*	—
Galdieria sulphuraria	—	—
Gracilaria tenuistipitata	√*	—
Rhodella violacea	—	—
haptophytes		
Emiliania huxleyi	√*	×
raphidophytes		
Heterosigma akashiwo	√*	—
cryptophytes		
Guillardia theta	×	—
Rhodomonas salina	√	—
bacillariophytes		
Phaeodactylum tricornutum	×	√
Thalassiosira pseudonana	×	√
viridiplantae		
Ostreococcus tauri	×	√
Ostreococcus lucimarinus	×	√
Chlamydomonas reinhardii	×	×
Chlorokybus atmosphyticus	×	—
Physcomitrella patens	×	√
Pinus taeda	×	√
Oryza sativa	×	√
Arabidopsis thaliana	×	√

Although chloroplast-encoded sensor kinases are unknown in green algae and in land plants, the sequencing of their nuclear genomes has now revealed many genes encoding two-component proteins. In the model higher plant Arabidopsis thaliana, there are 54 such genes and at least 16 of them encode putative histidine kinases. The possibility of one of these histidine kinase gene products being targeted to the chloroplast has been examined (Forsberg et al. 2001), but none was thought to possess a chloroplast-targeting sequence (Oelmüller et al. 2001; Lopez-Juez & Pyke 2005; Wagner & Pfannschmidt 2006). However, we recently identified a nuclear-encoded sensor kinase in Arabidopsis chloroplasts (Puthiyaveetil et al. 2008). Other investigators might have missed this chloroplast sensor kinase (CSK), because its gene is annotated 'unknown protein' and not even included as one of the 16 putative histidine kinase genes of Arabidopsis. The CSK gene, in contrast to the ycf26 sensor of non-green algae, has a wide phylogenetic distribution with recognizable homologues in all lineages of green algae and plants, and also in some red algae and diatoms (table 1; Puthiyaveetil et al. 2008).

As it stands today, chloroplasts seem to have at least two sensor kinases, one plastid-encoded and the other nuclear-encoded (table 1; Duplessis et al. 2007; Puthiyaveetil et al. 2008). These two CSKs, although quite different in some sequence features, show some commonalities in their functional design. The chloroplast-encoded sensor kinase of non-greens, ycf26, seems to be a transmembrane sensor as most of its examples have two to three predicted transmembrane helices. The predicted topology (Figure 2) of this protein includes one or two transmembrane helices at the N-terminus and a lumenal loop of 120–130 amino acids followed by another transmembrane helix. The predicted stromal-exposed domain of the enzyme has an HAMP linker domain and a PAS domain followed by a kinase domain (figure 2). HAMP domains are known to act as linkers that connect the periplasmic sensor domain with the cytoplasmic or stromal surface-exposed kinase domain of transmembrane histidine kinases (Aravind & Ponting 1999). PAS domains are well known for their role as internal sensors of the redox state (Taylor & Zhulin 1999). The kinase domain of ycf26 is typical of histidine sensor kinases, as it possesses all known sequence motifs for dimerization and transphosphorylation reactions (figure 3). The ycf26 sensor is seen in its minimal form in the raphidophycean alga, H. akashiwo (Duplessis et al. 2007) and in the cryptophycean alga, Rhodomonas salina. In these two species, ycf26 is likely to be a soluble stromal protein with only the PAS sensor domain and the kinase domain. It appears that in most ycf26 proteins the predicted lumenal loop and the internal PAS domain act as two separate sensor domains (Morrison et al. 2005).

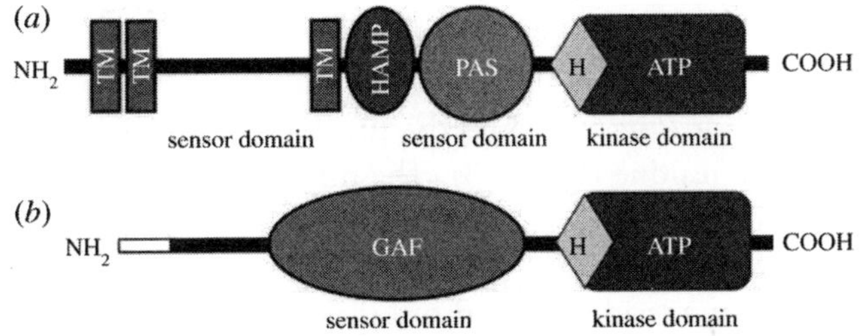

Figure 2. Predicted domain composition of chloroplast sensor kinases, (a) ycf26 and (b) CSK. Individual domains are labelled as follows: TM, transmembrane; HAMP, domain present in histidine kinases, adenylyl cyclases, methyl-accepting proteins and phosphatases; PAS, domain named after three proteins it occurs in, per, arnt and sim; GAF, domain first described for vertebrate cGMP-specific phosphodiesterase, a cyanobacterial adenylate cyclase and the bacterial formate hydrogen lyase transcription activator FhlA. The rectangle at the amino terminus of CSK represents the chloroplast transit peptide.

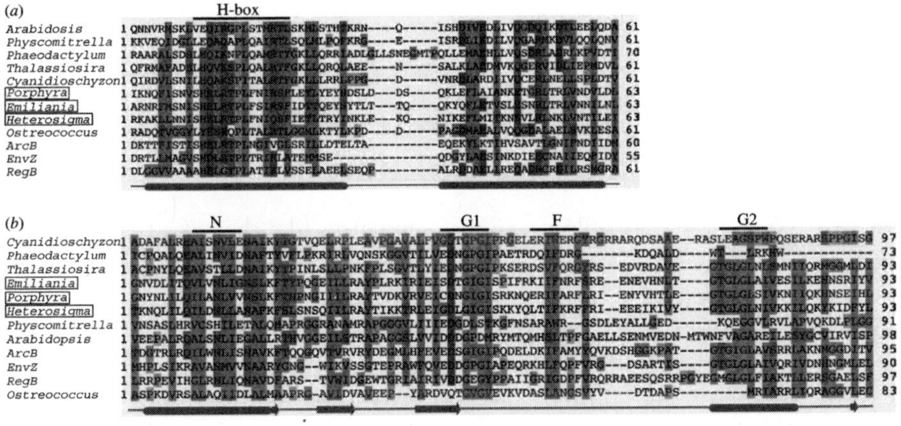

Figure 3. Chloroplast sensor kinases show molecular evolution in their kinase domains. (a) Multiple sequence alignment of ycf26 and CSK HisKA domain (dimerization and phosphoacceptor domain as defined by SMART database) from representative species with the three canonical histidine kinases from bacteria, ArcB, EnvZ and RegB. Species names of ycf26 sequences are boxed. Predicted secondary structures are shown at the bottom. Alpha helices are shown as cylinders, beta sheets as thick arrows and the line connecting them as loops and turns. The site of autophosphorylation, H-box, located in the first helix of HisKA domain, is indicated on the top of the alignment. (b) TheATP-binding domain is conserved in all chloroplast sensor kinases.ATPbinding domain of representative ycf26, CSK sequences are aligned with those of three canonical histidine kinases, ArcB, RegB and EnvZ. Species names of ycf26 sequences are boxed. N, G1, F and G2 motifs of theATP-binding domain are shown on top and the predicted secondary structures are shown at the bottom. Alignment was generated with CLUSTALW (Chenna et al. 2003) and edited with JALVIEW (Clamp et al. 2004). Secondary structures are predicted with JNET (Cole et al. 2008).

CSK, in contrast to ycf26, has no predicted membrane-spanning region (figure 2) and seems to be a soluble stromal protein. The sensor domain in CSK is a putative GAF domain, which, as a PAS domain, is implicated in redox sensing (Kumar et al. 2007; Vuillet et al. 2007). The kinase domain in different CSKs shows variation with respect to the conserved histidine residue in the H-box (Figure 3). CSKs from red algae and diatoms have the conserved histidine seen in bacterial sensor kinases, while the green algal CSKs have a tyrosine in place of the histidine residue (Figure 3). The moss Physcomitrella patens (Rensing et al. 2008) and all higher plants have a glutamate residue instead of the conserved histidine in their CSKs (Figure 3). While there are precedents for replacement of histidine residues in sensor kinases (Yeh & Lagarias 1998; Moussatche & Klee 2004), this is the first report, to our knowledge, of such a replacement occurring more than once in the evolution of a sensor kinase. This amino acid substitution is reflected in an altered phosphoryl group chemistry in the Arabidopsis CSK (Puthiyaveetil et al. 2008). Another interesting feature of CSK is that the cytosolically synthesized CSK precursor protein and the mature chloroplast CSK protein have the same molecular mass (Puthiyaveetil et al. 2008). This means that the transit peptide is retained in the mature protein, where it may play a role in CSK function.

Component 2. Chloroplast Response Regulators: from the Known to the Unknown

The first report of a chloroplast response regulator, like that of a CSK, came as a simple consequence of systematically sequencing chloroplast genomes. The first chloroplast response regulator was described in red algae and has a high sequence similarity to the bacterial OmpR response regulator transcription factor (Kessler et al. 1992). This chloroplast response regulator is generally known as ycf27, as 'OmpR-like protein', or simply as 'OmpR'. It has also been called 'transcriptional regulatory gene 1' (trg1) in H. akashiwo (Jacobs et al. 1999). ycf27 is found in many non-green algal lineages (table 2; Duplessis et al. 2007). However, among more than 75 sequenced chloroplast genomes of green algae and land plants, only the charophyte Chlorokybus atmosphyticus has the ycf27 gene (Duplessis et al. 2007). The predicted domain architecture of ycf27 conforms to known domain features of response regulators. The receiver domain is typical in having three sequence motifs (figure 4), including the invariant aspartate residue that receives phosphate from the histidine residue of a sensor kinase. The effector domain of ycf27 is a winged helix-turn-helix motif. This motif is a variation of the well-known DNA-binding helix-turn-helix domain. In addition to binding DNA, winged helix-turn-helix motifs also sometimes interact with RNA polymerase to influence gene transcription (Martinez-Hackert & Stock 1997).

Table 2. Distribution of chloroplast response regulators and response regulator-like proteins in photosynthetic eukaryotes. (A tick (O) indicates the presence and a cross (!) indicates the absence of ycf27/ycf29/TCP34. A dash (—) indicates that the complete genome sequence for that taxon is not available, so the presence or absence of ycf27/ycf29/ TCP34 is unknown. An asterisk (*) denotes that the occurrence is based on earlier reports (Weber et al. 2006; Duplessis et al. 2007). Chloroplast response regulators are mostly chloroplast gene products, but if the chloroplast response regulators exist as nuclear gene products, the nuclear location of the corresponding gene is indicated by a superscript N (^{N}).)

taxonomic group/organism	ycf27	ycf29	TCP34
glaucophytes			
Cyanophora paradoxa	√*	√	—
rhodophytes			
Porphyra purpurea	√*	√	—
Porphyra yezoensis	√*	√	—
Porphyridium aerugineum	√*	—	—
Cyanidioschyzon merolae	√*	√	×
Cyanidium caldarium	√*	√	—
Galdieria sulphuraria	√	—	—
Gracilaria tenuistipitata	√*	√	—
Rhodella violacea	√*	—	—
haptophytes			
Emiliania huxleyi	√*	√^{N}	×
raphidophytes			
Heterosigma akashiwo	√*	×	—
cryptophytes			
Guillardia theta	√*	√	—
Rhodomonas salina	√	√	—
bacillariophytes			
Phaeodactylum tricornutum	√^{N}	√^{N}	×
Thalassiosira pseudonana	√^{N}	√^{N}	×
viridiplantae			
Ostreococcus tauri	×	×	√^{N}
Ostreococcus lucimarinus	×	×	√^{N}
Chlamydomonas reinhardii	×	×	×
Chlorokybus atmosphyticus	√*	×	—
Physcomitrella patens	×	×	√^{N}
Pinus taeda	—	—	—
Oryza sativa	×	×	√^{*N}
Arabidopsis thaliana	×	×	√^{*N}

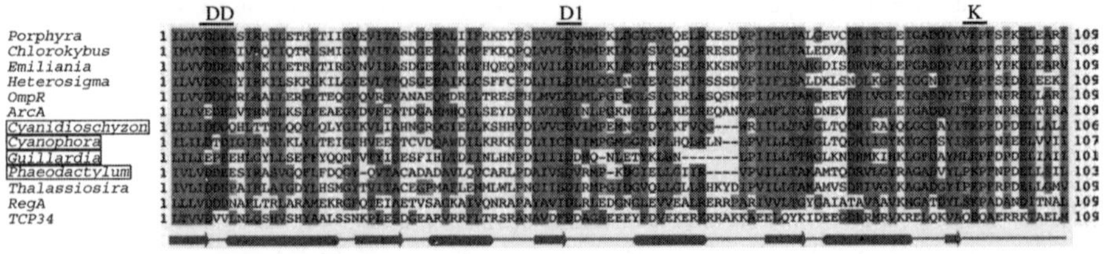

Figure 4. Conserved sequence features of chloroplast response regulators. The receiver domain of the representatives ycf27, ycf29 and the Arabidopsis TCP34 are aligned with those of three canonical response regulators in bacteria ArcA, RegA and OmpR. Species names of ycf29 sequences are boxed. Predicted secondary structures are shown at the bottom. Conserved sequence motifs of the receiver domain, DD, D1 and K, are indicated on top of the alignment. The fully conserved D1 motif, which lies in the loop connecting the third beta sheet and the third helix, is the conserved aspartate residue that receives the phosphate group from the phosphohistidine of the sensor kinase. Arabidopsis TCP34 shows a moderately conserved receiver domain with the 'K motif' still being present, but not entirely aligned with those from other response regulators.

A second chloroplast response regulator identified as a by-product of chloroplast genome sequencing is ycf29. This chloroplast response regulator shows high sequence similarity to the NarL response regulator transcription factor of Escherichia coli (Maris et al. 2002). ycf29 is also known as tctD in G. tenuistipitata or as NarL-like transcriptional regulator. ycf29 occurs together with ycf27 in some, but not all, non-green algae (table 2). Like ycf27, ycf29 appears to be limited in its phylogenetic distribution to non-green algae. The ycf29 protein has very similar properties to the ycf27 response regulator (Figure 4), with a small difference in their effector domains. ycf29 does not have the extended wing structure in its DNA-binding helix-turn-helix domain.

Most chloroplasts in the green lineages contain no gene for any response regulator. This conclusion does not preclude the existence of response regulators in their chloroplasts as products of nuclear genes, as shown for CSKs (Puthiyaveetil et al. 2008). Homologues of ycf27 or ycf29 cannot be readily identified in sequenced nuclear genomes of green algae and plants. Yet it is premature to conclude that nuclear-encoded ycf27 and ycf29 homologues are absent from chloroplasts of green algae and land plants. One possibility is that the chloroplast response regulators in greens are modified so as to accommodate input from a modified histidine kinase such as CSK. The coevolution of sensor kinases and response regulators in a cognate pair is well documented (Koretke et al. 2000; Skerker et al. 2008). One or more modified response regulators might thus have evaded identification by conventional sequence-similarity searches. Along these lines, a nuclear-encoded, modified response regulator protein, TCP34, has been identified in higher plant chloroplasts (Weber et al. 2006). This protein seems to be conserved in all sequenced plant genomes. Our analysis (table 2) identifies a homologue of TCP34 in the moss, P. patens. Physcomitrella also has a paralogue of TCP34 (Andrew Cuming, University of Leeds, personal communication). We also identify (table 2) single homologues of TCP34 in the nuclear genome of chlorophycean algae, Ostreococcus tauri and Ostreococcus lucimarinus, but, interestingly, not in Chlamydomonas reinhardii. The moss and algal TCP34 homologues are localized in the chloroplast, according to subcellular prediction programmes (results not shown).

The TCP34 protein combines a moderately conserved receiver domain (figure 4) with a tetratricopeptide repeat (TPR) as an effector domain (Weber et al. 2006). It also seems that a part of the receiver domain of this protein may be coopted as a DNA-binding protein with a helix-turn-helix motif. It has also been demonstrated that TCP34 is phosphorylated, as was discovered in the course of a search for a specific chloroplast protein kinase (Weber et al. 2006).

The Cyanobacterial Pedigree of Chloroplast Two-Component Systems

Eukaryotes acquired two-component systems from the bacterial ancestors of chloroplasts and mitochondria through endosymbiotic, lateral gene transfer (Koretke et al. 2000). Sequence-similarity searches reveal homologues of the chloroplast two-component systems in extant cyanobacterial lineages, demonstrating the ancient origin of these systems from the cyanobacterial ancestor of chloroplasts. In Synechocystis sp. PCC 6803, histidine kinases 33 and 2 (hik33 and hik2) are homologues of ycf26 and CSK, respectively (Ashby & Houmard 2006; Puthiyaveetil et al. 2008). hik33 is also known as 'drug sensory protein A' (dspA; Hsiao et al. 2004; Morrison et al. 2005) in Synechocystis sp. PCC 6803 and as NblS in Synechococcus elongatus PCC 7942. Homologues of the chloroplast response regulators ycf27 and ycf29 are the response regulators 26 and 1 (rre26 and rre1) in Synechocystis sp. PCC 6803 (Ashby & Houmard 2006; Duplessis et al. 2007; Puthiyaveetil et al. 2008). TCP34, the nuclear-encoded response regulator-like protein of higher plant chloroplasts, does not appear to have cyanobacterial counterparts. The phylogenetic signature of this protein seems to have been lost during its evolution. TCP34 might represent a eukaryotic innovation in which the receiver domain of a response regulator is fused to a TPR, the latter motif being usually found in eukaryotes (Weber et al. 2006).

The presence of cyanobacterial homologues strengthens the genealogy of the chloroplast two-component systems, but can this information shed light on their structure and function? It is well known that the two-component systems usually work in cognate pairs to govern a functional response, even though cross-talk can occur in certain cases. In Synechocystis sp. PCC 6803, various functional and mutagenesis studies, together with high-throughput two-hybrid screening, have identified cognate pairs or interaction partners in different two-component systems (Murata & Dmitry 2006; Sato et al. 2007; Kappell 2008). At least two cognate pairs are important from a plastid perspective. These are the cognate pairs formed between hik33 and rre26, and between hik2 and rre1 (Paithoonrangsarid et al. 2004; Murata & Dmitry 2006; Sato et al. 2007; Kappell 2008). ycf26 and ycf27 form one cognate pair, while CSK and ycf29 together form the other pair. Although some chloroplasts still keep these ancestral combinations between sensors and response regulators, most chloroplasts have deviated from this pattern, for reasons that we do not yet fully understand. These deviations have resulted in a striking pattern of lineage-specific retention or loss of the chloroplast two-component systems (figure 5).

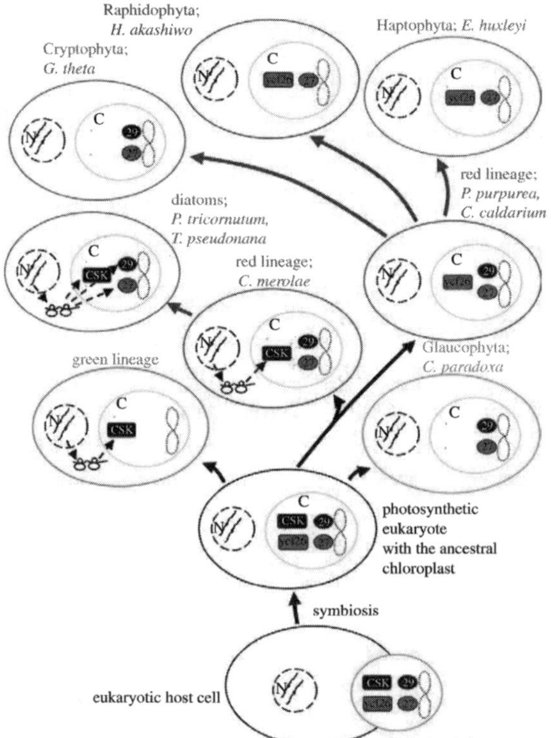

Figure 5. Lineage-specific distribution of the chloroplast two-component systems. Cyanobacteria (cyan) and chloroplasts (yellow) are represented as ovals. Names of lineages and their representative cells are coloured to correspond with their dominant pigments. Chloroplasts are additionally marked 'C' and the nucleus is represented by dashed circles and marked 'N'. When genes encoding either or both component of the two-component systems are moved to nucleus, their synthesis in cytoplasm and import back to chloroplasts is indicated by ribosomes and dashed arrows. Lineages leading to diatoms, cryptophytes, raphidophytes and haptophytes have involved secondary endosymbiosis with a red algal symbiont and are represented by thick blue arrows from the red lineage. The numbers 27 and 29 denote the response regulators ycf27 and ycf29, respectively.

Bricolage. Lineage-Specific Retention and Loss of Chloroplast Two-Component Systems

In photosynthetic eukaryotes, the chloroplast two-component systems as we understand them today consist of two sensors, two response regulators and a response regulator-like protein. The distribution of these proteins seems to follow a lineage-specific pattern (figure 5). In the lineage of rhodophytes leading to P. purpurea, Cyanidium caldarium and G. tenuistipitata, the chloroplast two-component systems consist of the ycf26 sensor and two response regulators, ycf27 and ycf29, all chloroplast-encoded (figure 5). In the raphidophyte H. akashiwo and the haptophyte Emiliania huxleyi, the same combination is seen except that the ycf29 response regulator is missing (figure 5). In the ancient red alga C. merolae, the response regulators are ycf27 and ycf29 proteins, both encoded in the chloroplast, while the sensor ycf26 is lost and presumably replaced by the nuclear-encoded CSK (figure 5). The glaucophyte Cyanophora paradoxa and the cryptophyte Guillardia theta resemble Cyanidioschyzon in their contingent of the chloroplast two-component systems (figure 5), but a nuclear-encoded CSK has yet to be demonstrated in their chloroplasts. Chloroplasts of bacillariophytes such as P. tricornutum and T. pseudonana do not encode any two-component system; nevertheless, they seem to contain the chloroplast two-component systems as products of nuclear genes, since homologues of CSK, ycf27 and ycf29 genes can be identified in their nuclear genomes (tables 1 and 2).

Chloroplasts in the green lineage (figure 5), with the single exception of C. atmosphyticus, seem not to encode the two-component systems. Homologues of CSK are readily identifiable in the sequenced nuclear genomes of higher plants, in the moss P. patens and in the prasinophycean alga, O. tauri (Puthiyaveetil et al. 2008). By contrast, chloroplast response regulators in green algae and land plants, again with the exception of C. atmosphyticus, remain unidentified. Whether the nuclear-encoded TCP34 protein can be counted as a genuine chloroplast response regulator remains to be seen.

Distribution of the chloroplast two-component systems appears to be a phylogenetic patchwork (figure 5). Nevertheless, a pattern emerges with regard to the location of genes encoding the chloroplast two-component systems. These genes are shown to move from chloroplast to nuclear genomes as we proceed from non-greens to greens (figure 5). Sequencing of more chloroplast and nuclear genomes may uncover yet further combinations of sensors and response regulators in chloroplasts. Genome sequencing will also answer questions such as the following: are there nuclear-encoded ycf26 homologues in algae and plants? Will a CSK homologue be found encoded in a chloroplast?

The phylogenetic distribution of the chloroplast two-component systems (figure 5) poses other unanswered questions. For example, why are non-greens reluctant to give up their chloroplast two-component genes to the nucleus, while greens are only too willing to do so? The answer to this question may lie in subtle differences between non-greens and greens in the structure and function of the chloroplast two-component systems.

The Missing Link between Photosynthesis and Gene Expression

If two-component systems have survived in chloroplasts, the obvious question that comes to mind is 'what do they do?' Available data point to their indispensable role in linking photosynthetic activity to the expression of genes that encode for the photosynthetic machinery. The chloroplast two-component systems fulfill this vital role by acting as redox sensors and redox response regulators (Allen 1993a,b) inside the chloroplast (figure 6). CSKs, acting as redox sensors, monitor the flux of electron transport in the thylakoid membrane. The flux of electron transport lies at the heart of photosynthetic activity. Sensor kinases pass this information on to response regulators, which, when required, selectively turn photosystem genes on and off (figure 6). This feedback regulation brings about redox homeostasis by ensuring that photosynthetic activity in chloroplasts is rapidly and unconditionally optimized to new environmental conditions that have changed unpredictably. This proposition on the action of the chloroplast two-component systems is in agreement with their known functional properties.

Little is known concerning the action of the ycf26 sensor in non-greens. This is partly because knockout mutant lines are not available in these algae (Duplessis et al. 2007). However, if the perceived redox-sensing sequence features of this kinase are correct, then its stromally positioned PAS domain will act as a redox sensor domain, on its own or with the help of redox-responsive prosthetic groups (figure 2) such as flavins. As if to compensate for the lack of information on the action of ycf26 in non-greens, the role of hik33 in cyanobacteria is known in more detail. In cyanobacteria, hik33 is known to regulate photosynthetic gene expression in response to nutrient stress and high light intensities (Hsiao et al. 2004). hik33 is also known in cyanobacteria as a cold sensor (Mikami et al. 2003b), as a sensor of osmotic stress (Mikami et al. 2002; Paithoonrangsarid et al. 2004; Shoumskaya et al. 2005) and as a sensor involved in the expression of oxidative stress-inducing genes (Kanesaki et al. 2007). This multifunctional role could easily qualify hik33 as a global regulator in cyanobacteria. With its lumenal loop and PAS domain acting as two separate sensor domains, hik33 will be able to combine or integrate multiple signals and act as a multi-sensor (Morrison et al. 2005). Environmental factors such as low temperature, nutrient stress and osmotic stress may slow down photosynthetic electron transport, generating redox signals that will themselves feed into the regulatory loop of hik33. Additionally, the redox-sensing functional role of ycf26 is underscored by the observation that this sensor is seen in its minimal form, with only the PAS sensor domain and the kinase domain being present, in the raphidophycean alga H. akashiwo and in the cryptophycean alga R. salina.

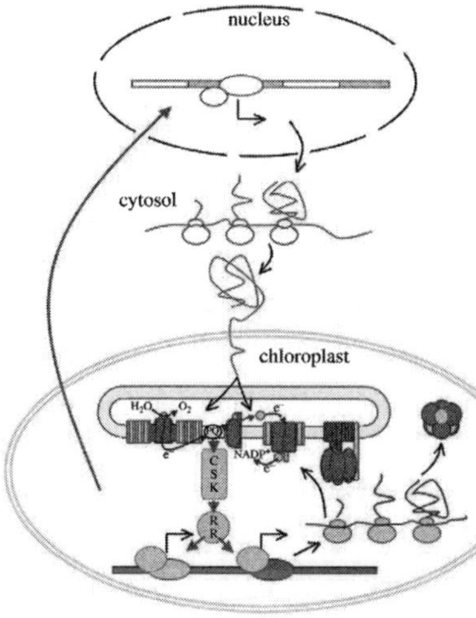

Figure 6. A model explaining the operation of a CSK-based two-component system in chloroplasts. CSK selectively switches on and off chloroplast genes in response to perturbations in the photosynthetic electron transport chain (depicted as electron transport from H2O to NADPC). CSK acts as a redox sensor and reports on electron flow. A response regulator protein (RR), the identity of which is yet to be revealed in green algae and plants, mediates CSK's control over gene expression. Chloroplast genes under CSK control are those which encode core components (shown in green) of the electron transfer complexes. The nuclear-encoded peripheral components are shown in yellow. The red arrow from the chloroplast to the nucleus represents plastidto-nuclear signalling.

For CSK, T-DNA insertion mutants are available in Arabidopsis, and their phenotype supports a role for CSK in coupling photosynthesis with chloroplast gene expression (Puthiyaveetil et al. 2008). This functional role of CSK was unravelled by the use of wavelengths of light preferentially absorbed by photosystem I (PS I) and photosystem II (PS II; Puthiyaveetil & Allen 2008a; Puthiyaveetil et al. 2008). Changes between lights of wavelength selectively absorbed by each photosystem affect the redox state of inter-photosystem electron carriers, including plastoquinone (PQ). The redox state of the PQ pool is a well-known regulatory signal affecting photosynthetic gene expression (Pfannschmidt et al. 1999a,b; Tullberg et al. 2000; Fey et al. 2005; Puthiyaveetil & Allen 2008b). Wavelength-dependent changes in chloroplast gene transcription are disrupted in CSK mutants, so CSK is proposed to act as a redox sensor that couples PQ redox state to gene expression in chloroplasts. In C. merolae, it has been shown that the mRNA of the CSK gene transiently accumulates after a switch to high light intensities (Minoda et al. 2005). CSK, like ycf26, shows redox-sensing sequence features, as it possesses a GAF domain at the N-terminus. GAF domains are known to sense redox signals (Kumar et al. 2007; Vuillet et al. 2007) and are structurally related to PAS domains (Ho et al. 2000). The cyanobacterial homologue of CSK, hik2, suggests functional properties that strengthen the redox-sensing functional role of CSK in chloroplasts. hik2 shows functional overlap with hik33 in sensing osmotic stress and low temperature (Mikami et al. 2003b). It has been further proposed that hik2 and hik33 are involved in the tolerance of PS II to environmental stress (Mikami et al. 2003a). Recent high-throughput yeast two-hybrid analysis has shown hik2 to interact with rppA (Sato et al. 2007), a redox response regulator in cyanobacteria (Li & Sherman 2000). This analysis also finds that hik2 interacts with the phycobilisome linker protein, apcE (Sato et al. 2007). These results are strongly indicative of a redox-sensing role for hik2 in cyanobacteria and, by extension, for CSK in chloroplasts.

The ycf27 response regulator in C. merolae binds upstream regions of the psbD/C operon and the psbA gene and is implicated in high-light acclimation (Minoda & Tanaka 2005). The cyanobacterial homologue of ycf27, rre26, is involved in the synthesis of factors responsible for coupling phycobilisomes to photosystems (Ashby & Mullineaux 1999). rre26 has recently been shown to bind the promoter region of the high-light-inducible hliB gene in Synechocystis PCC 6803 (Kappell & van Waasbergen 2007). Moreover, in cyanobacteria, the diverse physiological responses governed by the multi-sensor hik33 are mediated through the action of the response regulators rre26 and rre31 (Paithoonrangsarid et al. 2004; Murata & Dmitry 2006; Kappell 2008).

The ycf29 response regulator binds genes encoding the components of light-harvesting phycobilisome proteins such as cpcA and apcE (Minoda & Tanaka 2005). Transcriptional activation of these genes by ycf29 is believed to be part of an acclimatory response to decreased light intensity (Minoda & Tanaka 2005). In cyanobacteria, the action of hik2 is thought to be mediated through rre1, the ycf29 homologue (Paithoonrangsarid et al. 2004). TCP34, the response regulator-like protein of land plants, has been shown to bind psaA, rbcL, psbC and psbD gene core promoters in spinach chloroplasts (Weber et al. 2006).

It has been suggested that the distribution of ycf27 genes is somehow correlated with the retention or loss of phycobilisomes in various algal groups (Ashby et al. 2002). The loss of ycf27 genes from plastids has been linked with the loss of phycobilisomes. But the demonstration of ycf27 genes in the non-phycobilisome-containing genera (table 2) H. akashiwo, G. theta, E. huxleyi and C. atmosphyticus proves that no such link with phycobilisomes is required for the retention of ycf27 genes (Duplessis et al. 2007).

Our own analysis shows that the loss of the ycf29 response regulator gene is not linked with the loss of phycobilisomes as ycf29 genes are present in non-phycobilisome-containing species such as the diatom T. pseudonana and the cryptophyte G. theta (table 2). Nonetheless, there may exist some functional link between ycf29 and phycobilisomes. For instance, the ycf29 response regulator is involved in transcriptional activation of phycobiliproteins genes as part of an acclimatory response to low light intensity in C. merolae (Minoda & Tanaka 2005). The demonstration that ycf27 and ycf29 genes are retained in non-phycobilisome-containing algae is a clear indicator that these two response regulators have other target genes in chloroplasts.

Why are there Genes in the Cytoplasm?

The two-component systems connect photosynthesis to gene expression, but how exactly does this connection bring about homeostasis in chloroplast function? The answer lies in the chloroplast's resulting ability to control biogenesis of the electron transfer complexes participating in the light reactions of photosynthesis (Allen 2003). These electron transfer complexes are multi-subunit pigment–protein assemblages with their core protein components being encoded by the chloroplast genome and the peripheral protein components encoded by the nuclear genome. The biogenesis of these electron transfer complexes involves a hierarchical assembly in which the chloroplast-encoded core components are made and inserted into the thylakoid membrane first, followed by the assembly of the nuclear-encoded peripheral components (Choquet & Vallon 2000).

Some general principles have been recognized in the highly concerted assembly processes of electron transfer complexes in chloroplasts (Choquet & Vallon 2000). Nuclear-encoded subunits are found to be less essential to the stability of the complexes than the chloroplast-encoded core subunits. Among chloroplast-encoded subunits, there exists a sequential and ordered assembly that arises from a hierarchical organization in the expression of these subunits. In regulatory terms, this means that the translation of certain subunits is controlled by the state of assembly of the complex. To be precise, the rate of translation of some subunits depends on other subunits being present in the membrane. The former subunit is called a CES (control by epistasy of synthesis) protein and the latter is a dominant assembly factor (Choquet & Vallon 2000). The dominant assembly factors are less stable in the complexes and, in the absence of their assembly partners, are rapidly degraded by proteases. CES proteins on the contrary are more stable, but their synthesis is reduced in the absence of dominant assembly factors, by an autoregulatory translational loop. In Chlamydomonas, mutant studies have revealed that, in each complex of the thylakoid membrane, at least one subunit is under epistatic control of synthesis and is thus a CES protein. Cytochrome f in cytochrome b6f, psaA protein in PS I, D1 and CP47 in PS II, α subunit in ATP synthase and the large subunit of Rubisco are CES proteins (Choquet & Vallon 2000).

Control by epistasy of synthesis and biogenesis of electron transfer complexes, in general, offer some important insights into the gene content of chloroplast genomes and the genetic control mechanism of photosynthetic genes. Autoregulation of translation of CES proteins becomes possible only when their genes are present in the same location as where the assembly takes place (Choquet & Vallon 2000). CES properties certainly explain why genes encoding CES proteins should be present in chloroplasts, but they do not explain why genes encoding dominant assembly factors should also be present in chloroplasts. The retention of these latter genes, however, has to do with the nature of the biogenesis process itself. A process in which the chloroplast-encoded core subunits act as 'seeds' or 'nuclei' on which the nuclear-encoded peripheral subunits assemble to form functional electron transfer complexes.

The pivotal role of chloroplast-encoded subunits in assembly is further highlighted by the observation that some small subunits are encoded in the chloroplast genome and integral to assembly without forming part of a mature, functional complex (Boudreau et al. 1997). The pre-eminence of chloroplast-encoded subunits in the assembly process means that, if chloroplasts can rapidly control the availability of core subunits, they can 'make' or 'break' whole-electron transfer complexes. The precisely timed delivery of assembly-competent, chloroplast-encoded core subunits is essential not only for the assembly process itself, but also for the prompt binding of chlorophylls and other cofactors that are synthesized in the chloroplast. The deleterious effects of free chlorophylls and their biosynthetic intermediates are well known.

Control over the availability of chloroplast-encoded core subunits can be achieved by means of both transcriptional and translational regulation, although the former has been considered the principal mode of gene regulation. Since chloroplast-encoded core subunits consist of both dominant assembly factors and CES proteins, it is interesting to ask whether both classes of genes are under transcriptional control. If both categories of genes are under transcriptional control, will they differ in their levels of response? For example, will one class of genes respond faster than the other? The polycistronic nature of many chloroplast mRNAs suggests that both dominant assembly factor and CES protein genes could be equal targets of transcriptional regulation.

If rapid regulation of chloroplast-encoded genes is what tips the equilibrium of complex formation, how will chloroplasts know which chloroplast genes should be turned on and which turned off? The redox state of photosynthetic electron carriers is a source of information for deciding the composition of thylakoid membranes. However, this decision will require a continuous dialogue between electron carriers and chloroplast genes encoding core subunits. A minimal signal transduction chain in the form of a two-component system makes this dialogue possible. The chimeric nature of the electron transfer complexes further requires that the dialogue within the chloroplast be overheard by the nucleus, so that the availability of the nuclear-encoded peripheral subunits is also, eventually, brought under control. Plastid-to-nuclear signaling pathways (Gray et al. 2003; Fey et al. 2005; Lopez-Juez & Pyke 2005; Koussevitzky et al. 2007; Moulin et al. 2008) must have originated to achieve the balanced coordination that is required for the assembly of functional electron transfer complexes.

Box 1. The CoRR (co-location for redox regulation; Allen 2003) hypothesis states that genes have been retained in bioenergetic organelles to provide the direct and unconditional regulation of their expression by the redox state of their gene products, the core proteins of energy transduction in photosynthesis and respiration. Six out of the ten axioms and predictions of CoRR (Allen et al. 2005) apply directly to our present knowledge of the chloroplast two-component systems.

* The principle of selective value of redox control. For each gene under direct redox control, it is selectively advantageous for that gene to be retained and expressed only within the organelle.
* The principle of unselective gene transfer and protein import. Gene transfer from the organelle to the nucleus of the host cell is not selective for particular genes and there is no barrier to the successful import into the organelles of any precursor protein.
* The principle of selective value for nuclear location of genes. It is selectively advantageous to relocate those organellar genes that do not encode core subunits of electron transfer complexes, and for which expression need not be redox-regulated, to the nucleus of the host cell.
* The principle of primary involvement in energy transduction. Those genes for which direct redox control is vital, and for which gene products form the core subunits of the electron transfer complexes and participate in primary electron transfer reactions, are always contained in the organelles.
* The principle of secondary involvement in energy transduction. Genes for which products constitute the organelle genetic system, or for which products are associated with secondary events in energy transduction, may be retained in organelles in one group of organisms, but not in another.
* The principle of nuclear encoding of redox-signalling components. The redox signalling components, upon which the co-location of genes and gene products is based, are themselves not involved in primary electron transfer, so their genes have been relocated to the nucleus.

It has been suggested that the direct and rapid redox regulation of genes encoding core subunits of the electron transfer complexes by the process of redox chemistry is the reason why chloroplasts and mitochondria retain functional genomes (Allen 1993a; Allen 2003). As is evident from the biogenesis of electron transfer complexes, rapid regulation of chloroplast-encoded core subunits is required for the ordered assembly of electron transfer complexes. The property of the co-location of genes and gene products for redox regulation (CoRR) appears to be inseparably integrated into the assembly and biogenesis of electron transfer chains. This means that the gene content of chloroplast genomes is explained only when the CoRR hypothesis (Allen 2003; box 1) is viewed in the context of the biogenesis and assembly processes of electron transfer complexes. A synthesis of CoRR (Allen et al. 2005; box 1) and biogenetic principles might suggest that species-specific differences in assembly processes result in species-specific differences in the gene content of chloroplasts. An example of such difference is the observation that photosystem genes such as psaD, psaE and psaF are always chloroplast-encoded in red algae, but they are nuclear-encoded in green algae and plants.

Amplification, Gain Control and an Autonomous Clock

The emerging properties of the chloroplast two-component systems and their cyanobacterial homologues demonstrate their role in linking photosynthesis with gene expression. How does this general function account for the uneven distribution of their genes between chloroplast and nuclear genomes? One hypothesis is that the retention of two-component genes in chloroplasts confers a selective advantage to the organism bearing them, since they may then be able to mount a rapid and elaborate adaptive response to serious environmental assaults such as high light intensities (Allen 2003). This amplified response is thought to be mediated through a positive feedback loop in the two-component regulatory system, since its components are encoded in the operon that they control (Allen 1995; Allen & Nilsson 1997). Such a positive feedback loop has been shown to promote a transcriptional surge that jump-starts virulence in Salmonella typhimurium (Shin et al. 2006). Amplification of signals within chloroplasts may also explain the occurrence of chloroplast-encoded response regulator genes in more than one copy in some non-green algae. This proposal (Allen 1995) is also in agreement with the demonstrated high-light acclimatory roles of the chloroplast two-component systems and their cyanobacterial homologues (Hsiao et al. 2004; Kappell & van Waasbergen 2007).

Being part of the operon that it controls is also a characteristic feature of the transcriptional feedback loop in circadian oscillators. We propose that the circadian analogy can be applied to chloroplasts, and that the chloroplast-encoded two-component systems, acting as endogenous oscillators, generate a rhythmic pattern of chloroplast mRNA accumulation. A diurnal rhythm has been noted in photosynthetic electron transport (Okada & Horie 1979). The chloroplast two-component systems, as discussed earlier, connect the activity of photosynthetic electron transport chain to gene expression in chloroplasts. Thus, it is likely that an autoregulatory loop in the transcription of chloroplast-encoded two-component systems, when connected to the rhythmic activity of photosynthetic electron transport, will generate a rhythmic transcript accumulation pattern for chloroplast genes.

An endogenous rhythm of chloroplast transcript accumulation has been demonstrated (Doran & Cattolico 1997) and is thought to be driven by the activity of the photosynthetic electron transport chain. If our proposal is correct, chloroplasts in the green lineage may have lost this transcriptional loop as genes encoding the chloroplast two-component systems have moved to the nuclear genome. The probable existence of a post-translational feedback loop in the nuclear-encoded chloroplast two-component systems may still ensure an endogenous rhythm of transcript accumulation in chloroplasts, as has been observed in Arabidopsis (Puthiyaveetil & Allen 2008b). Additionally, the observation that, in CSK null mutants, rhythmicity of psaA transcript accumulation has been greatly attenuated (Puthiyaveetil et al. 2008) supports the role of a nuclear-encoded chloroplast two-component system in rhythmic transcriptional activity of chloroplast genes.

The endogenous rhythm in chloroplasts is distinct from the circadian rhythm in that the former is aperiodic in oscillation. The aperiodicity in the chloroplast's endogenous rhythm was demonstrated in algae, where the endogenous rhythm of transcript accumulation shows a temporally gated response to changes in photoperiod (Doran & Cattolico 1997). Doran & Cattolico (1997) showed that the ability of the algae to increase the amplitude of the transcriptional response in response to the changes in photoperiod was limited to the first 2 hours of the dark period. Likewise, in the red alga C. merolae, a similar chloroplast transcriptional response was noted in response to shifts from dark to light conditions (Minoda et al. 2005). Minoda et al. (2005) found that the transcription of photosystem genes and other chloroplast genes encoding core subunits of electron transfer complexes peaked 1 hour after dark to light switch, and then decreased for 6 hours. One can easily envisage how an endogenous oscillator made of the chloroplast two-component systems can generate such aperiodic transcriptional rhythms in chloroplasts.

There may be clear advantages for the endogenous rhythm of transcript accumulation in chloroplasts not being circadian. For example, chloroplasts are faced with aperiodic fluctuations in incident light quantity and quality, such as those caused by cloud cover or by transient shading by the leaf canopy. For aquatic algae, changes in position in the water column also constitute a major source of variability in light quantity and quality. In order to improve photosynthetic efficiency and to avoid free radical generation by inadvertent electron transfer reactions under such conditions, chloroplasts can rapidly tune the expression of genes encoding core subunits of the electron transfer complexes with the help of an aperiodic clock. As seen in the earlier section, the chloroplast's ability rapidly to regulate the expression of chloroplast-encoded core subunits gives them the upper hand in the assembly of electron transfer complexes, and hence in altering thylakoid composition. The chloroplast's endogenous rhythm in transcript accumulation is thus driven by the availability of light and can be best described as a dial, or light–dark, cycle.

Phase Differences as the Dialogue between Nuclear and Cytoplasmic Genetic Systems

It therefore appears that, as well as providing the redox-signalling device that connects photosynthesis with gene expression, the two-component systems have endowed chloroplasts with an autonomous clock. A common evolutionary origin of biological redox sensors and biological chronometers can be envisaged in the context of light–dark cycles acting upon photosynthesis: an ancestral redox sensor of photosynthetic electron transport may have diverged to form both a redox-signalling system entrained by light and, eventually, a parallel, free-running system recording the passage of time (Allen 1998).

Synchronization of the chloroplast clock with the nuclear–cytosolic clock may represent an ideal steady state under constant environmental conditions. However, an essential feature of all living cells is homeostasis—their capacity to adjust to a changing external environment and so maintain a constant internal environment. We suggest that phase differences between two chronometers underlie signalling between the chloroplast and the nucleus. A photosynthesis- and redox-entrained chloroplast clock may depart in phase from a photoreceptor-entrained and nuclear–cytosolic clock. We suggest that the regulatory coupling between nuclear and chloroplast genetic systems illustrated in figure 6 may result from, and correct, the phase differences between their two clocks.

Conserved Signalling Throughout a Major Evolutionary Transition

The two-component systems in chloroplasts persist from the ancestral cyanobacterial endosymbiont and thus epitomize a signalling system that could never have been put 'on hold' in evolution. Close coupling of photosynthesis with gene expression using a two-component system was vital before, during and after the transition of cyanobacterium to chloroplast. We now need to understand the precise nature of the redox signals sensed, the mechanism of signal perception by sensor kinases, the identity of their cognate response regulators and the interaction of these response regulators with their target genes. The two-component systems common to chloroplasts and cyanobacteria are keys to understanding the link between photosynthesis and gene expression, and may also illuminate the consequences of this link for eukaryotic cell evolution (Allen et al. 2008).

Acknowledgements

S.P. thanks Queen Mary, University of London for a research studentship. J.F.A. thanks the Royal Society for a Royal Society–Wolfson Research Merit Award and the Wellcome Trust for a Value in People Award.

References

1. Allen J.F. Control of gene-expression by redox potential and the requirement for chloroplast and mitochondrial genomes. J. Theor. Biol. 1993;165:609–631. doi:10.1006/jtbi.1993.1210
2. Allen J.F. Redox control of transcription—sensors, response regulators, activators and repressors. FEBS Lett. 1993;332:203–207. doi:10.1016/0014-5793-(93)80631-4
3. Allen J.F. Thylakoid protein-phosphorylation, state-1-state-2 transitions, and photosystem stoichiometry adjustment—redox control at multiple levels of gene-expression. Physiol. Plantarum. 1995;93:196–205. doi:10.1034/j.1399–3054.1995.930128.x
4. Allen J.F. Light, time and micro-organisms. In: Caddick M.X., Baumberg S., Hodgson D.A., Phillips-Jones M.K., editors. Microbial responses to light and time. Cambridge University Press; Cambridge, UK: 1998. pp. 1–31.
5. Allen J.F. The function of genomes in bioenergetic organelles. Phil. Trans. R. Soc. Lond. B. 2003;358:19–38. doi:10.1098/rstb.2002.1191

6. Allen J.F., Nilsson A. Redox signalling and the structural basis of regulation of photosynthesis by protein phosphorylation. Physiol. Plantarum. 1997;100:863–868. doi:10.1111/j.1399–3054.1997.tb00012.x
7. Allen J.F., Puthiyaveetil S., Strom J., Allen C.A. Energy transduction anchors genes in organelles. Bioessays. 2005;27:426–435. doi:10.1002/bies.20194
8. Allen J.F., Allen C.A., Puthiyaveetil S. Redox switches and evolutionary transitions. In: Allen J.F., Gantt E., Golbeck J.H., Osmond B., editors. Photosynthesis. Energy from the sun. Proceedings of the 14th international congress on photosynthesis. Springer; Heidelberg, Germany: 2008. pp. 1161–1166.
9. Aravind L., Ponting C.P. The cytoplasmic helical linker domain of receptor histidine kinase and methyl-accepting proteins is common to many prokaryotic signalling proteins. FEMS Microbiol. Lett. 1999;176:111–116. doi:10.1111/j.1574-6968.1999.tb13650.x
10. Ashby M.K., Houmard J. Cyanobacterial two-component proteins: structure, diversity, distribution, and evolution. Microbiol. Mol. Biol. Rev. 2006;70:472–509. doi:10.1128/MMBR.00046-05
11. Ashby M.K., Mullineaux C.W. Cyanobacterial ycf27 gene products regulate energy transfer from phycobilisomes to photosystems I and II. FEMS Microbiol. Lett. 1999;181:253–260. doi:10.1111/j.1574-6968.1999.tb08852.x
12. Ashby M., Houmard J., Mullineaux C.W. The ycf27 genes from cyanobacteria and eukaryotic algae: distribution and implications for chloroplast evolution. FEMS Microbiol. Lett. 2002;214:25–30. doi:10.1016/S0378-1097-(02)00834-0
13. Boudreau E., Takahashi Y., Lemieux C., Turmel M., Rochaix J.D. The chloroplast ycf3 and ycf4 open reading frames of Chlamydomonas reinhardtii are required for the accumulation of the photosystem I complex. EMBO J. 1997;16:6095–6104. doi:10.1093/emboj/16.20.6095
14. Chenna R., Sugawara H., Koike T., Lopez R., Gibson T.J., Higgins D.G., Thompson J.D. Multiple sequence alignment with the Clustal series of programs. Nucleic Acids Res. 2003;31:3497–3500. doi:10.1093/nar/gkg500
15. Choquet Y., Vallon O. Synthesis, assembly and degradation of thylakoid membrane proteins. Biochimie. 2000;82:615–634. doi:10.1016/S0300-9084-(00)00609-X
16. Clamp M., Cuff J., Searle S.M., Barton G.J. The Jalview Java alignment editor. Bioinformatics. 2004;20:426–427. doi:10.1093/bioinformatics/btg430
17. Cole C., Barber J.D., Barton G.J. The Jpred 3 secondary structure prediction server. Nucleic Acids Res. 2008;36:W197–W201. doi:10.1093/nar/gkn238

18. Doran E., Cattolico R.A. Photoregulation of chloroplast gene transcription in the chromophytic alga Heterosigma carterae. Plant Physiol. 1997;115:773–781.

19. Duplessis M.R., Karol K.G., Adman E.T., Choi L.Y., Jacobs M.A., Cattolico R.A. Chloroplast His-to-Asp signal transduction: a potential mechanism for plastid gene regulation in Heterosigma akashiwo (Raphidophyceae). BMC Evol. Biol. 2007;7:70. doi:10.1186/1471-2148-7-70

20. Fey V., Wagner R., Brautigam K., Wirtz M., Hell R., Dietzmann A., Leister D., Oelmuller R., Pfannschmidt T. Retrograde plastid redox signals in the expression of nuclear genes for chloroplast proteins of Arabidopsis thaliana. J. Biol. Chem. 2005;280:5318–5328. doi:10.1074/jbc.M406358200

21. Forsberg J., Rosenquist M., Fraysse L., Allen J.F. Redox signalling in chloroplasts and mitochondria: genomic and biochemical evidence for two-component regulatory systems in bioenergetic organelles. Biochem. Soc. Trans. 2001;29:403–407. doi:10.1042/BST0290403

22. Gray J.C., Sullivan J.A., Wang J.H., Jerome C.A., MacLean D. Coordination of plastid and nuclear gene expression. Phil. Trans. R. Soc. B. 2003;135–144-:144–145.

23. Ho Y.S., Burden L.M., Hurley J.H. Structure of the GAF domain, a ubiquitous signaling motif and a new class of cyclic GMP receptor. EMBO J. 2000;19:5288–5299. doi:10.1093/emboj/19.20.5288

24. Hsiao H.Y., He Q., Van Waasbergen L.G., Grossman A.R. Control of photosynthetic and high-light-responsive genes by the histidine kinase DspA: negative and positive regulation and interactions between signal transduction pathways. J. Bacteriol. 2004;186:3882–3888. doi:10.1128/JB.186.12.3882-3888.2004

25. Jacobs M.A., Connell L., Cattolico R.A. A conserved His-Asp signal response regulator-like gene in Heterosigma akashiwo chloroplasts. Plant Mol. Biol. 1999;41:645–655. doi:10.1023/A:1006394925182

26. Kanesaki Y., Yamamoto H., Paithoonrangsarid K., Shoumskaya M., Suzuki I., Hayashi H., Murata N. Histidine kinases play important roles in the perception and signal transduction of hydrogen peroxide in the cyanobacterium Synechocystis sp. PCC 6803. Plant J. 2007;49:313–324. doi:10.1111/j.1365-313-X.2006.02959.x

27. Kappell A.D. Biology department. University of Texas; Arlington, TX: 2008. The control of gene expression by high light stress in cyanobacteria through the apparent two-component NblS-RpaB signal transduction pair; p. 82.

28. Kappell A.D., van Waasbergen L.G. The response regulator RpaB binds the high light regulatory 1 sequence upstream of the high-light-inducible hliB

gene from the cyanobacterium Synechocystis PCC 6803. Arch. Microbiol. 2007;187:337–342. doi:10.1007/s00203-007-0213-1

29. Kessler U., Maid U., Zetsche K. An equivalent to bacterial ompR genes is encoded on the plastid genome of red algae. Plant Mol. Biol. 1992;18:777–780. doi:10.1007/BF00020019

30. Koretke K.K., Lupas A.N., Warren P.V., Rosenberg M., Brown J.R. Evolution of two-component signal transduction. Mol. Biol. Evol. 2000;17:1956–1970.

31. Koussevitzky S., Nott A., Mockler T.C., Hong F., Sachetto-Martins G., Surpin M., Lim J., Mittler R., Chory J. Signals from chloroplasts converge to regulate nuclear gene expression. Science. 2007;316:715–719. doi:10.1126/science.1140516

32. Kumar A., Toledo J.C., Patel R.P., Lancaster J.R., Jr, Steyn A.J. Mycobacterium tuberculosis DosS is a redox sensor and DosT is a hypoxia sensor. Proc. Natl Acad. Sci. USA. 2007;104:11 568–11 573. doi:10.1073/pnas.0705054104

33. Li H., Sherman L.A. A redox-responsive regulator of photosynthesis gene expression in the cyanobacterium Synechocystis sp. strain PCC 6803. J. Bacteriol. 2000;182:4268–4277. doi:10.1128/JB.182.15.4268-4277.2000

34. Lopez-Juez E., Pyke K.A. Plastids unleashed: their development and their integration in plant development. Int. J. Dev. Biol. 2005;49:557–577. doi:10.1387/ijdb.051997el

35. Maris A.E., Sawaya M.R., Kaczor-Grzeskowiak M., Jarvis M.R., Bearson S.M., Kopka M.L., Schroder I., Gunsalus R.P., Dickerson R.E. Dimerization allows DNA target site recognition by the NarL response regulator. Nat. Struct. Biol. 2002;9:771–778. doi:10.1038/nsb845

36. Martinez-Hackert E., Stock A.M. Structural relationships in the OmpR family of winged-helix transcription factors. J. Mol. Biol. 1997;269:301–312. doi:10.1006/jmbi.1997.1065

37. Mikami K., Kanesaki Y., Suzuki I., Murata N. The histidine kinase Hik33 perceives osmotic stress and cold stress in Synechocystis sp. PCC 6803. Mol. Microbiol. 2002;46:905–915. doi:10.1046/j.1365-2958.2002.03202.x

38. Mikami K., Dulai S., Sulpice R., Takahashi S., Ferjani A., Suzuki I., Murata N. Histidine kinases, Hik2, Hik16 and Hik33, in Synechocystis sp. PCC 6803 are Involved in the tolerance of photosystem II to environmental stress. Plant Cell Physiol. 2003;44:S82.

39. Mikami K., Suzuki I., Murata N. Sensors of abiotic stress in Synechocystis. Topics in Current Genetics. 2003;4:103–119.

40. Minoda A., Tanaka K. Roles of the transcription factors encoded in the plastid genome of Cyanidioschyzon merolae. In: Est A.v.d., Bruce D., editors. Photosynthesis: fundamental aspects to global perspectives. Alliance Communications Group; Lawrence, KS: 2005. pp. 728–729.

41. Minoda A., Nagasawa K., Hanaoka M., Horiuchi M., Takahashi H., Tanaka K. Microarray profiling of plastid gene expression in a unicellular red alga, Cyanidioschyzon merolae. Plant Mol. Biol. 2005;59:375–385. doi:10.1007/s11103-005-0182-1

42. Morrison S.S., Mullineaux C.W., Ashby M.K. The influence of acetyl phosphate on DspA signalling in the cyanobacterium Synechocystis sp. PCC6803. BMC Microbiol. 2005;5:47. doi:10.1186/1471-2180-5-47

43. Moulin M., McCormac A.C., Terry M.J., Smith A.G. Tetrapyrrole profiling in Arabidopsis seedlings reveals that retrograde plastid nuclear signaling is not due to Mg-protoporphyrin IX accumulation. Proc. Natl Acad. Sci. USA. 2008;105:15 178–15 183. doi:10.1073/pnas.0803054105

44. Moussatche P., Klee H.J. Autophosphorylation activity of the Arabidopsis ethylene receptor multigene family. J. Biol. Chem. 2004;279:48734–48741. doi:10.1074/jbc.M403100200

45. Murata N., Dmitry A.L. Histidine kinase Hik33 is an important participant in cold-signal transduction in cyanobacteria. Physiol. Plantarum. 2006;126:17–27. doi:10.1111/j.1399-3054.2006.00608.x

46. Nixon B.T., Ronson C.W., Ausubel F.M. Two-component regulatory systems responsive to environmental stimuli share strongly conserved domains with the nitrogen assimilation regulatory genes ntrB and ntrC. Proc. Natl Acad. Sci. USA. 1986;83:7850–7854. doi:10.1073/pnas.83.20.7850

47. Oelmüller, R., Peskan, T., Westermann, M., Sherameti, I., Chandok, M., Sopory, S.K., Wöstemeyer, A., Kusnetsov, V., Bezhani, S. & Pfannschmidt, T. 2001 Novel aspects in photosynthesis gene regulation. In Signal transduction in plants (eds S. S. K. & O. R.), pp. 259–277. Dordrecht, The Netherlands: Kluwer.

48. Okada M., Horie H. Diurnal rhythm in the hill reaction in cell-free extracts of the green alga Bryopsis maxima. Plant Cell Physiol. 1979;20:1403–1406.

49. Paithoonrangsarid K., et al. Five histidine kinases perceive osmotic stress and regulate distinct sets of genes in Synechocystis. J. Biol. Chem. 2004;279:53 078–53 086. doi:10.1074/jbc.M410162200

50. Pfannschmidt T., Nilsson A., Allen J.F. Photosynthetic control of chloroplast gene expression. Nature. 1999;397:625–628. doi:10.1038/17624

51. Pfannschmidt T., Nilsson A., Tullberg A., Link G., Allen J.F. Direct transcriptional control of the chloroplast genes psbA and psaAB adjusts photosynthesis to light energy distribution in plants. IUBMB Life. 1999;48:271–276.

52. Puthiyaveetil S., Allen J.F. A bacterial-type sensor kinase couples electron transport to gene expression in chloroplasts. In: Allen J.F., Gantt E., Golbeck J.H., Osmond B., editors. Photosynthesis. Energy from the sun. Proceedings of the 14th international congress on photosynthesis. Springer; Heidelberg, Germany: 2008. pp. 1187–1192.

53. Puthiyaveetil S., Allen J.F. Transients in chloroplast gene transcription. Biochem. Biophys. Res. Commun. 2008;368:871–874. doi:10.1016/j.bbrc.2008.01.167

54. Puthiyaveetil S., et al. The ancestral symbiont sensor kinase CSK links photosynthesis with gene expression in chloroplasts. Proc. Natl Acad. Sci. USA. 2008;105:10 061–10 066. doi:10.1073/pnas.0803928105

55. Rensing S.A., et al. The Physcomitrella genome reveals evolutionary insights into the conquest of land by plants. Science. 2008;319:64–69. doi:10.1126/science.1150646

56. Sato S., Shimoda Y., Muraki A., Kohara M., Nakamura Y., Tabata S. A large-scale protein–protein interaction analysis in Synechocystis sp. PCC6803. DNA Res. 2007;14:207–216. doi:10.1093/dnares/dsm021

57. Shin D., Lee E.J., Huang H., Groisman E.A. A positive feedback loop promotes transcription surge that jump-starts Salmonella virulence circuit. Science. 2006;314:1607–1609. doi:10.1126/science.1134930

58. Shoumskaya M.A., Paithoonrangsarid K., Kanesaki Y., Los D.A., Zinchenko V.V., Tanticharoen M., Suzuki I., Murata N. Identical Hik-Rre systems are involved in perception and transduction of salt signals and hyperosmotic signals but regulate the expression of individual genes to different extents in Synechocystis. J. Biol. Chem. 2005;280:21 531–21 538. doi:10.1074/jbc.M412174200

59. Skerker J.M., Perchuk B.S., Siryaporn A., Lubin E.A., Ashenberg O., Goulian M., Laub1 M.T. Rewiring the specificity of two-component signal transduction systems. Cell. 2008;133:1043–1054. doi:10.1016/j.cell.2008.04.040

60. Stock A., Koshland D.E., Stock J. Homologies between the Salmonella-Typhimurium chey protein and proteins involved in the regulation of chemotaxis, membrane-protein synthesis, and sporulation. Proc. Natl Acad. Sci. USA. 1985;82:7989–7993. doi:10.1073/pnas.82.23.7989

61. Stock J.B., Ninfa A.J., Stock A.M. Protein phosphorylation and regulation of adaptive responses in bacteria. Microbiol. Rev. 1989;53:450–490.

62. Stock A.M., Robinson V.L., Goudreau P.N. Two-component signal transduction. Annu. Rev. Biochem. 2000;69:183–215. doi:10.1146/annurev.biochem.69.1.183

63. Taylor B.L., Zhulin I.B. PAS domains: internal sensors of oxygen, redox potential, and light. Microbiol. Mol. Biol. Rev. 1999;63:479–506.

64. Tullberg A., Alexciev K., Pfannschmidt T., Allen J.F. Photosynthetic electron flow regulates transcription of the psaB gene in pea (Pisum sativum L.) chloroplasts through the redox state of the plastoquinone pool. Plant Cell Physiol. 2000;41:1045–1054. doi:10.1093/pcp/pcd031

65. Vuillet L., et al. Evolution of a bacteriophytochrome from light to redox sensor. EMBO J. 2007;26:3322–3331. doi:10.1038/sj.emboj.7601770

66. Wagner R., Pfannschmidt T. Eukaryotic transcription factors in plastids: bioinformatic assessment and implications for the evolution of gene expression machineries in plants. Gene. 2006;381:62–70. doi:10.1016/j.gene.2006.06.022

67. Weber P., Fulgosi H., Piven I., Muller L., Krupinska K., Duong V.H., Herrmann R.G., Sokolenko A. TCP34, a nuclear-encoded response regulator-like TPR protein of higher plant chloroplasts. J. Mol. Biol. 2006;357:535–549. doi:10.1016/j.jmb.2005.12.079

68. Yeh K.C., Lagarias J.C. Eukaryotic phytochromes: light-regulated serine/threonine protein kinases with histidine kinase ancestry. Proc. Natl Acad. Sci. USA. 1998;95:13 976–13 981. doi:10.1073/pnas.95.23.13976

Ecological Selection Pressures for C_4 Photosynthesis in the Grasses

Colin P. Osborne and Robert P. Freckleton

ABSTRACT

Grasses using the C_4 photosynthetic pathway dominate grasslands and savannahs of warm regions, and account for half of the species in this ecologically and economically important plant family. The C_4 pathway increases the potential for high rates of photosynthesis, particularly at high irradiance, and raises water-use efficiency compared with the C_3 type. It is therefore classically viewed as an adaptation to open, arid conditions. Here, we test this adaptive hypothesis using the comparative method, analysing habitat data for 117 genera of grasses, representing 15 C_4 lineages. The evidence from our three complementary analyses is consistent with the hypothesis that evolutionary selection for C_4 photosynthesis requires open environments, but we find an equal likelihood of C_4 evolutionary origins in mesic, arid and saline habitats. However, once the pathway has arisen, evolutionary transitions into arid *habitats occur at higher rates in C_4 than C_3 clades. Extant C_4 genera therefore occupy a wider range of drier habitats than their C3 counterparts because the C_4 pathway represents a pre-adaptation to arid conditions. Our analyses warn against evolutionary inferences based solely upon the high occurrence of extant C_4 species in dry habitats, and provide a novel interpretation of this classic ecological association.*

Keywords: C_4 photosynthesis, adaptation, water-use efficiency, aridity, shade, phylogeny

Introduction

The majority of terrestrial plant species use the C_3 photosynthetic pathway. However, the efficiency of this process is compromised by photorespiration, and its rate is strongly limited by CO_2 diffusion from the atmosphere. Photorespiration increases at low CO_2 concentrations and high temperatures, and CO_2 limitation is accentuated by the reduction of stomatal aperture under arid conditions (Björkman 1971; Osmond et al. 1982). The evolution of C_4 photosynthesis has solved each of these problems via a suite of physiological and anatomical adaptations that concentrate CO_2 at the site of carbon fixation, minimize photorespiration and raise the affinity of photosynthesis for CO_2 at low mesophyll concentrations (Björkman 1971; Osmond et al. 1982). As a consequence, C_4 plants have the potential to achieve higher rates of photosynthesis than their C_3 counterparts, particularly at high irradiance (Black et al. 1969). Since C_4 photosynthesis draws mesophyll CO2 down to lower concentrations than the C_3 type, it also allows stomatal conductance to be reduced, leading to greater water-use efficiency than the C_3 pathway under the same environmental conditions (Downes 1969). The C_4 pathway is therefore classically viewed as an adaptation to declining levels of atmospheric CO_2 (Ehleringer et al. 1991), and hot, open, arid environments (Björkman 1971; Loomis et al. 1971).

Approximately half of the world's grass species use C4 photosynthesis (Sage et al. 1999a), and these plants dominate grassland and savannah ecosystems in warm climate regions (Sage et al. 1999b). They also include economically important food crops such as maize and sugarcane, and biofuel crops such as switchgrass and Miscanthus. Recent phylogenetic data suggest that the C4 pathway evolved in 9–18 independent clades of grasses during the past 32 million years (Myr) (Christin et al. 2008; Vicentini et al. 2008). However, only the earliest of these evolutionary origins coincided with the major decline in CO2 that occurred during the Oligocene (32–25 Myr ago; Pagani et al. 2005; Christin et al. 2008; Roalson 2008; Vicentini et al. 2008). One phylogenetic analysis suggests that the evolution of the C4 pathway became more likely after the CO2 decrease (Christin et al. 2008), and complementary studies suggest that the C4 origination events were clustered in time (Vicentini et al. 2008), and occurred in grass clades that were already adapted to warm climates (Edwards & Still 2008). However, adaptive hypotheses about the suite of local ecological factors that are selected for C4 photosynthesis remain largely untested (Roalson 2008). Chief among these are the hypothesized roles of water deficits caused by aridity or salinity, and the formation of open habitats via disturbance (Sage 2001).

The C4 photosynthetic pathway offers grasses the potential to achieve higher rates of leaf carbon fixation with a similar or lower expenditure of water than C3 species (Loomis et al. 1971; Gifford 1974). It also maximizes dry matter production when water is available in limited pulses (Williams et al. 1998), and allows the conservation of water in a drying soil (Kalapos et al. 1996). These physiological benefits are moderated by a trade-off between the photosynthetic rate and the intrinsic water-use efficiency of C4 leaves (Meinzer 2003). However, they are consistent with the common occurrence of C4 grass species in seasonally arid ecosystems, deserts and on saline soils (Sage et al. 1999b). Compelling evidence for the ecological sorting of C4 species into drier habitats than C3 species was provided by a recent comparative study of the largely exotic Hawaiian grass flora (Edwards & Still 2008).

Despite their prevalence in dry habitats, C4 grasses also occupy a diverse range of mesic, shaded and flooded ecological niches, and the primary importance of aridity for the ecological success of these species has therefore been challenged (Ehleringer et al. 1997; Sage et al. 1999b; Keeley & Rundel 2003). Large-scale spatial patterns also highlight a more complex relationship with climate than predicted by water-use efficiency alone, with the biomass of C4 grasses relative to other plant functional types increasing, rather than decreasing, with rainfall across the Great Plains of North America (Paruelo & Lauenroth 1996). In fact, the potential for C4 photosynthesis to drive high rates of productivity means that there are sound theoretical reasons to expect a selective advantage for the pathway in moist soil environments, whenever high temperatures are coupled with moderate-to-high light availability (Long 1999; Sage et al. 1999b; Keeley & Rundel 2003; Sage 2004).

Spatial correlations with environmental variables suggest that some of the observed variation in the ecological niche of C4 grasses may be explained by the contrasts in the tolerance of aridity between different phylogenetic groups (Hartley 1950; Taub 2000). Unravelling the confounding effects of physiology and phylogeny will therefore be crucial if we are to make realistic predictions about the future impacts of increasing aridity on community composition in subtropical grasslands (Christensen et al. 2007), and move towards a greater understanding of the role of palaeoclimate change in driving the expansion of C4 grassland ecosystems in the geological past (Osborne 2008).

The aim of this study is to investigate the ecological selection pressures for C4 photosynthesis in the grasses, using the comparative method to test the alternative hypotheses of adaptation (Harvey & Pagel 1991). Drawing upon a recently published phylogeny (Christin et al. 2008), we have compiled a global habitat dataset for 117 genera of grasses, sampling each of the major clades and 15 independent C4 lineages. Analyses of these data address two key questions. First, which ecological factors have selected for the C4 pathway, in particular, is it an adaptation to aridity? And secondly, to what extent is variation in the ecological niches of different C4 plant groups explained by phylogenetic history? Our results are consistent with the hypothesis that selection for C4 photosynthesis occurred in open habitats but was independent of water availability, whereas subsequent evolutionary transitions into arid habitats were faster in C4 than C3 clades.

Material and Methods

Phylogenetic Framework

Phylogenetic relationships were based on the calibrated consensus tree of Christin et al. (2008). Species sampling for this tree was designed to include all postulated origins of the C_4 photosynthetic pathway within the grasses, and to minimize the distance between the stem group and crown group nodes. The topology was obtained by Bayesian inference using the chloroplast DNA markers rbcL and ndhF, and calibrated using Bayesian molecular dating, with minimum ages for six nodes based on fossil evidence (Christin et al. 2008). Branch lengths are therefore proportional to time elapsed. The grass phylogeny was kindly provided by Pascal-Antoine Christin (University of Lausanne).

Since the complete phylogenetic analysis spanned the entire order Poales, we first extracted the 187 species belonging to the grass family Poaceae. The tree indicated that a number of genera were polyphyletic (e.g. Panicum, Merxmuellera), and these were removed as it was not possible to generate unequivocal trait data for these. One genus that appeared to be paraphyletic (Brachiaria) was combined together with its sister (Urochloa) to form a monophyletic clade. This procedure resulted in a phylogeny of 129 grass genera.

Ecological Data

The photosynthetic type (C_3 or C_4) within each genus was assigned following Sage et al. (1999a). However, a number of genera could not be categorically assigned a photosynthetic type, since they contained C_3, C_4 and C_3–C_4 intermediate species (Neurachne, Alloteropsis and Steinchisma). Rather than excluding these genera from the analysis, we assigned photosynthetic type based on the majority of species (Neurachne and Alloteropsis=C_4 and Steinchisma=C_3), and tested the sensitivity of our analyses to this assumption by examining the effects of a reversal in the photosynthetic type for these genera.

Habitat and diversity data were then derived from the information compiled by Watson & Dallwitz (1992 onwards). For each genus, we recorded the number of species and type of habitats occupied, including information on water requirements (e.g. hydrophyte, xerophyte), tolerance of saline conditions (halophyte and glycophyte) and the occupation of shaded habitats (shaded and open). Water requirements were assigned a numerical score, giving equal weighting to the extremes (hydrophyte=5, helophyte=4, mesophyte=3 and xerophyte=1), and resulting in a continuous sequence of values for each genus. The habitat types occupied by each genus were then characterized using the mean and range of these values. Two further binary traits recorded the presence or absence of shade species, and the presence or absence of xerophytes. Since halophytes tolerate physiological drought imposed via high osmotic pressure, we also included genera containing halophytes in the 'xerophyte' category. However, all of the halophytic genera except one (Spartina) contained xerophytes. Habitat data were not available for all clades, and our final dataset included a total of 117 genera, sampling 15 out of the 17 hypothesized origins of C4 photosynthesis in the grasses (Christin et al. 2008).

Phylogenetic Comparative Analysis

In the first set of analyses we aimed to determine whether photosynthetic pathway is associated with several continuous ecological traits. Photosynthetic pathway was coded as a binary categorical variable (C_3 versus C_4). The number of species within a genus, and the mean and range of genus water requirements were coded as continuous variables. To test whether these were correlated with photosynthetic pathway, we used a generalized linear model in which the continuous variable was the dependent variable and the photosynthetic pathway a categorical predictor. In order to control for phylogenetic dependence we simultaneously estimated Pagel's λ (Pagel 1999) using the approach described in Freckleton et al. (2002). This parameter measures, and controls for, the degree to which the residual variation shows phylogenetic non-independence according to the predictions of a simple Brownian model of trait evolution. According to this, a value of λ=0 indicates that there is no phylogenetic dependence in the data, while λ=1 indicates that the residuals show strong phylogenetic dependence.

Modelling Evolutionary Pathways

In the second set of analyses, our objective was to model the transitions between C_3 and C_4 photosynthetic pathways and to determine whether these are associated with transitions between habitat types, specifically shaded versus open habitats, and xeric versus mesic ones. We modelled the evolutionary transitions using approaches described in Pagel (1994, 1999) and Pagel & Meade (2006). In brief, this method is based on a continuous-time Markov model, which models the transitions of discrete characters between states. For a pair of binary traits, there are four possible states (state 1=00, state 2=01, state 3=10, state 4=11) and eight parameters, which are the instantaneous rates of change between the states (denoted by q_{ij}, measuring the rate of change from state i to j), assuming that instantaneously only a single change in one character may occur. The model was fitted using the reversible jump Markov chain Monte Carlo methods described in Pagel & Meade (2006) using the package BayesTraits (http://www.evolution.rdg.ac.uk/BayesTraits.html), and parameters were sampled from their posterior distributions.

In the first analysis, we wished to test whether transitions between C3 and C4 pathways were dependent on habitat openness. Thus, each genus was coded as either exclusively confined to open habitats (0) or sometimes/always occupying shaded habitats (1), and as being C3 (0) or C4 (1). We fitted the full model allowing for all single-step transitions between the states. In order to test the hypotheses concerning the rates of evolution between the states, we conducted three comparisons: firstly, we asked whether the rate of transition between C3 and C4 differed between open and shaded habitats (by contrasting rates q13 and q24). Secondly, we asked whether the rate of transition from open to shaded habitats differed between C3 and C4 lineages (by contrasting q12 and q34). And finally, we asked whether the transition from shaded to open habitats differed between C3 and C4 lineages (by contrasting q21 and q43).

In the second analysis, we tested whether the transitions between C3 and C4 pathways were accompanied by changes in the aridity of occupied habitat. Each genus was coded as being either exclusively confined to waterlogged/mesic habitats (0) or sometimes/always occupying xeric/saline habitats (1), and again we fitted a full model including eight parameters. From the posterior distribution of parameter estimates, we compared the distributions of the estimates of rates of transition from C3 to C4 in xeric and mesic habitats. Again, we used the fitted parameters to test three hypotheses: firstly, we asked whether the rate of transition between C3 and C4 pathways differed in mesic and xeric habitats (by contrasting q13 and q24). Secondly, we asked whether the rate of transition from mesic to xeric habitats differed between C3 and C4 lineages (by contrasting q12 and q34). And finally, we asked whether the transition from xeric to mesic habitats differed between C3 and C4 lineages (by contrasting q21 and q43).

To contrast qij and qkl, for each model in the posterior distribution we calculated the difference qij–qkl. For the whole set of models in the posterior distribution, we then examined the distribution of values of these differences to determine whether there were systematic deviations from zero. These differences are presented in the supplementary information together with the estimated parameters for all models.

The possibility of evolutionary reversals from the C4 pathway to the C3 type remains a key area of uncertainty in phylogenetic models. Phylogenetic analyses of the numerous C3 and C4 clades in the subfamily Panicoideae suggest that the hypotheses of multiple evolutionary origins and/or reversions are equally parsimonious (Giussani et al. 2001) and, in the genus Alloteropsis, a C4 to C3 reversal is the single most parsimonious interpretation (Ibrahim et al. 2009). Although the convergent evolution of amino acid sequences in a C4-specific enzyme does provide compelling evidence for multiple C4 origins in this grass subfamily (Christin et al. 2007), phylogenetic analyses still indicate a high likelihood of reversion events in the Panicoideae (Vicentini et al. 2008).

However, one issue of concern in such analysis is that, when analysing the evolution of a binary trait, if one of the trait states has a higher speciation rate, reconstructions can appear to support the enhanced rates of reversals from rare to common states (Maddison 2006), and this problem affects the method used here. We note below that we find evidence consistent with higher rates of diversification in C4 grass clades, raising the possibility of a non-random distribution of extinction probabilities across C3 and C4 lineages.

Clearly, the issue of reversible transitions between photosynthetic pathways is contentious and must be considered in ecological models of C4 grass evolution. We therefore conducted two sets of analysis to consider the sensitivity of our results to this. In the first instance, we conducted the analysis as described above, including the possibility of reversions. We then re-analysed the data, prohibiting reversals from C4 to C3. This constrained model included six rather than eight parameters. We asked two further questions using the full, eight-parameter models; if they are possible, do C4 to C3 reversals depend on shading or aridity (q31 versus q42)?

Results

Comparative Analysis

Species number is significantly higher within C_4 than C_3 genera (table 1; figure 1a), and the range of habitat water requirements within each genus is significantly greater for the C_4 than the C_3 type (table 1; figure1b). Species number is 33 per cent greater in C_4 compared with C_3 genera (figure 1a), while the range of habitat water requirements almost doubles (increasing by 85%; figure 1b). Neither shows significant phylogenetic dependence (λ=0; table 1). However, there is a significant linear association between species number and the range of habitat water requirements ($F_{1,90}$=26.32, p=1.7×10^{-6}). The range of habitats occupied within each genus explains about a quarter of its species number (R^2=0.22). Critically, the introduction of photosynthetic type as a categorical predictor does not significantly improve the fit of this statistical model to the data ($F_{2,90}$=1.88, p=0.17). This means that the observed association between species number and photosynthetic type may be entirely due to habitat diversity rather than a direct effect of C_4 photosynthesis per se. In other words, C_4 genera occupy a wider range of habitats and this, in turn, is associated with a larger number of species per genus.

Table 1. Results of generalized linear models testing for an association between photosynthetic pathway (C_3 or C_4) and species number or habitat characteristics. ('Species number' indicates the total number of species within each genus. 'Water range' and 'water mean' refer to the range and mean of habitat water categories, taken across all of the species within each genus. The results show the F-ratio, degrees of freedom (d.f.) and significance level (p-value) for photosynthetic pathway as a categorical predictor in each model. Pagel's λ estimates the degree of phylogenetic dependence in the data, ranging from 0 (no dependence) to 1 (strong dependence).)

variable	*F*-ratio	d.f.	*p*-value	λ
species number	6.95	1, 115	9.5×10^{-3}	0.00
water range	7.78	1, 90	6.4×10^{-3}	0.00
water mean	6.76	1, 90	1.1×10^{-2}	0.83

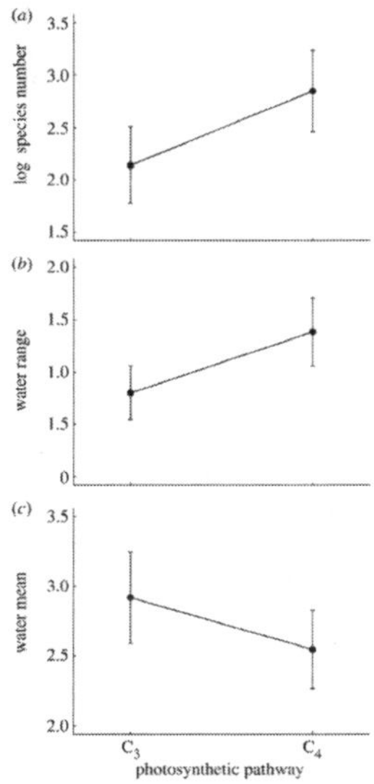

Figure 1. Species number and habitat water requirements in extant C_3 and C_4 genera. The plots show mean ±95% C.I. for (a) species number, (b) range of water requirements tolerated and (c) mean water requirements for each photosynthetic type.

The mean habitat water requirement is significantly lower in C4 than C3 genera (table 1; figure 1c), and shows a strong, statistically significant phylogenetic dependence ($\lambda \rightarrow 1$; table 1). Therefore, C4 genera occupy a wider range of drier habitats than their C3 counterparts, but different clades of grasses differ markedly in their habitat water requirements. These results are insensitive to the assumptions made about photosynthetic pathway in the genera Neurachne, Alloteropsis and Steinchisma.

Evolutionary Transitions

Figure 2 summarizes the rates of evolutionary transitions between states, considering the phylogenetic tree as a whole, and all of the postulated origins of C_4 photosynthesis. All of these results are insensitive to the assumptions made about photosynthetic pathway in the genera Neurachne, Alloteropsis and Steinchisma.

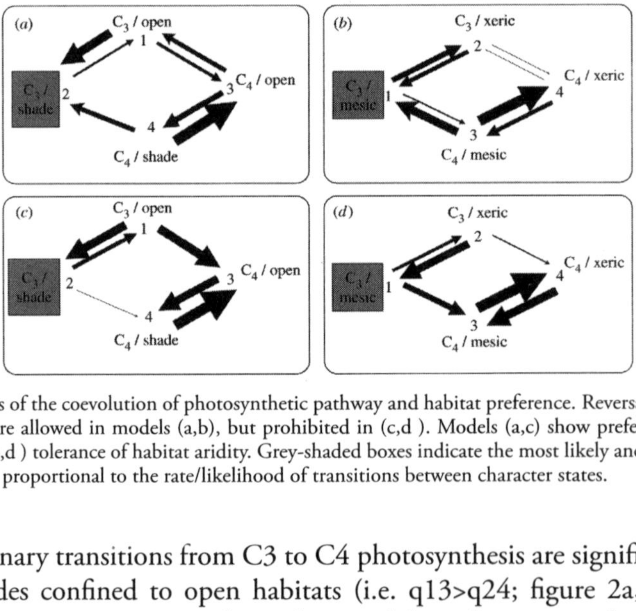

Figure 2. Models of the coevolution of photosynthetic pathway and habitat preference. Reversals from C_4 to C_3 photosynthesis are allowed in models (a,b), but prohibited in (c,d). Models (a,c) show preference for habitat openness, and (b,d) tolerance of habitat aridity. Grey-shaded boxes indicate the most likely ancestral condition, and arrow size is proportional to the rate/likelihood of transitions between character states.

Evolutionary transitions from C3 to C4 photosynthesis are significantly faster in grass clades confined to open habitats (i.e. q13>q24; figure 2a,c), and this result is robust to assumptions about the possibility of reversions from C4 to C3 photosynthesis (figure 2a versus figure 2c). The same analysis shows that grass clades occupying shaded habitats are significantly more likely to become confined to open habitats if they are C4 than C3 (i.e. q43>q21; figure 2a,c). However, the rate of evolutionary transitions from open to shaded habitats is independent of photosynthetic type, and C3 and C4 species are therefore equally likely to adapt to shade (i.e. q12=q34; figure 2a,c). Again, these results are robust to the assumptions made about C4 to C3 reversions (figure 2a versus figure 2c). If C4 to C3 reversals are possible, they occur at the same rate (are equally likely) in open and shaded habitats (i.e. q31=q42; figure 2a).

The likelihoods of ancestral character states at each node in the phylogeny are shown in figure 3. The model indicates with a high posterior probability that the last common ancestor of the Poaceae was a C3 shade species (figure 3, node A). It also illustrates the most likely evolutionary pathway to C4 photosynthesis, whereby a transition into open habitats was a necessary pre-condition for the origin of the C4 pathway. For example, the model shows with high likelihood that the last common ancestors of the C4 clades Chloridoideae (figure 4, node B) and x=10 Paniceae (figure 4, node C) were confined to open habitats. However, the open habitat reconstructions for last common ancestors of the C4 clades Andropogoneae (figure 4, node D) and the 'main clade' of x=9 Paniceae (figure 4, node E) have lower associated probabilities.

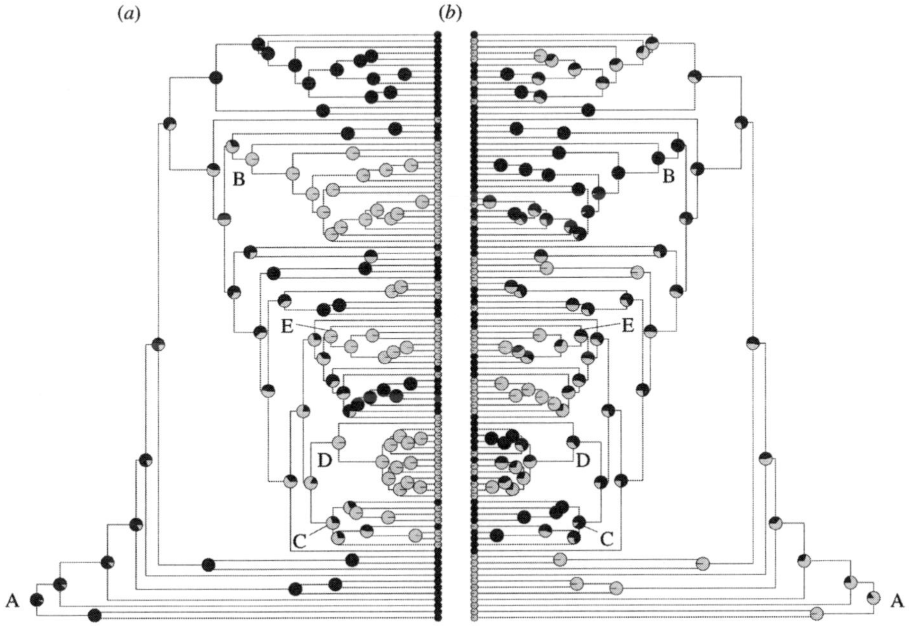

Figure 3. Likelihood of alternative ancestral states for nodes in the phylogenetic tree, showing (a) photosynthetic pathway (yellow circles, C_4; blue circles, C_3) and (b) preference for habitat openness (yellow circles, shade; blue circles, open habitat). Ancestral values were computed for individual traits using the likelihood method of Pagel (1994) and phylogenies drawn using the ace and plot.phylo functions in APE (Paradis et al. 2004).

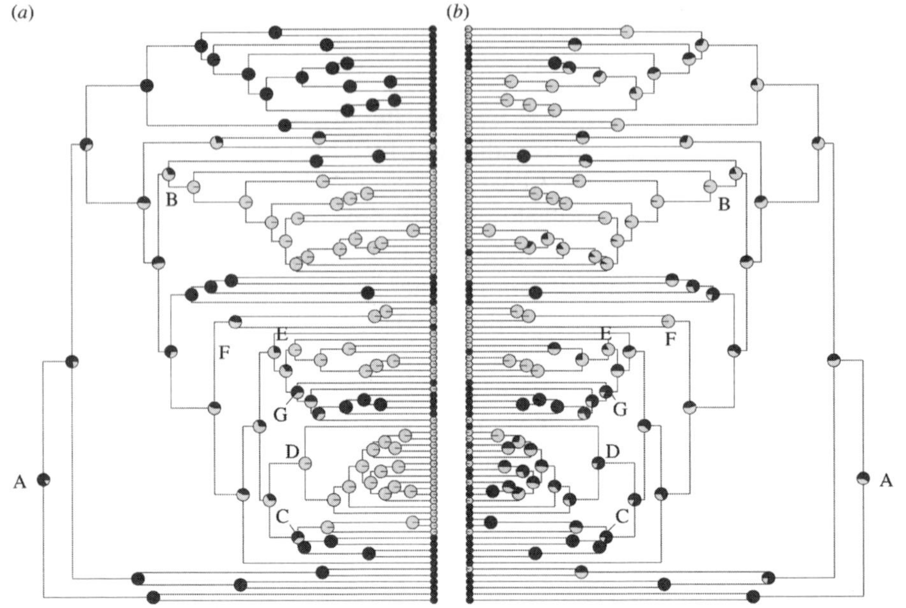

Figure 4. Likelihood of alternative ancestral states for nodes in the phylogenetic tree, showing (a) photosynthetic pathway (yellow circles, C_4; blue circles, C_3) and (b) preference for habitat aridity (yellow circles, xeric; blue circles, mesic). Ancestral values were computed for individual traits using the likelihood method of Pagel (1994) and phylogenies drawn using the ace and plot.phylo functions in APE (Paradis et al. 2004).

Unexpectedly, evolutionary transitions from C3 to C4 photosynthesis occur at the same rate (are equally likely) in grass clades that contain xerophytic or halophytic species, and those confined to mesic or waterlogged habitats (i.e. q13=q24; figure 3b,d). However, the rate/likelihood of evolutionary transitions from mesic to xeric habitats is significantly higher in C4 than in C3 grass clades (i.e. q34>q12; figure 3b,d). By contrast, species are equally likely to become confined to mesic or waterlogged habitats if they are C3 or C4 (i.e. the rate of evolutionary transition from xeric to mesic habitats is independent of photosynthetic type, q21=q43; figure 3b,d). As in the previous analysis, these results are robust to the assumptions made about the possibility of C4 to C3 reversions (figure 3b versus figure 3d). If C4 to C3 reversals are possible, they depend significantly on habitat water availability, and evolutionary reversion is significantly faster/more likely in mesic or waterlogged habitats than xeric ones (i.e. q31>q42; figure 2b).

The second model of ancestral character states is shown in figure 4 and indicates that the most likely common ancestor of the Poaceae was a C3 species confined to mesic habitats (node A). It also illustrates important contrasts between clades in the habitat where the C4 pathway evolved. For example, the model shows with a high probability (greater than 80%) that the last common ancestors of the C4 clades Chloridoideae (figure 4, node B), 'Arundinelleae' (figure 4, node F) and the main clade of x=9 Paniceae (figure 4, node E) occupied xeric habitats, whereas ancestors of the Andropogoneae (figure 4, node D), the x=9 Paniceae clade containing Echinochloa and Alloteropsis (figure 4, node G) and x=10 Paniceae (figure 4, node C) were more likely confined to mesic habitats (probability greater than 80%). This contrast in the ancestral state of independent C4 clades illustrates how the phylogenetic correlation in mean habitat water requirements may arise (table 1).

Discussion

Ecological Selection

Our three complementary analyses provide robust statistical support for the following adaptive hypothesis of C_4 pathway evolution in the grasses. Selection for C_4 photosynthesis occurs in open habitats, but may take place in mesic, arid or saline conditions. Once the pathway has evolved, C_4 lineages adapt to arid and saline habitats at a faster rate than C_3 lineages, and are more likely to become confined to open environments; C_4 photosynthesis in the grasses therefore represents a pre-adaptation (exaptation) to xeric conditions. However, evolutionary transitions into shaded and mesic habitats are independent of photosynthetic type. If reversals from the C_4 to C_3 type occur, they do so in mesic or waterlogged habitats, irrespective of the habitat light regime. The net result of these evolutionary processes is that extant C_4 genera occupy a drier range of habitats than their C_3 counterparts. This association of photosynthetic pathway with aridity in extant genera may interact with temperature, but we were unable to test this with our dataset.

Seasonal aridity, fire, the activity of large mammalian herbivores and edaphic factors increase the availability of open habitats through the reduction of woody plant cover (Sankaran et al. 2008). Our data are therefore consistent with the hypothesis that these factors raise the likelihood of C4 pathway evolution in the grasses (Sage 2001). The strong statistical dependence of C4 pathway evolution on habitat openness is also consistent with the environmental responses of photosynthesis in extant C3 and C4 grasses: temperature and irradiance are greater in open than shaded environments, especially in the period after a disturbance event (Knapp 1984), which enhances the advantage of C4 photosynthesis for CO2 fixation over the C3 type (Black et al. 1969; Björkman 1971). Our finding that shade adaptation is independent of photosynthetic type is therefore surprising, especially since C4 grasses are virtually absent from the deep shade of forest floor environments (Sage 2001). However, the shade beneath trees in tropical woodlands and savannahs is associated with high soil moisture and nutrient contents, and the tolerance of low irradiance gives grasses the opportunity to exploit these soil resource patches (Ludwig et al. 2001).

The analysis of evolutionary transitions across the whole grass phylogeny provides no statistical evidence for an overall dependence of C4 pathway evolution on aridity. However, it does not exclude the possibilities that (i) arid or saline conditions may select for C4 photosynthesis in some grass clades (e.g. Chloridoideae) and not others (e.g. Andropogoneae) or (ii) high evaporative demand and soil drying between episodic rainfall events (Williams et al. 1998) or after fire (Knapp 1984) may be important selection pressures for C4 photosynthesis in mesic habitats. A previous comparative analysis suggested that the C4 pathway has evolved in grass clades of warm climate regions (Edwards & Still 2008), where high rates of evaporation and shallow rooting systems may lead to leaf water deficits of −1.5 MPa, even when the soil is wet (Le Roux & Bariac 1998). Although these adaptive interpretations are possible, they are not necessary, because our finding that C4 photosynthesis is a pre-adaptation to arid conditions is strongly supported across the whole phylogenetic tree. It is consistent with the well-known association between photosynthetic pathway and leaf water consumption (e.g. Black et al. 1969; Downes 1969). However, it warns against adaptive inferences based solely upon correlations in extant species between photosynthetic pathway and habitat aridity, such as those observed in our data (table 1) and by previous authors (Edwards & Still 2008).

Diversity and Data Quality

The association between species number and the range of habitats occupied by each genus could arise for a number of reasons. First, the origin of C_4 photosynthesis may represent a 'key innovation' (Hunter 1998) that stimulates evolutionary diversification by increasing the rate of transition into xeric niches compared with the C_3 type. In this case, ecological selection is implicated in both the origins of C_4 photosynthesis and subsequent diversification within C_4 grass clades. However, it is important to note that, while the number of species and range of habitats may on average be larger within each C_4 than C_3 genus, this does not mean that C_4 grasses occupy a wider range of habitats overall. A second possible explanation for the observed correlation is sampling bias. If the sample of C_4 grasses is biased towards large genera, then the wider habitat range could be a statistical artefact arising from the greater probability of encountering species from different habitats in large samples. Testing these alternative explanations will require phylogenetic measures of diversification rates, rather than the genus-based approach used here. This is because different genera may have begun to diverge at different times, and genus size depends crucially on the attention paid to each group by taxonomists.

The habitat data used in our analysis are simple, qualitative characterizations of the ecology of each genus. However, despite the basic nature of this information, we still found strong associations between photosynthetic pathway and habitat, with highly significant statistical support. The qualitative agreement between the three different analyses lends further confidence to our findings. While it is possible that the phylogeny may have biased sampling via the selection of species whose phylogenetic position is important, but whose ecology is atypical, this should have been counteracted by the explicit consideration of branch lengths in our analysis. A final sampling issue arises from our use of binary habitat traits, which potentially underestimate habitat diversity in large genera. However, the strong positive correlation between the range of water requirements and species number in each genus suggests that this did not bias our findings.

Our analysis suggested that the distribution of traits is consistent with the possibility of reversions from C4 to C3 types. This echoes the findings in other analyses (Ibrahim et al. 2009; Vicentini et al. 2008); however, we should be cautious about this conclusion at this stage. As noted previously, if we analyse traits that shape the phylogeny via speciation (or extinction) rates, then the outcome of the analyses can be misleading. The problem described by Maddison (2006) would arise in our dataset if the rate of speciation were greater in species with one photosynthetic pathway than the other, and the result in figure 1a indicates that this may have been the case, subject to the caveats above.

Conclusions

We have sought statistical evidence for an adaptive hypothesis of C_4 pathway evolution in the grasses. Our analyses are consistent with the hypothesis that selection for C_4 photosynthesis requires open environments, but indicate that the high occurrence of C_4 clades in dry habitats arises because the pathway is a pre-adaptation to xeric conditions. These results provide a novel interpretation of the classic association of C_4 plants with arid environments.

Acknowledgements

The authors thank the Royal Society for funding via University Research Fellowships, Andrew Meade for advice on the use of BayesTraits, and David Beerling and Ian Woodward for their comments on the manuscript.

References

1. Björkman O. Comparative photosynthetic CO_2 exchange in higher plants. In: Hatch M.D., Osmond C.B., Slatyer R.O., editors. Photosynthesis and photorespiration. Wiley-Interscience; New York, NY: 1971. pp. 18–32.
2. Black C.C., Chen T.M., Brown R.H. Biochemical basis for plant competition. Weed Sci. 1969;17:338–344.
3. Christensen J.H., et al. Regional climate projections. In: Solomon S., Qin D., Manning M., Chen Z., Marquis M., Averyt K.B., Tignor M., Miller H.L., editors. Climate change 2007: the physical science basis. Contribution of working group I to the fourth assessment report of the intergovernmental panel on climate change. Cambridge University Press; Cambridge, UK: 2007. pp. 847–940.
4. Christin P.A., Salamin N., Savolainen V., Duvall M.R., Besnard G. C_4 photosynthesis evolved in grasses via parallel adaptive genetic changes. Curr. Biol. 2007;17:1241–1247. doi:10.1016/j.cub.2007.06.036
5. Christin P.A., Besnard G., Samaritani E., Duvall M.R., Hodkinson T.R., Savolainen V., Salamin N. Oligocene CO_2 decline promoted C4 photosynthesis in grasses. Curr. Biol. 2008;18:37–43. doi:10.1016/j.cub.2007.11.058
6. Downes R.W. Differences in transpiration rates between tropical and temperate grasses under controlled conditions. Planta. 1969;88:261–273. doi:10.1007/BF00385069
7. Edwards E.J., Still C.J. Climate, phylogeny and the ecological distribution of C_4 grasses. Ecol. Lett. 2008;11:266–276. doi:10.1111/j.1461-0248.2007.01144.x
8. Ehleringer J.R., Sage R.F., Flanagan L.B., Pearcy R.W. Climate change and the evolution of C_4 photosynthesis. Trends Ecol. Evol. 1991;6:95–99. doi:10.1016/0169-5347(91)90183-X
9. Ehleringer J.R., Cerling T.E., Helliker B.R. C_4 photosynthesis, atmospheric CO_2, and climate. Oecologia. 1997;112:285–299. doi:10.1007/s004420050311
10. Freckleton R.P., Harvey P.H., Pagel M. Phylogenetic analysis and comparative data: a test and review of the evidence. Am. Nat. 2002;160:712–726. doi:10.1086/343873
11. Gifford R.M. A comparison of potential photosynthesis, productivity and yield of plant species with differing photosynthetic metabolism. Aust. J. Plant Physiol. 1974;1:107–117. doi:10.1071/pp9740107
12. Giussani L.M., Cota-Sanchez J.H., Zuloaga F.O., Kellogg E.A. A molecular phylogeny of the grass subfamily Panicoideae (Poaceae) shows multiple origins of C_4 photosynthesis. Am. J. Bot. 2001;88:1993–2012. doi:10.2307/3558427

13. Hartley W. The global distribution of tribes of the Gramineae in relation to historical and environmental factors. Aust. J. Agric. Res. 1950;1:355–373. doi:10.1071/AR9500355

14. Harvey P.H., Pagel M.D. Oxford University Press; Oxford, UK: 1991. The comparative method in evolutionary biology.

15. Hunter J.P. Key innovations and the ecology of macroevolution. Trends Ecol. Evol. 1998;13:31–36. doi:10.1016/S0169-5347(97)01273-1

16. Ibrahim D.G., Burke T., Ripley B.S., Osborne C.P. A molecular phylogeny of the genus Alloteropsis (Panicoideae, Poaceae) suggests an evolutionary reversion from C_4 to C3 photosynthesis. Ann. Bot. 2009;103:127–136. doi:10.1093/aob/mcn204

17. Kalapos T., van den Boogaard R., Lambers H. Effect of soil drying on growth, biomass allocation and leaf gas exchange of two annual grass species. Plant Soil. 1996;185:137–149. doi:10.1007/BF02257570

18. Keeley J.E., Rundel P.W. Evolution of CAM and C_4 carbon-concentrating mechanisms. Int. J. Plant Sci. 2003;164:S55–S77. doi:10.1086/374192

19. Knapp A.K. Post-burn differences in solar radiation, leaf temperature and water stress influencing production in a lowland tallgrass prairie. Am. J. Bot. 1984;71:220–227. doi:10.2307/2443749

20. Le Roux X., Bariac T. Seasonal variations in soil, grass and shrub water status in a West African humid savanna. Oecologia. 1998;113:456–466. doi:10.1007/s004420050398

21. Long S.P. Environmental responses. In: Sage R.F., Monson R.K., editors. C4 plant biology. Academic Press; San Diego, CA: 1999. pp. 215–283.

22. Loomis R.S., Williams W.A., Hall A.E. Agricultural productivity. Annu. Rev. Plant Physiol. 1971;22:431–468. doi:10.1146/annurev.pp.22.060171.002243

23. Ludwig F., de Kroon H., Prins H.H.T., Berendse F. Effects of nutrients and shade on tree-grass interactions in an East African savanna. J. Veg. Sci. 2001;12:579–588. doi:10.2307/3237009

24. Maddison W.P. Confounding asymmetries in evolutionary diversification and character change. Evolution. 2006;60:1743–1746. doi:10.1554/05-666.1

25. Meinzer F.C. Functional convergence in plant responses to the environment. Oecologia. 2003;134:1–11. doi:10.1007/s00442-002-1088-0

26. Osborne C.P. Atmosphere, ecology and evolution: what drove the Miocene expansion of C4 grasslands? J. Ecol. 2008;96:35–45.

27. Osmond C.B., Winter K., Ziegler H. Functional significance of different pathways of CO_2 fixation in photosynthesis. In: Lange O.L., Nobel P.S., Osmond

C.B., Ziegler H., editors. Encyclopedia of plant physiology, new series volume 12b physiological plant ecology II. Water relations and carbon assimilation. Springer-Verlag; Berlin, Germany: 1982. pp. 479–547.

28. Pagani M., Zachos J., Freeman K.H., Tipple B., Boharty S. Marked decline in atmospheric carbon dioxide concentrations during the Paleogene. Science. 2005;309:600–603. doi:10.1126/science.1110063

29. Pagel M. Detecting correlated evolution on phylogenies: a general method for the comparative analysis of discrete characters. Proc. R. Soc. B. 1994;255:37–45. doi:10.1098/rspb.1994.0006

30. Pagel M. Inferring the historical patterns of biological evolution. Nature. 1999;401:877–884. doi:10.1038/44766

31. Pagel M., Meade A. Bayesian analysis of correlated evolution of discrete characters by Reversible-Jump Markov Chain Monte Carlo. Am. Nat. 2006;167:808–825. doi:10.1086/503444

32. Paradis E., Claude J., Strimmer K. APE: analysis of phylogenetics and evolution in R language. Bioinformatics. 2004;20:289–290. doi:10.1093/bioinformatics/btg412

33. Paruelo J.M., Lauenroth W.K. Relative abundance of plant functional types in grasslands and shrublands of North America. Ecol. Appl. 1996;6:1212–1224. doi:10.2307/2269602

34. Roalson E.H. C4 photosynthesis: differentiating causation and coincidence. Curr. Biol. 2008;18:R167–R168. doi:10.1016/j.cub.2007.12.015

35. Sage R.F. Environmental and evolutionary preconditions for the origin and diversification of the C_4 photosynthetic syndrome. Plant Biol. 2001;3:202–213. doi:10.1055/s-2001-15206

36. Sage R.F. The evolution of C4 photosynthesis. New Phytol. 2004;161:341–370. doi:10.1111/j.1469-8137.2004.00974.x

37. Sage R.F., Li M., Monson R.K. The taxonomic distribution of C_4 photosynthesis. In: Sage R.F., Monson R.K., editors. C_4 plant biology. Academic Press; San Diego, CA: 1999a. pp. 551–584.

38. Sage R.F., Wedin D.A., Li M. The biogeography of C_4 photosynthesis: patterns and controlling factors. In: Sage R.F., Monson R.K., editors. C_4 plant biology. Academic Press; San Diego, CA: 1999b. pp. 313–373.

39. Sankaran M., Ratnam J., Hanan N. Woody cover in African savannas: the role of resources, fire and herbivory. Glob. Ecol. Biogeogr. 2008;17:236–245. doi:10.1111/j.1466-8238.2007.00360.x

40. Taub D.R. Climate and the U.S. distribution of C_4 grass subfamilies and decarboxylation variants of C_4 photosynthesis. Am. J. Bot. 2000;87:1211–1215. doi:10.2307/2656659

41. Vicentini A., Barber J.C., Aliscioni A.S., Giussani A.M., Kellogg E.A. The age of the grasses and clusters of origins of C_4 photosynthesis. Glob. Chang. Biol. 2008;14:2963–2977. doi:10.1111/j.1365-2486.2008.01688.x

42. Watson, L. & Dallwitz, M. J. 1992 onwards. The grass genera of the world: descriptions, illustrations, identification, and information retrieval; including synonyms, morphology, anatomy, physiology, phytochemistry, cytology, classification, pathogens, world and local distribution, and references. Version: 28th November 2005. http://delta-intkey.com.

43. Williams K.J., Wilsey B.J., McNaughton S.J., Banyikwa F.F. Temporally variable rainfall does not limit yields of Serengeti grasses. Oikos. 1998;81:463–470. doi:10.2307/3546768

Modeling the Fitness Consequences of a Cyanophage-Encoded Photosynthesis Gene

Jason G. Bragg and Sallie W. Chisholm

ABSTRACT

Background

Phages infecting marine picocyanobacteria often carry a psbA gene, which encodes a homolog to the photosynthetic reaction center protein, D1. Host encoded D1 decays during phage infection in the light. Phage encoded D1 may help to maintain photosynthesis during the lytic cycle, which in turn could bolster the production of deoxynucleoside triphosphates (dNTPs) for phage genome replication.

Methodology / Principal Findings

To explore the consequences to a phage of encoding and expressing psbA, we derive a simple model of infection for a cyanophage/host pair — cyanophage

P-SSP7 and Prochlorococcus MED4—for which pertinent laboratory data are available. We first use the model to describe phage genome replication and the kinetics of psbA expression by host and phage. We then examine the contribution of phage psbA expression to phage genome replication under constant low irradiance (25 μE m^{-2} s^{-1}). We predict that while phage psbA expression could lead to an increase in the number of phage genomes produced during a lytic cycle of between 2.5 and 4.5% (depending on parameter values), this advantage can be nearly negated by the cost of psbA in elongating the phage genome. Under higher irradiance conditions that promote D1 degradation, however, phage psbA confers a greater advantage to phage genome replication.

Conclusions / Significance

These analyses illustrate how psbA may benefit phage in the dynamic ocean surface mixed layer.

Introduction

The marine picocyanobacteria Prochlorococcus and Synechococcus are numerically dominant phytoplankton in nutrient-poor open ocean ecosystems, and are an important contributor to photosynthesis in the oceans [1–3]. They are infected by cyanophages including members of the families Podoviridae, Myoviridae and Siphoviridae [4], which can be abundant in regions where these cells dominate (e.g. [5–8]). Several genomes of these marine cyanophages have been sequenced, revealing gene content and organization broadly similar to confamilial phages [9–11]. For example, cyanophage P-SSP7, which infects Prochlorococcus MED4, has many genomic similarities to the T7 phage that infects Escherichia coli [11].

The genomes of marine Synechococcus and Prochlorococcus cyanophages often contain genes that are absent from the genomes of morphologically related phages that do not infect marine cyanobacteria [11]. A striking example of this is the psbA photosynthesis gene [12], [13]. This gene is found in the genomes of a large proportion of cyanophages known to infect marine picocyanobacteria [14], [15], suggesting that it confers a fitness advantage. The product of the psbA gene in the host cell, the D1 protein, forms part of the photosystem II reaction center, and turns over relatively rapidly during photosynthesis [16]. Over the course of phage infection, host-encoded D1 proteins decline following the inhibition of host transcription and the decay of host psbA transcripts [17], while phage-encoded D1 proteins increase [17]. It is hypothesized that the latter replace damaged host D1 proteins, and help to maintain photosynthesis throughout the lytic cycle. This, in turn, could increase the relative fitness of phage that carry the psbA gene [12], [18]. In some cyanophages, reproduction (e.g. [19]) and genome replication [17] are severely limited in the dark, indicating that photosynthesis can be important for phage genome replication, which potentially limits the production of phage progeny. During the cyanophage P-SSP7 lytic cycle, psbA is transcribed contemporaneously with several metabolism genes that have probable roles in dNTP synthesis (e.g. ribonucleotide reductase), as well as genome replication enzymes [20]. This adds weight to the suggestion that the psbA gene helps phage P-SSP7 to acquire resources to make dNTPs during infection.

Models of phage infection in well established phage/host systems, such as T7/E. coli [21], have provided significant insights into factors affecting phage reproduction and fitness [22]–[24]. Inspired by these works, we have developed an intracellular model of infection of Prochlorococcus MED4 by the Podovirus P-SSP7. The model concentrates on processes of phage genome replication, the production of dNTPs, and the expression of psbA by host and phage, and can find good agreement with experimental measurements collected over the cyanophage P-SSP7 lytic cycle [17], [20]. We use the model to ask basic questions about the advantages to the phage of carrying this gene that are not yet tractable experimentally: How much can phage psbA expression benefit phage genome replication? To what degree is this contingent on environmental conditions, particularly the ambient light environment?

Methods

Model Development

Approach

After the genome of cyanophage P-SSP7 enters a host cell, phage genes are expressed, the phage genome is replicated, new phage particles are assembled, and the host cell is lysed — all over a period of about 8 hours [20]. Many of these processes are carried out using products of phage-encoded genes, which are expressed at different times during the cycle of infection [20]. Our model links phage genome replication to the production of deoxynucleoside triphosphates (dNTPs), which in turn is linked to photosynthesis and the kinetics of host and phage psbA expression (Fig. 1). It incorporates elements of previous models of phage genome replication [21], [25], and D1 protein kinetics [26]. More specifically, we model phage genome replication within a host cell as a function of the availability of dNTPs, which are supplied by (i) scavenging from the degraded host genome, and (ii) a pathway for the synthesis of new deoxynucleotides. We assume that the supply of dNTPs from each of these sources can depend on photosynthesis. In turn, photosynthesis is modeled as a function of the number of functional photosystem II (PSII) subunits, which become non-functional when their D1 core proteins are damaged, and regain their function when the damaged D1 is excised from the photosystem and replaced with the protein product of either a host or phage psbA gene (Fig. 1).

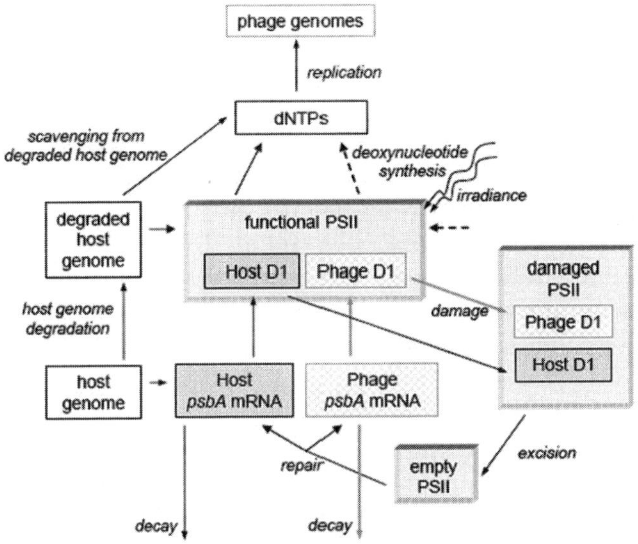

Figure 1. Schematic diagram of model. Phage genomes are made using dNTPs from two possible sources. First, dNTPs can be made by scavenging deoxynucleotides from the host genome. This process can occur in the dark, but is bolstered by photosynthesis. Second, dNTPs can be newly synthesized by a process that is dependent on the products of photosynthesis (dashed lines). Photosynthesis is dependent on functional PSII subunits, which contain the D1 protein. During exposure to light, D1 proteins can become damaged, and are excised from PSII subunits, and replaced with D1 proteins from either host or phage encoded psbA mRNAs.

Our model incorporates processes that are carried out by phage genes that begin to be expressed at different times following infection [20]. We therefore need a way to represent relatively abrupt increases in the velocity of processes that are carried out by different proteins (generically, P), at different, specific times following infection. We do this using Hill functions [27], $P_x = \left| \dfrac{t^n}{t^n + t_x^n} \right.$ or , where t is the time since infection, tx is the time at which process Px reaches half its maximum rate (the 'timing parameter'), and n is a parameter that controls the abruptness of the increase. The Hill function is a sigmoidal curve that increases from zero to one with increasing values of t, and can provide a reasonable description of the expression of some relevant phage genes, at least in terms of mRNA abundances. Below, we represent Hill functions in our equations using 'P' followed by a subscript that corresponds to the process that is being represented.

Phage Genome Replication

We assume that protein products of phage genes, such as DNA polymerase, are essential for phage genome replication. Following the expression of these genes, genome replication occurs as a function of the availability of dNTPs, N (Fig. 1). We model the change in phage genomes in a host cell (G_P) over time (following [21], [25]), as

$$\frac{d(G_P)}{dt} = \frac{\mathbf{1}}{L_P} \frac{V_r N}{L_P N + + K_{mr}} P_r. \tag{1}$$

Here LP is the length of the phage genome (in base pairs). The term Pr is a Hill function that represents the time-dependent expression of phage genome replication genes. Vr represents the maximum rate of DNA elongation per functional unit of phage genome replication machinery (hereafter, polymerase), multiplied by the maximum abundance of polymerases. Kmr is the value of N at which elongation by a polymerase reaches half its maximum rate. We set GP(0) = 1, to reflect infection by a single virion.

Phage Acquisition of dNTPs

We assume that cyanophage P-SSP7 can acquire dNTPs from two possible sources during infection of Prochlorococcus MED4 (Fig. 1). The first is scavenging from the host genome, which is degraded during infection [20]. The second involves the synthesis of new deoxynucleotides [11].

We model degradation of the host's genome (GH) as

$$\frac{d(G_H)}{dt} = -\frac{V_{G\mathbf{deg}} G_H}{G_H + K_{mGH}} P_{G\mathbf{deg}}, \tag{2}$$

where P_{Gdeg} is a Hill function representing the expression of genes that degrade the host genome (with time parameter t_{Gdeg}), and V_{Gdeg} is the maximum rate of degradation of the host genome. The term $\frac{V_{G\deg} G_H}{G_H + K_{mGH}} P_{G\deg}$, with small K_{mGH}, is approximately equal to V_{Gdeg} until the host genome is almost entirely degraded. This formulation is based on the observation that the decline in host genomes is approximately linear, following a delay of 4–5 h [20], as well as the necessity for degradation to cease as G_H approaches 0. We set $G_H(0) = 1$ to reflect a single host genome at the time of infection.

We assume the phage can then make dNTPs from the degraded host genome (GHdeg). We model the rate of production of dNTPs from this source (sG; dNTPs cell−1 h−1) as . Here 2LH is the number of deoxynucleotides in the host genome (2 per base pair, times LH base pairs per genome), and VN is the maximum rate at which the degraded genome can be converted to dNTPs. We set the parameter KN to a small value so that when genes for degrading the host genome are expressed and degraded host genome is available, the term $\frac{G_{H\deg}}{G_{H\deg}+K_N}P_{G\deg}$ is approximately equal to 1, and dNTPs are produced from degraded genomes at a rate of approximately 2LHVN.

We then consider the possibility that the production of dNTPs from degraded genomes (2LHVN) is limited by photosynthesis. We represent this in the model by letting 2LHVN = ε+μ, where ε represents the rate at which dNTPs can be made from degraded genomes during infection in the dark, and μ represents the photosynthesis-dependent production of dNTPs from host genomes. In turn, we assume that μ is limited by the abundance of functional photosystem II subunits or μ = zγFPSII, where FPSII is the number of functional photosystem II subunits per cell, γ is the rate of photosynthesis per functional PSII, and z is the efficiency with which products of photosynthesis are used in converting degraded genomes to dNTPs. Below, we develop a model for the proportion of PSII subunits that are functional (fPSII) during infection. We therefore let FPSII = UfPSII, where U is the total number of PSII subunits per cell. This means we have μ = zγUfPSII, or if we represent zγU by the parameter κ (in dNTPs cell−1 h−1), μ = κfPSII.

We note that the change over time in the proportion of the host genome in a degraded state is then given by

$$\frac{d\left(G_{H\deg}\right)}{dt}=\frac{V_{G\deg}G_H}{G_H+D_{mGH}}P_{G\deg}-\frac{V_N G_{H\deg}}{G_{H\deg}+K_N}P_{G\deg}. \tag{3}$$

We now consider the possibility that new deoxynucleotides (i.e., not from the host genome) are produced during infection as a source of dNTPs for phage genome replication (sP; dNTPs cell−1 h−1). This possibility is suggested by the observation that cyanophage P-SSP7 encodes [11] and transcribes [20] a ribonucleotide reductase gene, whose protein product likely functions in converting ribonucleotides to deoxynucleotides. We assume this source of dNTPs is dependent on photosynthesis, as well as the activity of genes that are encoded by the phage, and whose expression is described by a Hill function, PS. Once these phage genes are expressed, the rate of supply of dNTPs from this source is assumed to be proportional to the rate of photosynthesis, which is limited by the abundance of functional PSII subunits, or sP = vγFPSIIPS, where v is the efficiency with which products of photosynthesis are converted to dNTPs. We then represent vγU using a single parameter, λ (in dNTPs cell−1 h−1), such that sP = λfPSIIPS. Presently we lack detailed mechanistic information about this potential extra source of dNTPs, which would be useful for refining the model. For example, if photosynthesis powers the conversion of a finite cellular resource to dNTPs, the depletion of this resource ought to be modeled.

After accounting for the incorporation of free dNTPs into genomes, we have a rate equation for dNTPs per cell (N):

$$\frac{d(N)}{dt} = s_G + s_P - 2\frac{V_r N}{N + K_{mr}} P_r . \qquad (4)$$

We next model the proportion of PSII subunits that are functional (fPSII), to insert in both sG and sP. Functional PSII subunits are lost when their D1 proteins become damaged. Following excision of the damaged D1 protein, PSII subunits become functional upon receiving a new D1 protein. Our approach is similar to that of [26], in modeling the proportions of PSII subunits that (i) are functional (fPSII), (ii) contain damaged D1 proteins (dPSII), and (iii) have had damaged D1 proteins excised ('empty' PSII subunits, or xPSII) (Fig. 1). For functional and damaged subunits, we track PSII subunits containing host- versus phage-encoded D1 proteins separately. For example, for functional PSII subunits, fPSII = fPSIIH+fPSIIP, where fPSIIH and fPSIIP contain host D1 and phage D1, respectively. We also assume that during the course of infection, the total number of PSII subunits (U) in a cell is constant, and that fPSIIH+fPSIIP+dPSIIH+dPSIIP+xPSII = 1. This yields the following system of equations:

$$\frac{d(f_{PSIIH})}{dt} = -k_{D1dam} f_{PSIIH} + k_{\tau D1} R_{HpsbA} x_{PSII} \qquad (5)$$

$$\frac{d(d_{PSIIH})}{dt} = k_{D1dam} f_{PSIIH} - k_{exc} d_{PSIIH} \qquad (6)$$

$$\frac{d(f_{PSIIH})}{dt} = -k_{D1dam} f_{PSIIH} + k_{\tau D1} R_{PpsbA} x_{PSII} \qquad (7)$$

$$\frac{d(f_{PSIIH})}{dt} = k_{D1dam} f_{PSIIH} - k_{exc} d_{PSIIP} \qquad (8)$$

where $x_{PSII} = 1-(f_{PSIIH}+f_{PSIIP}+d_{PSIIH}+d_{PSIIP})$. Here k_{D1dam} is the rate at which D1 proteins in functional PSII subunits are damaged by irradiance, k_{exc} is the rate at which damaged D1 proteins are excised from PSII subunits, and $k_{\tau D1}$ is the rate at which damaged PSII subunits are repaired using psbA mRNA transcripts. R_{HpsbA} and R_{PpsbA} are the abundances of host and phage psbA transcripts, respectively. This formulation assumes that D1 proteins are represented only in functional and damaged PSII subunits, and that psbA transcripts are limiting to repair.

The expression of host and phage psbA transcripts are modeled as follows:

$$\frac{d\left(R_{HpsbA}\right)}{dt} = k_{HpsbA}G_H\left(1-P_{Rpol}\right) - d_{RpsbA}R_{HpsbA} \tag{9}$$

$$\frac{d\left(R_{PpsbA}\right)}{dt} = k_{PpsbA}P_{PpsbA} - d_{RpsbA}R_{PpsbA}. \tag{10}$$

Here dRpsbA is the decay rate of psbA mRNA transcripts. kHpsbA and kPpsbA are the maximum rates of transcription of host and phage psbA mRNAs, respectively. Host psbA is transcribed until either the host genome is gone, or until host RNA polymerase is inhibited. The inhibition of host RNA polymerase by a phage protein is represented using the term (1-PRpol), where PRpol is a Hill function. PPpsbA is a Hill function representing the commencement of transcription of phage psbA at a time of approximately tPpsbA.

The above formulation includes assumptions that can be tested experimentally, and improved in future versions of the model. We assume, for example, that host and phage psbA transcripts have identical rates of decay (dRpsbA). We also assume that empty PSII subunits can be repaired at identical rates using products of host and phage psbA genes, that PSII subunits containing host and phage D1 proteins have similar rates of damage (kD1dam) and excision (kexc), and that functional PSII subunits containing host and phage D1 have similar rates of photosynthesis. In reality, these properties of host and phage psbA transcripts or D1 proteins could be different. For example, it has been suggested that phage D1 might be more resistant to photodamage than host D1 [28]. Furthermore, we assume that the total number of photosystem II subunits and the maximum rates of excision and repair are constant over the course of infection, while in reality, these values may decay as a function of time. It would be useful to measure these properties of infected cells experimentally, and revise the model if necessary. More broadly, our model clearly uses a highly simplified representation of photosynthesis, an extremely complex process influenced by a large number of factors [29]. Our goal was to abstract this complexity with a focus on the potential advantage to phage of supplementing the supply of D1 during infection.

Results

Model Validation

Approach

The parameterization of the model is described in detail below, and parameter values are listed in Table 1. Our general approach was as follows: We began by considering parameters that govern the abundance of host and phage psbA transcripts, and then estimated parameters for the abundance of host and phage D1 proteins. In the experiments that are the basis for the model, cells were grown under continuous light [17]. We therefore assumed the abundances of host psbA mRNAs and the proportions of functional, damaged and empty PSII subunits were in steady state prior to infection, and set equations (5), (6) and (9) equal to 0. This imposed relationships between parameters and initial conditions of some variables (Table 1), reducing the number of free parameters. Finally, we used data for genome replication in the light and dark [17] to estimate parameters for the dependence of dNTP acquisition on photosynthesis. Data from Lindell et al. [20] were used to estimate parameters for the degradation of host genomes and for the timing of expression of phage genes involved in genome replication and dNTP production.

Lindell et al. [17], [20] studied populations of cells that were infected with phage, while our model is based on infection of a single host cell. In comparing model predictions to these experimental data, we assume that our model represents infection of an average cell. To estimate the number of phage genomes per host cell at different times after infection, we normalized by the number of phages measured at 1 h post-infection. Given that our estimates of phage genome replication depend on this normalization, we place our emphasis on the proportional advantage or disadvantage conferred by phage psbA, rather than the absolute number of genomes. Lindell et al. [17] used a low multiplicity of infection (0.1 phage for every host cell) for the experiment in which phage genome replication was measured. Under these conditions, most infected cells would have been infected by a single virion. When using data for intracellular levels of host psbA transcripts, D1 proteins, and genomes, we assumed that 50% of cells were infected. A higher multiplicity of infection (3 phage per host cell) was used in the experiments from which these data were collected, and 50% represents the maximum level of infection that has been observed for this phage [17]. We also assumed that measurements of D1 protein abundance made by [17] detected both functional and damaged D1 proteins.

We integrated equations (1)–(10) using ode45, a MATLAB® (The MathWorks, Natick, MA) variable time step numerical ODE solver, which implements a medium order Runge-Kutta scheme.

Table 1. Model parameters and initial conditions.

Parameter	Description	Units	Value
G_P	phage genomes	genomes cell^{-1}	1^{*}
G_H	host genomes	genomes cell^{-1}	1^{*}
G_{Hdeg}	degraded host genomes	genomes cell^{-1}	0
N	dNTPs	dNTPs cell^{-1}	0
f_{PSIIH}, f_{PSIIP}	proportion of PSII subunits that are functional and contain host, phage D1	dimensionless	$\frac{1-x_{PSII}}{1+\frac{k_{D1dam}}{k_{exc}}}=0.46$, 0^{*}
d_{PSIIH}, d_{PSIIP}	proportion of PSII subunits that are damaged and contain host, phage D1	dimensionless	$\frac{1-x_{PSII}}{1+\frac{k_{exc}}{k_{D1dam}}}=0.04$, 0^{*}
x_{PSII}	proportion of PSII subunits that are empty	dimensionless	$0.5^{*,a}$
R_{HpsbA}, R_{PpsbA}	*psbA* transcripts	dimensionless	1, 0^{*}
L_H, L_P	genome length of host, phage	bp genome^{-1}	1657990, 44970
V_r	max velocity of phage DNA elongation	bp h^{-1} cell^{-1}	1332000
K_{mr}	half-saturation for DNA replication	dNTP cell^{-1}	1224
t_r	timing parameter for phage genome replication	h	2
V_{Gdeg}	max velocity of host genome degradation	genomes cell^{-1} h^{-1}	0.35
K_{mGH}	half-saturation for host genome degradation	genomes cell^{-1}	0.000001
t_{Gdeg}	timing parameter for host genome degradation	h	5
ε	production of dNTPs from degraded host genome in the dark	dNTP h^{-1} cell^{-1}	127665
κ	production of dNTPs from degraded host genome in the light	dNTP h^{-1} cell^{-1}	0
K_N	half saturation for dNTP production from degraded host genome	genomes cell^{-1}	0.000001
λ	production of dNTPs in the light	dNTP h^{-1} cell^{-1}	1027800
t_S	timing of dNTP synthesis from source s_P	h	4
k_{D1dam}	damage to functional D1 proteins	h^{-1}	0.35^{a}
k_{exc}	excision of damaged D1 proteins	h^{-1}	4^{a}
k_{rD1}	repair of empty PSII subunits	h^{-1}	$\frac{k_{exc}d_{PSIIH}}{R_{HpsbA}x_{PSII}}=0.32$
d_{RpsbA}	*psbA* transcript decay	h^{-1}	0.27^{b}
k_{HpsbA}	host *psbA* transcription	h^{-1} (genomes cell^{-1})$^{-1}$	0.27^{b}
k_{PpsbA}	phage *psbA* transcription	h^{-1}	0.016
t_{Rpol}	timing parameter for inhibition of host RNA polymerase	h	1
t_{PpsbA}	timing parameter for transcription of phage *psbA*	h	1.3
n	Hill parameter	dimensionless	5

*Initial condition.
[a]Values were systematically varied in exploring the kinetics of D1 protein degradation, excision and repair. All combinations of the following values were used:
k_{D1dam} = [0.01, 0.025, 0.05, 0.06, 0.07, 0.08, 0.09, 0.1, 0.125, 0.15, 0.175, 0.2, 0.25, 0.3, 0.35, 0.4, 0.5].
k_{exc} = [0.5, 0.75, 1, 1.25, 1.5, 1.75, 2, 2.5, 3, 3.5, 4, 4.5, 5, 7.5, 10].
$x_{PSII}(0)$ = [0.05, 0.1, 0.15, 0.2, 0.25, 0.3, 0.4, 0.5, 0.6, 0.7, 0.8, 0.9].
[b]This estimate is based on microarray measurements of host *psbA* mRNA expression. Measurements of host *psbA* transcript abundances that were made using RT-PCR [17] suggested a greater value of d_{RpsbA}. We therefore present additional analyses based on a value of d_{RpsbA} = 0.72 in Text S1.

Expression of Photosynthesis Genes

Experimental evidence shows that following infection by the cyanophage P-SSP7, the abundance of psbA transcripts in the host cell declines [17]. We assumed that host transcription was largely inhibited ($1-P_{Rpol}\approx 0$) by 1 hour after infection (Fig. 2A, Table 1), and calculated the decay constant (d_{RpsbA}) using experimental observations [17]. We set the initial value of host psbA mRNA to $R_{HpsbA}(0) = 1$, and normalized the abundance of host psbA transcripts to this initial (maximum) value.

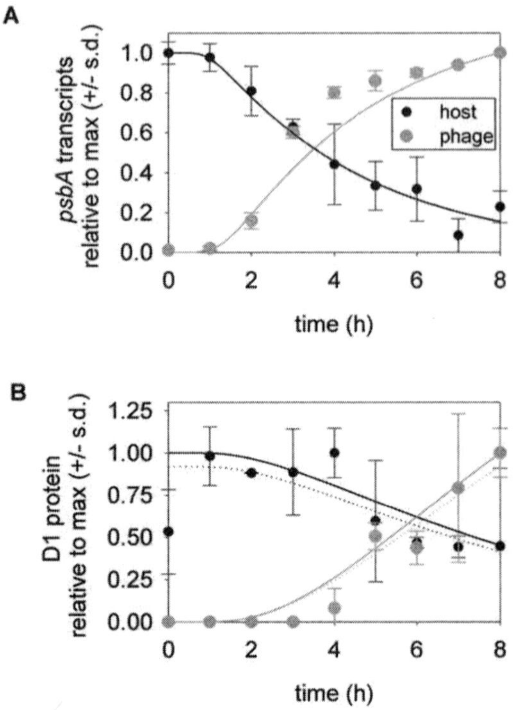

Figure 2. Measured (data points) and modeled (lines) levels of host and phage psbA transcripts (A) and D1 protein product (B) during the lytic cycle of infection of Prochlorococcus MED4 by cyanophage P-SSP7. For modeled levels of D1 protein, solid lines represent the sum of functional and damaged D1, and dotted lines represent functional D1 only. Data are from [17]. Data for host expression levels were transformed assuming that 50% of cells were infected [17].

Following the decay of host psbA transcripts, Lindell et al. [17] observed a drop in the level of host D1 proteins, such that host D1 abundance had decreased to approximately 45% of its maximum value (measured 1 hour after infection) after 8 hours of infection (Fig. 2B). In parameterizing the dynamics of host D1 proteins, we first assumed cells were in steady state prior to infection, which constrained parameters according to RHpsbA(0)kτD1xPSII(0) = kexcdPSIIH(0) = kD1damfPSIIH(0). This means only xPSII(0) (and correspondingly, kτD1; see Table 1) was free to vary for given pair of kexc and kD1dam values (since RHpsbA(0) = 1, and xPSII(0)+dPSIIH(0)+fPSIIH(0) = 1). We did not have independent estimates of the parameters kD1dam, kexc and xPSII(0) for Prochlorococcus under the conditions of the experiment [17], and values of kD1dam and kexc may vary substantially among organisms and growth conditions [26]. We therefore used measurements from a study of Prochlorococcus PSII function and D1 protein abundance under transient exposure to high irradiance [30] to estimate possible ranges of parameters kD1dam, kexc and xPSII(0). We then solved our model of D1 dynamics 3060 times, comprising all combinations of 17 values of kD1dam, 15 values of kexc, and 12 values of xPSII(0) (see Table 1). Out of these 3060 simulations, 126 resulted in a drop in the abundance of host D1 proteins after 8 hours of infection that was similar to the value measured in the laboratory [17]. From here onward, we present analyses that focus on one set of parameters (kD1dam = 0.35 and kexc = 4, with xPSII(0) = 0.5), but we did perform all subsequent analyses using all 126 combinations of parameters, to confirm that our conclusions are robust across this range of parameter values. The model can provide a reasonable description of the drop in host D1 proteins during infection, as illustrated in Fig. 2B (black line and symbols). However, we note that the model does not predict several features of the experimental observations, and in particular, the low level of host D1 at 0 hours, and the sudden drop in host D1 between 4 hours and 5 hours after infection. We are not aware of any mechanisms that might account for these observations, so have not attempted to replicate them with the present model.

We next modeled the abundance of phage psbA transcripts using the decay constant (dRpsbA) calculated above for host psbA transcripts. Our model describes the shape of the experimentally derived curve of phage psbA mRNA abundance reasonably well (Fig. 2A), though modeled levels of mRNAs approached an asymptotic level more slowly than observed [17].

The empirical observation that phage D1 proteins accumulated to approximately 10% of all D1 after 8 hours of infection [17] was predicted by the model when phage psbA transcription was set to 5.9% of the rate at which host psbA was transcribed prior to infection (kPpsbA = 0.016; Fig. 2B), and with the same values of kD1dam and kexc that were used for host D1. The model predicted the increase in the level of phage D1 slightly sooner than it was observed experimentally. This could be due to a time delay for the translation of D1, or may simply reflect experimental variability.

Degradation of the Host Genome

Experimental evidence showed that host genomes are mostly degraded between 4 and 8 hours after infection [20] and the loss of host genomes was approximately linear. The model provides a good description of these observations with $t_{Gdeg} = 5$ and $V_{Gdeg} = 0.35$, and with K_{mGH} set to a small value (0.000001) (data not shown).

Genome Replication

To study phage genome replication in the model, we first simulate infection in the dark, setting the photosynthesis-dependent production of dNTPs equal to zero (setting $\lambda = 0$ and $\kappa = 0$). We then needed to estimate values for DNA replication kinetic parameters (V_r and K_{mr}), the timing of phage DNA replication machinery (t_r) and the production of dNTPs using degraded host genomes in the absence of photosynthesis (ε). Phage T7 has a rate of DNA elongation of approximately 1,332,000 (h^{-1} polymerase^{-1}) [21], [31]. In the absence of data for cyanophage P-SSP7, we set V_r = 1,332,000 (bp h^{-1} cell^{-1}). We did not multiply this value by the number of phage polymerases in the host cells since (i) we do not have data on phage polymerase abundance, and (ii) this value of V_r is already sufficiently large to be non-limiting to genome replication (see below). We also estimated K_{mr} based on the corresponding value for deoxynucleotide incorporation by T7 phage enzymes [21], [32], adjusted according to the size of a Prochlorococcus MED4 cell, which is assumed to be a sphere with diameter 0.6 µm. We assumed that phage genome replication enzymes were expressed approximately 2 hours post-infection (t_r = 2), based on observations of phage DNA polymerase transcript abundance in [20]. We then found that ε = 127,665 could provide a reasonable description of genome replication in the dark, if all dNTPs used in phage genome replication in the dark were derived from the host genome (Fig. 3).

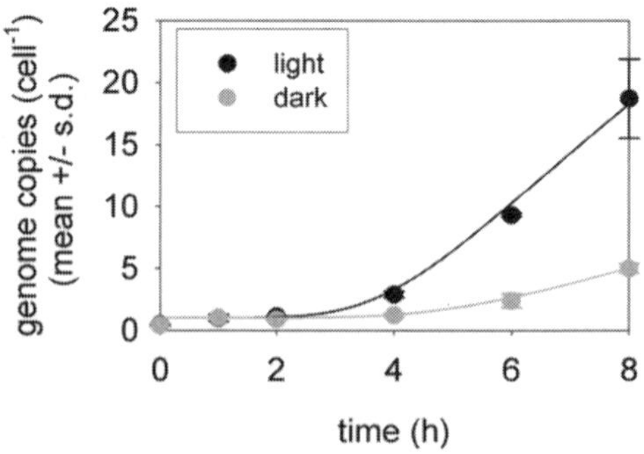

Figure 3. Measured (data points) and modeled (lines) genome copies of cyanophage P-SSP7 during the lytic cycle under light (25 µE m^{-2} s^{-1}) and dark conditions. Genome copies were measured as genomes per ml of culture [17], and were transformed to a per cell basis for comparison to the model.

Our model includes the possibility that photosynthesis increases the production of dNTPs from degraded host genomes, and the possibility that photosynthesis promotes the synthesis of new deoxynucleotides. However, since we do not know the relative importance of these possible sources of dNTPs, we analyze their potential contribution to dNTP production during infection in the light separately. Here we present analyses that assume extra dNTPs made in the light were derived from the synthesis of new deoxynucleotides (i.e., $\lambda>0$ and $\mu = 0$). However, we confirmed that similar results are obtained if we assume instead that the extra dNTPs made in the light were derived from the degraded host genome.

For infection in the light, we use the same values of parameters Vr, Kmr, tr and ε, and found that λ = 1,027,800 (dNTP h−1 cell−1) gave a reasonable description of phage genome replication (Fig. 3). With this parameterization, dNTP availability is strongly limiting to genome replication: a 10% increase in dNTP production by photosynthesis (λ) results in a 7.3% increase in genome replication, whereas a 10% increase in the maximum velocity of genome replication (Vr) results in almost no increase in genome replication.

In Silico Knockout of Phage psbA

The major goal of this study is to consider the fitness consequences to a phage of encoding and expressing the psbA gene. Having described the kinetics of infection reasonably well with our model (Figs 2 and 3), we can now turn off transcription of phage psbA ($k_{PpsbA} = 0$), and study how this affects the predicted number of phage genomes in infected cells after 8 h of infection. Using the parameter values presented in Table 1, we predict that a phage unable to express psbA would produce 2.81% fewer genomes after 8 h of infection. However, if a phage did not encode psbA, its genome would be shorter, by approximately 1080 bp. Taking this into account, a phage that did not encode psbA would produce only 0.55% fewer genomes after 8 h of infection than a phage that encodes and expresses psbA. In the dark, where there is presumably no advantage to expressing psbA, a phage without this gene is predicted to produce 1.97% more genomes than a phage with it.

Ideally, we would like to consider the consequences of psbA to phage genome replication under different and changing levels of irradiance. However, many of the parameters used in our model are likely to change as a function of irradiance, in ways that can be difficult to predict (see [29]). Therefore we limit ourselves to one specific case, where cells are moved from 25 μE m−2 s−1 to 50 μE m−2 s−1 one hour after infection has begun, when the capacity of the cells to respond to the changing light may be largely compromised by infection. This means we can use the same initial conditions and parameter values as in our previous simulations (Table 1), except for two parameters that will be affected directly by the increased irradiance (kD1dam and λ). We assume that the rate of damage to functional PSII subunits (kD1dam) increases proportionally with irradiance (i.e., kD1dam is doubled; [26]), and that the rate of photosynthesis of Prochlorococcus MED4 is greater at 50 μE m−2 s−1 than at 25 μE m−2 s−1 by a factor of approximately 1.75 [33], [34].

We found that in the case where irradiance increases from 25 μE m−2 s−1 to 50 μE m−2 s−1 one hour after infection, a phage that does not express psbA is predicted to produce 4.31% fewer genomes than a phage that does. Here, a phage that does not express or encode psbA is predicted to produce 2.10% fewer genomes than a phage that does encode and express psbA. We therefore predict that psbA will have a greater impact on phage genome replication under this switch to a higher level of irradiance, such as could occur in the surface mixed layer of the oceans. We note that this prediction also holds if photosynthesis (and λ) increases by a factor of either 1.5 or 2 under the switch to higher irradiance, rather than by a factor of 1.75. Further, we performed analyses similar to the above using a range of different values of kD1dam, kexc and xPSII(0). Across these simulations, expressing psbA usually led to a modest increase in genome replication in continuous light (of between 2.5 and 4.5%), though this increase was typically smaller (between 0.3 and 2.3%) when the cost of encoding psbA was considered. The predicted advantage of expressing and encoding psbA was typically greater when a switch to higher light was simulated during infection, though the precise size of the advantage conferred by psbA varied.

Cyanophage dNTP Diets

In addition to psbA, marine cyanophages encode a variety of genes that potentially help them acquire dNTPs (e.g., ribonucleotide reductase, transaldolase; see [35]). It is therefore interesting to ask more broadly: Under what set of circumstances can a phage increase its total genome replication by encoding an additional gene or module of genes that help it acquire extra dNTPs?

Consider a phage that encodes genes that allow it to access dNTPs from a single source, s1 (e.g. scavenging from the host genome). Now, a mutant acquires an extra gene or module of genes that allow it to access an extra source of dNTPs, s2. If genes needed to access this second source of dNTPs elongate the wild type genome, we want to know when the mutant will make more genomes than the wild type by some time post-infection, or when

GM(t)>GW(t)(10)

where $G_M(t)$ and $G_W(t)$ are the numbers of genomes in cells infected by mutant and wild type phages (respectively) at time t.

We explore this question using our original model as a starting point, but with modifications that allow it to be studied analytically. We assume that the supply of dNTPs is highly limiting to genome replication (N<<Kmr), such that we can rewrite equation (1) as $\frac{d(G_p)}{dt} \approx \frac{1}{L_p} V_R N$, where VR is the rate of DNA elongation (here in bp dNTP−1 h−1 cell−1). We also assume that the dNTP revenue from each phage-encoded source can be expressed as a function of time, such that a phage

using s1 and s2 has $\frac{d(N)}{dt} = s_1(t) + s_2(t) - 2V_R N$. In reality, processes of phage-encoded dNTP acquisition (s1 and s2) and genome replication may begin at different times post-infection. For simplicity, we assume these processes all begin at the same time (tb hours after infection), and let time t = 0 in this model refer to this time tb when these processes begin. Solving for N(t), and then GP(t) yields

$$G_P(t) = 1 + \frac{1}{2L_P}\left[N_b\left(1-\exp(-2V_R t)\right) + s_1(t) * \left(1-\exp(-2V_R t)\right) + s_2(t) * \left(1-\exp(-2V_R t)\right)\right] \quad (11)$$

where N_b is the number of dNTPs in the cell at time t_b, there is one phage genome in the cell at time t_b, and '*' represents a convolution product. If we assume N_b is very small ($N_b \approx 0$) and sub (11) into (10), we get

$$\frac{1}{2(L_R + L_1 + L_2)}[\Psi_1 + \Psi_2] > \frac{1}{2(L_R + L_1)}[\Psi_1] \quad (12)$$

where Ψ_1 and Ψ_2 represent dNTPs derived from s_1 and s_2 (respectively), L_1 and L_2 represent the length of genes needed to encode s1 and s2 (respectively) and LR represents the length of the rest of the genome. This can be expressed as $\frac{\Psi_2}{\Psi_1} > \frac{L_2}{(L_R + L_1)}$, meaning that a new module of genes will increase phage genome replication if it leads to a proportional increase in dNTP production that is greater than the proportional increase in genome length that it causes. Alternatively, we could say that for a new module of genes to increase phage genome replication, it must have a ratio of dNTPs contributed / cost in genome length, $\frac{\Psi_2}{L_2}$, that exceeds a threshold, $\frac{\Psi_1}{(L_R + L_1)}$.

While this model is oversimplified, and has required assumptions that limit its applicability, it nevertheless may help us to understand some of the variability among cyanophages in methods they use to acquire dNTPs. For example, it suggests that if two similar phages acquire dNTPs by scavenging from the genomes of their hosts (i.e., their s1), but one phage infects a host with a smaller genome from which fewer dNTPs can be produced (smaller Ψ1), this phage might be more likely to exploit an additional source of dNTPs, if given the opportunity. This may be one factor that helps to explain why cyanophages infecting Prochlorococcus, which has a very small genome, might encode genes that help acquire dNTPs from other sources (see [36] for discussion of related issues).

Further, it can be shown that if the new source of dNTPs, s2, is highly profitable, the phage may no longer be advantaged by encoding s1. We would expect this to be the case when $\frac{1}{2(L_R + L_2)}[\Psi_2] > \frac{1}{2(L_R + L_1 + L_2)}[\Psi_2 + \Psi_2]$, or when $\frac{\Psi_2}{(L_R + L_2)} > \frac{\Psi_1}{L_1}$. This illustrates a way in which one source of dNTPs could replace another in the genome of a phage, over evolutionary time.

This analysis has strong parallels with diet theory models that predict when a foraging animal should incorporate an encountered prey item into its diet, based on the energetic gain from the prey item, balanced against the cost in terms of time of pursuing it [37], [38]. It thus adds to an impressive list of circumstances in which phage strategies can be understood using analogies to theory developed for foraging animals (e.g. [39], [40]).

Discussion

The goal of this simple modeling exercise was to predict the advantage conferred to a cyanophage of carrying and expressing the psbA gene. More specifically, we consider the hypothesis that phage psbA expression augments the photosynthetic apparatus of the host during infection, following the decay of host psbA transcripts, and we do not consider possible alternative or additional advantages of phage psbA. We have intentionally oversimplified the complex processes of infection, photosynthesis and dNTP synthesis in an effort to match the model to the scope and resolution of the available data. The modeled predictions serve as hypotheses to be tested when the means to knock out specific genes in these cyanophage genomes are eventually developed.

First, we predict that under low continuous irradiance, phage psbA expression increases phage genome replication, and potentially phage fitness, relative to a 'mutant' that does not contain this gene. This advantage is substantially reduced, however, if one accounts for the cost to the phage of elongation of the cyanophage P-SSP7 genome by psbA. Second, we predict that the slight advantage conferred by phage-encoded psbA may be greater under conditions of light stress, such as an increase in irradiance during infection. This is due to the more rapid decay of host D1 proteins at higher irradiance, and could contribute substantially to the advantage conferred by psbA to cyanophage P-SSP7 in the dominant habitat of this particular Prochlorococcus host — the surface mixed layer of the ocean. Finally, the model predicts that during infection in the dark, where there is presumably no advantage to expressing psbA, encoding psbA would result in a net decrease in genome replication of approximately 2%. Taken together, these results illustrate how the benefits of psbA to cyanophage genome replication may vary substantially among infections that occur at different times over the diel cycle, or for cells that are subject to different conditions of irradiance due to mixing [18]. These are all testable hypotheses.

It is clear that the selective advantage of psbA to phage will be determined by the benefit it confers during all conditions under which infection occurs, weighted by their frequency of occurrence. Therefore to fully understand the fitness consequences of carrying the psbA gene to phage, we will need to better understand how psbA influences genome replication over a much broader range of conditions, including at different times over the diel cycle where properties of host photosynthesis will change dynamically [29] and hosts will contain different numbers of genomes [41] and free dNTPs. To connect these predictions for genome replication to fitness, we will also need a better understanding of when genome replication limits phage burst size (e.g. see [24], [42], [43]) and of the interactions between burst size and other factors, such as the timing of cell lysis, and the availability and quality of hosts (e.g. [39], [40], [44]–[47]).

We have learned recently that marine cyanophage encode a number of genes that are absent in the genomes of non-marine phages and share homology with genes involved in microbial metabolism [11]. As we attempt to understand both the evolutionary significance of these genes and the distribution of phage genes in the ocean [35], it will be useful to have theoretical tools. To begin building such tools, here we have developed a model exploring the selective advantage of one specific gene of host origin that is commonly encoded by marine cyanophages, as well as more general tradeoffs between acquiring dNTPs and elongating the genome. We hope these models will form the basis for a more powerful and predictive modeling framework, and contribute substantially to our understanding of phage dynamics in marine microbial communities.

Acknowledgements

We thank S. Abedon, D. Campbell, D. Lindell, M. Follows and S. Dutkiewicz for valuable comments on earlier versions of this manuscript. We thank M. Sullivan, D. Lindell, D. Endy and members of the Chisholm lab for helpful discussions of this work.

Author Contributions

Conceived and designed the experiments: JGB. Analyzed the data: JGB. Wrote the paper: JGB SWC.

References

1. Partensky F, Hess WR, Vaulot D (1999) Prochlorococcus, a marine photosynthetic prokaryote of global significance. Microbiology and Molecular Biology Reviews 63: 106–127.

2. Bouman HA, Ulloa O, Scanlan DJ, Zwirglmaier K, Li WKW, et al. (2006) Oceanographic basis of the global surface distribution of Prochlorococcus ecotypes. Science 312: 918–921.
3. Johnson ZI, Zinser ER, Coe A, McNulty NP, Woodward EMS, et al. (2006) Niche partitioning among Prochlorococcus ecotypes along ocean-scale environmental gradients. Science 311: 1737–1740.
4. Mann NH (2003) Phages of the marine cyanobacterial picophytoplankton. FEMS Microbiology Reviews 27: 17–34.
5. Waterbury JB, Valois FW (1993) Resistance to co-occurring phages enables marine Synechococcus communities to coexist with cyanophages abundant in seawater. Applied and Environmental Microbiology 59: 3393–3399.
6. Suttle CA, Chan AM (1994) Dynamics and distribution of cyanophages and their effect on marine Synechococcus spp. Applied and Environmental Microbiology 60: 3167–3174.
7. Sullivan MB, Waterbury JB, Chisholm SW (2003) Cyanophages infecting the oceanic cyanobacterium Prochlorococcus. Nature 424: 1047–1051.
8. DeLong EF, Preston CM, Mincer T, Rich V, Hallam SJ, et al. (2006) Community genomics among stratified microbial assemblages in the ocean's interior. Science 311: 496–503.
9. Chen F, Lu J (2002) Genomic sequence and evolution of marine cyanophage P60: a new insight on lytic and lysogenic phages. Applied and Environmental Microbiology 68: 2589–2594.
10. Mann NH, Clokie MRJ, Millard A, Cook A, Wilson WH, et al. (2005) The genome of S-PM2, a "photosynthetic" T4-type bacteriophage that infects marine Synechococcus strains. Journal of Bacteriology 187: 3188–3200.
11. Sullivan MB, Coleman ML, Weigele PR, Rohwer F, Chisholm SW (2005) Three Prochlorococcus cyanophage genomes: signature features and ecological interpretations. PloS Biology 3: e144.
12. Mann NH, Cook A, Millard A, Bailey S, Clokie M (2003) Marine ecosystems: bacterial photosynthesis genes in a virus. Nature 424: 741–741.
13. Lindell D, Sullivan MB, Johnson ZI, Tolonen AC, Rohwer F, et al. (2004) Transfer of photosynthesis genes to and from Prochlorococcus viruses. Proceedings of the National Academy of Sciences of the United States of America 101: 11013–11018.
14. Millard A, Clokie MRJ, Shub DA, Mann NH (2004) Genetic organization of the psbAD region in phages infecting marine Synechococcus strains. Proceedings of the National Academy of Sciences of the United States of America 101: 11007–11012.

15. Sullivan MB, Lindell D, Lee JA, Thompson LR, Bielawski JP, et al. (2006) Prevalence and evolution of core photosystem II genes in marine cyanobacterial viruses and their hosts. PLoS Biology 4: 1344–1357.

16. Ohad I, Kyle DJ, Arntzen CJ (1984) Membrane protein damage and repair: removal and replacement of inactivated 32-kilodalton polypeptides in chloroplast membranes. Journal of Cell Biology 99: 481–485.

17. Lindell D, Jaffe JD, Johnson ZI, Church GM, Chisholm SW (2005) Photosynthesis genes in marine viruses yield proteins during host infection. Nature 438: 86–89.

18. Bailey S, Clokie MRJ, Millard A, Mann NH (2004) Cyanophage infection and photoinhibition in marine cyanobacteria. Research in Microbiology 155: 720–725.

19. Mackenzie JJ, Haselkorn R (1972) Photosynthesis and the development of blue-green algal virus SM-1. Virology 49: 517–521.

20. Lindell D, Jaffe JD, Coleman ML, Futschik ME, Axmann IM, et al. (2007) Genome-wide expression dynamics of a marine virus and its host reveal features of co-evolution. Nature 449: 83–86.

21. Endy D, Kong D, Yin J (1997) Intracellular kinetics of a growing virus: a genetically structured simulation for bacteriophage T7. Biotechnology and Bioengineering 55: 375–389.

22. Endy D, You L, Yin J, Molineux IJ (2000) Computation, prediction, and experimental tests of fitness for bacteriophage T7 mutants with permuted genomes. Proceedings of the National Academy of Sciences of the United States of America 97: 5375–5380.

23. You L, Yin J (2002) Dependence of epistasis on environment and mutation severity as revealed by in silico mutagenesis of phage T7. Genetics 160: 1273–1281.

24. You L, Suthers PF, Yin J (2002) Effects of Escherichia coli physiology on growth of phage T7 in vivo and in silico. Journal of Bacteriology 184: 1888–1894.

25. Buchholtz F, Schneider FW (1987) Computer simulation of T3 / T7 phage infection using lag times. Biophysical Chemistry 26: 171–179.

26. Tyystjärvi E, Mäenpää P, Aro E-M (1994) Mathematical modelling of photoinhibition and Photosystem II repair cycle. I. Photoinhibition and D1 protein degradation in vitro and in the absence of chloroplast protein synthesis in vivo. Photosynthesis Research 41: 439–449.

27. Murray JD (1989) Mathematical biology. Berlin: Springer-Verlag.

28. Sharon I, Tzahor S, Williamson S, Shmoish M, Man-Aharonovich D, et al. (2007) Viral photosynthetic reaction center genes and transcripts in the marine environment. ISME Journal 1: 492–501.

29. Falkowski PG, Raven JA (2007) Aquatic photosynthesis. Princeton, , NJ: Princeton University Press.

30. Six C, Finkel ZV, Irwin AJ, Campbell DA (2008) Light variability illuminates niche-partitioning among marine picocyanobacteria. PLoS One 2: e1341.

31. Rabkin SD, Richardson CC (1990) In vivo analysis of the initiation of bacteriophage T7 DNA replication. Virology 174: 585–592.

32. Donlin MJ, Johnson KA (1994) Mutants affecting nucleotide recognition by T7 DNA polymerase. Biochemistry 33: 14908–14917.

33. Moore LR, Chisholm SW (1999) Photophysiology of the marine cyanobacterium Prochlorococcus: ecotypic differences among cultured isolates. Limnology and Oceanography 44: 628–638.

34. Partensky F, Hoepffner N, Li WKW, Ulloa O, Vaulot D (1993) Photoacclimation of Prochlorococcus sp. (Prochlorophyta) strains isolated from the north Atlantic and the Mediterranean sea. Plant Physiology 101: 285–296.

35. Breibart M, Thompson LR, Suttle CA, Sullivan MB (2007) Exploring the vast diversity of marine viruses. Oceanography 20: 135–139.

36. Brown CM, Lawrence JE, Campbell DA (2006) Are phytoplankton population density maxima predictable through analysis of host and viral genomic DNA content? Journal of the Marine Biological Association of the United Kingdom 86: 491–498.

37. MacArthur RH, Pianka ER (1966) On optimal use of a patchy environment. American Naturalist 100: 603–609.

38. Schoener TW (1971) Theory of feeding strategies. Annual Review of Ecology and Systematics 2: 369–404.

39. Wang I-N, Dykhuizen DE, Slobodkin LB (1996) The evolution of phage lysis timing. Evolutionary Ecology 10: 545–558.

40. Bull JJ (2006) Optimality models of phage life history and parallels in disease evolution. Journal of Theoretical Biology 241: 928–938.

41. Vaulot D, Marie D, Olson RJ, Chisholm SW (1995) Growth of Prochlorococcus, a photosynthetic prokaryote, in the equatorial pacific ocean. Science 268: 1480–1482.

42. Kim H, Yin J (2004) Energy-efficient growth of phage Q beta in Escherichia coli. Biotechnology and Bioengineering 88: 148–156.

43. You L, Yin J (2006) Evolutionary design on a budget: robustness and optimality of bacteriophage T7. IEE Proceedings Systems Biology 153: 46–52.

44. Abedon ST (1989) Selection for bacteriophage latent period length by bacterial density: a theoretical examination. Microbial Ecology 18: 79–88.

45. Abedon ST, Herschler TD, Stopar D (2001) Bacteriophage latent period evolution as a response to resource availability. Applied and Environmental Microbiology 67: 4233–4241.

46. Abedon ST, Hyman P, Thomas C (2003) Bacteriophage latent-period evolution as a response to bacteria availability: An experimental examination. Applied and Environmental Microbiology 69: 7499–7506.

47. Wang I-N (2006) Lysis timing and bacteriophage fitness. Genetics 172: 17–26.

An Evaluation of the Effects of Exogenous Ethephon, an Ethylene Releasing Compound, on Photosynthesis of Mustard (Brassica juncea) Cultivars that Differ in Photosynthetic Capacity

N. A. Khan

ABSTRACT

Background

The stimulatory effect of CO_2 on ethylene evolution in plants is known, but the extent to which ethylene controls photosynthesis is not clear. Studies on the effects of ethylene on CO_2 metabolism have shown conflicting results. Increase *or inhibition of photosynthesis by ethylene has been reported. To understand the physiological processes responsible for ethylene-mediated changes in photosynthesis, stomatal and mesophyll effects on photosynthesis and ethylene biosynthesis in response to ethephon treatment in mustard (Brassica juncea) cultivars differing in photosynthetic capacity were studied.*

Results

The effects of ethephon on photosynthetic rate (P_N), stomatal conductance (g_S), carbonic anhydrase (CA) activity, 1-aminocyclopropane carboxylic acid synthase (ACS) activity and ethylene evolution were similar in both the cultivars. Increasing ethephon concentration up to 1.5 mM increased P_N, g_S and CA maximally, whereas 3.0 mM ethephon proved inhibitory. ACS activity and ethylene evolution increased with increasing concentrations of ethephon. The corresponding changes in g_s and CA activity suggest that the changes in photosynthesis in response to ethephon were triggered by altered stomatal and mesophyll processes. Stomatal conductance changed in parallel with changes in mesophyll photosynthetic properties. In both the cultivars ACS activity and ethylene increased up to 3.0 mM ethephon, but 1.5 mM ethephon caused maximum effects on photosynthetic parameters.

Conclusion

These results suggest that ethephon affects foliar gas exchange responses. The changes in photosynthesis in response to ethephon were due to stomatal and mesophyll effects. The changes in g_S were a response maintaining stable intercellular CO_2 concentration (C_i) under the given treatment in both the cultivars. Also, the high photosynthetic capacity cultivar, Varuna responded less to ethephon than the low photosynthetic capacity cultivar, RH30. The photosynthetic capacity of RH30 increased with the increase in ethylene evolution due to 1.5 mM ethephon application.

Background

Photosynthesis is controlled by several intrinsic and extrinsic factors. Of these, plant hormones have received considerable attention in the past in photosynthetic responses of plants. Ethylene is a phytohormone that influences every aspect of plant growth and development [1]. It is synthesized by the activity of 1-aminocyclopropane carboxylic acid synthase (ACS). The response of plants to ethylene depends on the sensitivity of plants to the gas. Conflicting results on the effects of ethylene-releasing compounds on net photosynthetic rate (P_N) have been reported. It has been shown to increase P_N [2-7] or decrease it [8,9], but no definite reason has been assigned for this. It has been shown that the increase in P_N with ethylene-releasing compounds was due to the increase in chlorophyll per unit leaf area [10] or by greater light interception [11]. In my earlier report it has been shown that alteration in photosynthesis was due to the changes in ACS activity [12]. The goal of this work was to compare stomatal and mesophyll effects on P_N in response to ethephon treatment. For that, P_N, stomatal conductance (g_S) and carbonic anhydrase (CA) activity were recorded. To find a possible relationship of ethylene-mediated changes in foliar gas exchange parameters, activity of ACS and ethylene evolution were also determined. The work was carried out in two cultivars of mustard previously shown to have different photosynthetic capacity [12].

Results

The effects of ethephon on P_N, g_S and CA were found significant in both the cultivars (Figures 1, 2). Ethephon at 1.5 mM increased the characteristics maximally, increasing P_N by 31.8 and 41.8%, g_S by 15.0 and 17.1% and CA by 84.6 and 71.4% in Varuna and RH30, respectively. Higher concentration of ethephon (3.0 mM) decreased the characteristics in both the cultivars. The ratio of intercellular to ambient CO_2 concentration (C_i/C_a) was constant.

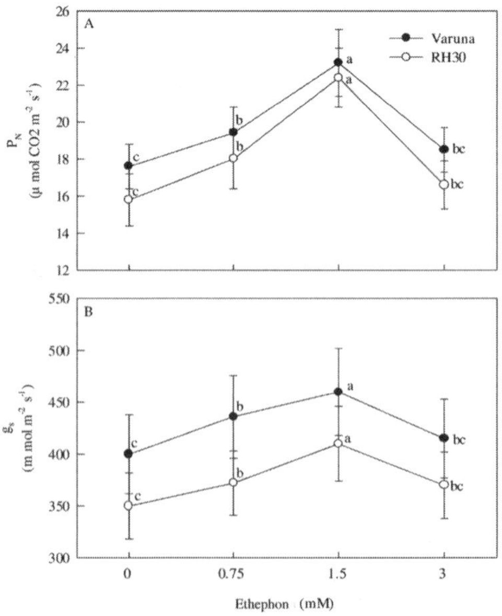

Figure 1. Effects of ethephon on photosynthesis and stomatal conductance in mustard. Effect of different concentrations of ethephon (2-chloroethyl phosphonic acid) applied at 30 d after sowing on net photosynthetic rate (P_N) (A) and stomatal conductance (g_S) (B) in high photosynthetic capacity cultivar Varuna and low photosynthetic capacity cultivar RH30 of mustard (Brassica juncea) at 15 d after the treatment. Each data point represent treatment mean ± SE. Values at each data point within the cultivar sharing the same letter are not significantly different at P < 0.05.

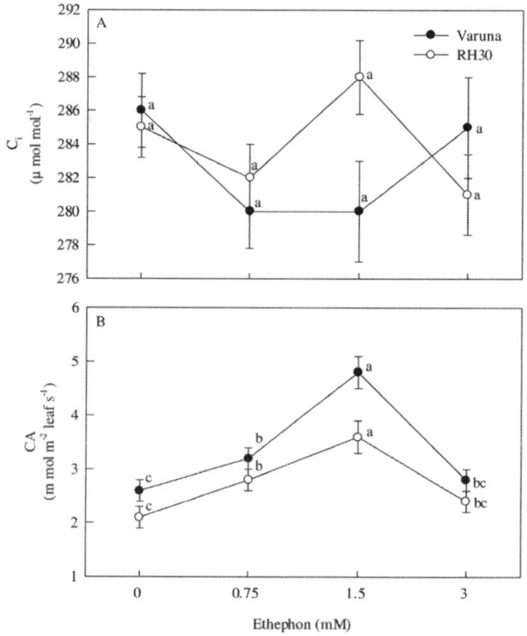

Figure 2. Effects of ethephon on intercellular CO_2 concentration and carbonic anhydrase activity in mustard. Effect of different concentrations of ethephon (2-chloroethyl phosphonic acid) applied at 30 d after sowing on intercellular CO_2 concentration (C_i) (A) and carbonic anhydrase (CA) activity (B) in high photosynthetic capacity cultivar Varuna and low photosynthetic capacity cultivar RH30 of mustard (Brassica juncea) at 15 d after the treatment. Each data point represent treatment mean ± SE. Values at each data point within the cultivar sharing the same letter are not significantly different at $P < 0.05$.

Ethephon application significantly affected ACS activity and ethylene evolution, and were greatest with 3.0 mM ethephon (Table 1). Low photosynthetic capacity cultivar, RH30 was more responsive to ethephon than the high photosynthetic capacity cultivar, Varuna. Application of 1.5 mM ethephon increased ethylene by 52.6% in Varuna and 75.0% in RH30. Increase in ethylene with 1.5 mM ethephon was associated with the increase in PN, gS and CA. Ethylene evolution with 3.0 mM ethephon proved inhibitory for photosynthetic parameters.

Table 1. Effects of different concentrations of ethephon (2-chloroethyl phosphonic acid) applied at 30 d after sowing on the activity of 1-aminocyclopropane carboxylic acid synthase (ACS; (ng ACC kg^{-1} leaf (FM) s^{-1}) and ethylene evolution (ng kg^{-1} leaf (FM) s^{-1}) in two cultivars of mustard (Brassica juncea) at 15 d after the treatment. Values ± SE. Data followed by the same letter within a column are significantly not different.

Ethephon Treatments (mM)	High photosynthetic capacity cultivar (Varuna)		Low photosynthetic capacity cultivar (RH 30)	
	ACS	Ethylene	ACS	Ethylene
0	62.2 ± 5.4c	3.8 ± 0.2c	40.8 ± 3.6c	2.4 ± 0.2c
0.75	64.6 ± 5.8c	4.4 ± 0.3bc	44.4 ± 4.0c	3.2 ± 0.2bc
1.5	68.2 ± 5.8b	5.8 ± 0.3b	52.8 ± 4.6b	4.2 ± 0.3b
3.0	73.5 ± 6.0a	7.0 ± 0.5a	62.3 ± 5.6a	6.0 ± 0.4a

Discussion

Maximum rates of photosynthesis were found with 1.5 mM ethephon. The increase in PN due to ethephon has been reported [4-6]. Increased gs and CA values in both the cultivars showed stomatal and mesophyll effects on photosynthesis. Mesophyll effects are characterized as a product of CO_2 binding capacity and the electron transport capacity. The carboxylation capacity determines the mesophyll effects [13,14]. Increase in CA activity at the site of CO_2 fixation exhibited the enhanced carboxylation reaction [15-17]. The changes in stomatal conductance due to ethephon were to maintain stable intercellular CO_2 concentration (C_i) under the given treatment. Thus, stomatal and mesophyll processes contributed to the increase in P_N in response to ethephon. The ethephon-induced effects on photosynthetic parameters were mediated by ethylene evolved due to ethephon treatment. Taylor and Gunderson [18] showed a relationship between ethylene-enhanced g_S and ethylene-enhanced P_N. Higher concentration of ethephon (3.0 mM) decreased the P_N and g_S. Such condition of inhibition of P_N by ethylene-releasing compound has been observed by Kays and Pallas [8] and Rajala and Peltonen-Sainio [9]. In all these studies ethylene has been attributed to the changes in P_N due to its effect on g_S. Mattoo and White [19] reported that ethylene affected CO_2 assimilation and the plant responded depending on the tissue concentration. On the similar lines, Dhawan et al. [20], Kao and Yang [21] and Grodzinski et al. [22] reasoned that decrease in CO_2 regulated P_N and was related to ethylene evolution. In the present study, low photosynthetic cultivar, RH30 responded more to ethephon than the high photosynthetic cultivar, Varuna. In control plants, lesser ethylene evolution in RH30 than Varuna was responsible for lesser P_N. As the ethylene evolution increased with ethephon application, the capacity of RH30 for P_N also increased resulting in higher per cent increase in P_N than the Varuna. An increase of 75% ethylene in RH30 due to 1.5 mM ethephon increased P_N by 41.8%, whereas 52.6% increase in ethylene in Varuna due to the same treatment increased P_N by 31.8%. Earlier strong positive correlation between ACS activity and P_N has been shown [12].

It therefore, appears possible that the threshold value for ethylene with 1.5 mM ethephon was comparable to that which elicits the ethylene-mediated hormonal responses, which differ with the cultivars inherent capacity of physiological processes. It is that there is some requirement of ethylene for optimum response. Low and high concentration represent the two ends of an optimum curve, promoting at low concentration and inhibiting at high.

Conclusions

This study shows that ethephon affects P_N in both high and low photosynthetic capacity cultivars, Varuna and RH30. In both the cultivars, changes in P_N were due to stomatal and mesophyll effects. Ethephon-induced P_N was attributed to ethylene evolution. The high photosynthetic capacity cultivar, Varuna responded less to ethephon than the low photosynthetic capacity cultivar, RH30. The low P_N of RH30 was due to low level of ethylene. The low photosynthetic capacity of RH30 could be enhanced to give higher P_N through increase in ethylene evolution. However, for both the cultivars there is a range of physiologically active concentration of ethylene beyond which it exerts inhibitory effects.

Methods

Two cultivars of mustard (Brassica juncea L. Czern & Coss.), namely Varuna (high photosynthetic capacity) and RH30 (low photosynthetic capacity) were grown from seeds in 10 m^2 field plots in complete randomized design with five replications. At seedling establishment a plant population of 12 plants m^{-2} was maintained and recommended plant cultivation procedures were adopted. A uniform recommended soil application of 18 g N, 3 g P and 3 g K m^{-2} was given at the time of sowing so as the nutrients were non-limiting.

At 30 d after sowing, 0, 0.75, 1.5 and 3.0 mM ethephon (2-chloroethyl phosphonic acid) was sprayed with a hand sprayer. Ethephon is a direct ethylene source when applied to plants and elicits response identical to those induced by ethylene gas [23,24]. Since ethephon on hydrolysis releases ethylene and phosphorus, therefore equivalent amount of phosphorus present in 3.0 mM ethephon was given to all treatments including control to nullify the effects of phosphorus.

At 45 d after sowing (15 d after ethephon treatment) PN, gS, Ci, CA activity, ACS activity and ethylene evolution were determined.

Measurement of Photosynthetic Parameters

P_N, g_S and C_i were measured using infrared gas analyzer (LiCOR 6200, Lincoln, NE) on fully expanded upper most leaves at saturating light intensity on four plants from each replicate. The atmospheric conditions during the experiment between 1100–1200 h were: photosynthetic active radiation about 1050 μmol m^{-2} s^{-1}, relative humidity 64% and temperature 23°C, atmospheric CO_2 concentration 360 μmol mol^{-1}.

Measurement of Carbonic Anhydrase Activity

The leaves used for photosynthesis measurement were selected for CA activity determination. CA was measured by the method of Dwivedi and Randhava [25].

Leaves were cut into small pieces in 10 mL of 0.2 M cystein at 4°C. The solution adhering to the leaf surface was removed and immediately transferred to a tube having 4 mL phosphate buffer (pH 6.8). A 4 mL of 0.2 M sodium bicarbonate in 0.002 M sodium hydroxide and 0.2 mL of 0.002% bromothymol blue was added to the tube. The tubes were kept at 4°C for 20 min after shaking. Liberated CO_2 during the catalytic action of enzyme on sodium bicarbonate was estimated by titrating the reaction mixture against 0.05 N hydrochloric acid.

Measurement of ACS Activity and Ethylene Evolution

Activity of ACS was measured adopting the methods of Avni et al. [26] and Woeste et al. [27]. Leaf tissue was grind in 100 mM N-2 hydroxyethylenepiperazine N-2 ethanesulfonic acid buffer (pH 8.0) containing 4 mM dithiothreitol, 2.5 mM pyridoxal phosphate and 25% polyvinylpolypyrrolidone. The preparation was homogenized and centrifuged at 12000 g for 15 min. One mL of the supernatant was placed in a 30 mL tube and 0.1 mL of 5 mM S-adenosyl methionine (AdoMet) was added. This was incubated for 1 h at 22°C. The 1-aminocyclopropane carboxylic acid formed was determined by its conversion to ethylene by the addition of 0.1 mL of 20 mM HgCl2 followed by 0.1 mL of 1:1 mixture of saturated NaOH/NaCl and incubated on ice for 10 min, and ethylene evolution was measured on a gas chromatograph. For control set AdoMet was not added. For ethylene evolution 5 mL of gas phase was removed with a syringe and ethylene was measured on a gas chromatograph (GC 5700, Nucon, New Delhi) equipped with 1.8 m Porapack N (80/100 mesh) column, a flame ionization detector and an integrator. Nitrogen was used as carrier gas. The flow rates of nitrogen, hydrogen and oxygen were 0.5, 0.5 and 5 mL s-1, respectively. The oven temperature was 100°C and detector was at 150°C. Ethylene identification was based on the retention time and quantified comparing with the peaks from standard ethylene concentrations.

Data Analysis

Data were analyzed statistically and standard error of the mean value was calculated. Analysis of variance was performed to identify the significant differences among treatments at $P < 0.05$ [28].

Abbrevations

Adomet – S-adenosyl methionine; ACS – 1-aminocyclopropane carboxylic acid synthase; CA – carbonic anhydrase; C_i – intercellular CO_2 concentration; g_S – stomatal conductance; P_N – net photosynthetic rate

Acknowledgements

The author gratefully acknowledges financial assistance by the University Grants Commission, New Delhi for the work and to the two anonymous reviewers for constructive criticism on the earlier version of the manuscript.

References

1. Abeles FB, Morgan PW, Saltveit ME: Ethylene in Plant Biology. London. UK: Academic Press; 1992.
2. Buhler B, Drumm H, Mohr H: Investigation on the role of ethylene in phytochrome-mediated photomorphogenesis. II. Enzyme levels and chlorophyll synthesis. Planta 1978, 142:119–122.
3. Grewal HS, Kolar JS: Response of Brassica juncea to chlorocholine chloride and ethrel sprays in association with nitrogen application. J Agric Sci 1990, 114:87–91.
4. Subrahmanyam D, Rathore VS: Influence of ethylene on carbon-14 labelled carbon dioxide assimilation and partitioning in mustard. Plant Physiol Biochem 1992, 30:81–86.
5. Pua EC, Chi GL: De novo shoot morphogenesis and plant growth of mustard (Brassica juncea) in vitro in relation to ethylene. Physiol Plant 1993, 88:467–474.
6. Khan NA, Lone NA, Samiullah : Response of mustard (Brassica juncea L.) to applied nitrogen with or without ethrel sprays under non-irrigated conditions. J Agron Crop Sci 2000, 184:63–66.
7. Khan NA, Singh S, Khan M, Samiullah : Interactive effect of nitrogen and plant growth regulators on biomass partitioning and seed yield of mustard. Brassica 2003, 5:64–71.
8. Kays SJ, Pallas JE Jr: Inhibition of photosynthesis by ethylene. Nature 1980, 385:51–52.
9. Rajala A, Peltonen-Sainio P: Plant growth regulator effects on spring cereal root and shoot growth. Agron J 2001, 93:936–943.
10. Grewal HS, Kolar JS, Cheema SS, Singh G: Studies on the use of growth regulators in relation to nitrogen for enhancing sink capacity and yield of gobhi-serson (Brassica napus). Indian J Plant Physiol 1993, 36:1–4.
11. Woodrow L, Grodzinski B: An evaluation of the effects of ethylene on carbon assimilation in Lycopersicon esculentum Mill. J Exp Bot 1989, 40:361–368.

12. Khan NA: Activity of 1-aminocyclopropane carboxylic acid synthase in two mustard (Brassica juncea L.) cultivars differing in photosynthetic capacity. Photosynthetica 2004, 42:477–480.
13. Pell EJ, Eckardt NA, Enyedi AJ: Timing of ozone stress and resulting status of ribulose bisphosphate carboxylase/oxygenase and associated net photosynthesis. New Phytol 1992, 120:397–405.
14. Eichelmann H, Laisk A: Ribulose-1,5-bisphosphate carboxylase/oxygenase content, assimilatory change and mesophyll conductance in leaves. Plant Physiol 1999, 119:179–189.
15. Badger RR, Price GD: The role of carbonic anhydrase in photosynthesis. Ann Rev Plant Physiol Plant Mol Biol 1994, 45:369–393.
16. Moroney JV, Bartlett SG, Samuelsson G: Carbonic anhydrase in plants and algae. Plant Cell Environ 2001, 24:141–153.
17. Khan NA, Javid S, Samiullah : Physiological role of carbonic anhydrase in CO_2 fixation and carbon partitioning. Physiol Mol Biol Plants 2004, 10:153–166.
18. Taylor GE Jr, Gunderson CA: The response of foliar gas exchange to exogenously applied ethylene. Plant Physiol 1986, 82:653–657.
19. Mattoo AK, White WB: Regulation of ethylene biosynthesis. In The Plant Hormone Ethylene. Edited by: Mattoo AK, Suttle JC. CRC Press, Boca Raton, London; 1991:21–42.
20. Dhawan KR, Bassi PK, Spencer MS: Effects of carbon dioxide on ethylene production and action in intact sunflower plants. Plant Physiol 1981, 68:831–834.
21. Kao CH, Yang SF: Light inhibition of the conversion of 1-aminocyclopropane-1-carboxylic acid to ethylene in leaves is mediated through carbon dioxide. Planta 1982, 155:261–266.
22. Grodzinski B, Boesel L, Horton RF: Ethylene release from leaves of Xanthium strumarium L. and Zea mays L. J Exp Bot 1982, 33:344–354.
23. Cooke AR, Randall DI: 2-Haloethanephosphonic acids as ethylene releasing agents for the induction of flowering in pineapples. Nature 1968, 218:974–975.
24. Edgerton LJ, Blanpied GD: Regulation of growth and fruit maturation with 2-chloroethanephosphonic acid. Nature 1968, 219:1064–1065.
25. Dwivedi RS, Randhava NS: Evaluation of a rapid test for the hidden hunger in plants. Plant Soil 1974, 40:445–451.
26. Avni A, Bailey BA, Mattoo AK, Anderson JD: Induction of ethylene biosynthesis in Nicotiana tabacum by a Trichoderma viride xylanase is correlated to

the accumulation of 1-aminocyclopropane carboxylic acid (ACC) synthase and ACC oxidase transcripts. Plant Physiol 1994, 106:1049–1055.

27. Woeste KE, Ye C, Kieber JJ: Two Arabidopsis mutants that overproduce ethylene are affected in the post-transcriptional regulation of 1-aminocyclopropane 1-carboxylic acid synthase. Plant Physiol 1999, 119:521–529.

28. Gomez KA, Gomez AA: Statistical Methods for Agricultural Research. New York: A Wiley-Interscience Publication; 1984.

High-Susceptibility of Photosynthesis to Photoinhibition in the Tropical Plant Ficus Microcarpa L. f. cv. Golden Leaves

Shunichi Takahashi, Ayumu Tamashiro, Yasuko Sakihama, Yasusi Yamamoto, Yoshinobu Kawamitsu and Hideo Yamasaki

ABSTRACT

Background

The tropical plant Ficus microcarpa L. f. cv. Golden Leaves (GL) is a high-light sensitive tropical fig tree in which sun-leaves are yellow and shade-leaves are green. We compared the response of photosynthetic activities to strong light between GL and its wild-type (WT, Ficus microcarpa L. f.).

Results

Field measurements of maximum photosystem II (PSII) efficiency (F_v/F_m) of intact sun-leaves in GL showed that photo synthetic activity was severely photoinhibited during the daytime (F_v/F_m = 0.46) and subsequently recovered in the evening (F_v/F_m = 0.76). In contrast, WT did not show any substantial changes of F_v/F_m values throughout the day (between 0.82 and 0.78). Light dependency of the CO_2 assimilation rate in detached shade-leaves of GL showed a response similar to that in WT, suggesting no substantial difference in photosynthetic performance between them. Several indicators of photoinhibition, including declines in PSII reaction center protein (D1) content, F_v/F_m value, and O_2 evolution and CO_2 assimilation rates, all indicated that GL is much more susceptible to photoinhibition than WT. Kinetics of PAM chlorophyll a fluorescence revealed that nonphotochemical quenching (NPQ) capacity of GL was lower than that of WT.

Conclusion

We conclude that the photosynthetic apparatus of GL is more highly susceptible to photoinhibition than that of WT.

Background

Exposure of leaves to strong light sometimes causes the reduction of photosynthetic efficiency, a phenomenon referred to as photoinhibition [1-3]. The susceptibility of plants to photoinhibition depends on the species and growth light-environments [4]. In general, shade plants or low-light grown plants are more susceptible to photoinhibition than sun plants or high-light grown plants [4]. Since photoinhibition has a potential to lower productivity and plant growth, avoidance of photoinhibition is critical for the fitness and survival of plants in natural habitats [2,5,6].

It is now widely accepted that harmful Reactive Oxygen Species (ROS) produced upon illumination are involved in the mechanism of photoinhibition [7]. Singlet-excited oxygen (1O2) can be generated by the interaction of O2 with triplet-excited chlorophyll (3Chl) formed in the PSII reaction center [8]. Superoxide radical (O2-) is unavoidably produced by the Mehler reaction via electron transfer to O2 at photosystem I. Dismutation of O2- results in the formation of hydrogen peroxide (H2O2) and the reaction between H2O2 and transition metal ions generates hydroxyl radical (•OH) which is the most reactive radical among ROS [7]. These ROS can oxidize molecules in chloroplasts including D1 protein in the PSII reaction center [9] and thiol enzymes in the Calvin-Benson cycle [10,11] to inhibit partial reactions of photosynthesis eventually leading to photoinhibition.

Recently, ROS have been shown to also bring about photoinhibition through the inhibition of de novo synthesis of D1 protein of PSII that is essential to recover from photoinhibition [12].

To protect photosynthetic machinery from ROS-mediated photoinhibition, chloroplasts contain a high amount of ascorbate in stroma at 20 to 300 mM [13]. In the stroma, ascorbate contributes to suppress the accumulation of photo-produced-H2O2 by acting as the electron donor for ascorbate peroxidase (APX) which detoxifies H2O2 to H2O [14]. Ascorbate can be also involved in the detoxification of 1O2, O2- and •OH via nonenzymatic reduction [13]. In these reactions, ascorbate is univalently oxidized to monodehydroascorbate (MDA). To maintain ascorbate concentration in the chloroplasts, MDA should be promptly regenerated to ascorbate during illumination. MDA can be reduced to ascorbate by NAD(P)H in a reaction catalyzed by monodehydroascorbate reductase (MDAR) [15] or directly by reduced ferredoxin [16]. MDA, an organic radical, can undergo spontaneous dismutation to produce ascorbate and dehydroascorbate (DHA) in the absence of MDA reduction. DHA can be reduced by glutathione either through nonenzymatic reaction at a slow rate or enzymatic reaction at a much higher rate [17]. Dehydroascorbate reductase (DHAR, EC 1.8.5.1) is considered to be involved in enzymatic DHA reduction in the chloroplasts [17]. Until now, however, direct in vivo evidence showing the physiological significance of DHAR has been yet not available.

We previously reported that leaves of Ficus microcarpa L. f. cv. Golden Leaves (GL), a tropical fig tree, lack heat-stable DHAR activity [18]. In GL, the canopy sun-leaves, which are always exposed to direct sunlight, show characteristic yellow whereas those in wild-type (WT, Ficus microcarpa L. f.) exhibit normal green even when exposed to direct sunlight. The mechanism for yellow leaves production in GL is unknown. In barley, it has been reported that leaves incubated with CO2-enriched air show yellow color due to photoinhibition of photosynthesis [19,20], a phenomenon apparently similar to that observed in GL. We hypothesized that GL possess an incomplete machinery of photosynthesis which is susceptible to photoinhibition. The aim of this study was to directly examine the hypothesis. The side-by-side comparisons shown in this study demonstrate that GL is much more susceptible to photoinhibition of photosynthesis than WT.

Results

Photoinhibition in the Field

GL is a cultivar of the tropical fig tree Ficus microcarpa L. f. (WT) that is natively distributed in subtropical/tropical regions of Asia and Oceania. A significant characteristic of GL is the presence of yellow leaves that contain high amounts of flavonoids and negligible amounts of chlorophyll and carotenoids [21]. The reduced content of photosynthetic pigments in the yellow leaves can be increased by shading them from high-light. In shade conditions, therefore, GL exhibits normal green leaves which cannot be morphologically distinguished from WT leaves (Fig. 1). We have suggested that high susceptibility of GL to photoinhibition could explain the phenomenon [21]. However, there was no direct evidence available to show the susceptibility of GL to photoinhibition.

Figure 1. Photograph of Ficus microcarpa L. f. (WT) and Ficus microcarpa L. f. cv. Golden Leaves (GL). Each branch was collected from canopy of mature trees grown in the field. Characteristic yellow leaves are observed only in sun-leaves of GL.

Figure 2 shows diurnal change of maximum PSII efficiency (Fv/Fm) in field grown attached sun-leaves collected from the canopy. Chlorophyll content of sun-leaves in GL (yellow) was 0.05 g m-2 and that in WT (green) was 0.44 g m-2. There was no significant difference between GL and WT in Fv/Fm values measured early in the morning (6 a.m.). The value of Fv/Fm in GL decreased from 0.77 (6 a.m.) to 0.46 (2 p.m.) and subsequently recovered to 0.74 in the evening (6 p.m.). In contrast to GL, Fv/Fm values in WT were almost constant throughout a day. Shade-leaves of GL (green), collected from the inside of tree, did not show any decrease in the Fv/Fm value (data not shown). These data obtained in the field conditions suggest that GL is highly susceptible to high-light.

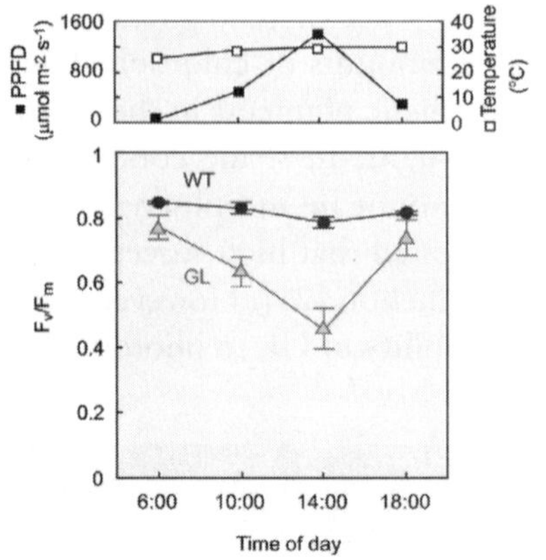

Figure 2. Photoinhibition of GL observed in the field-grown conditions. Top panel, a typical diurnal change in PPFD (▪) on the leaves and atmospheric temperature (□) in the field. Bottom panel, a diurnal change in maximum PSII efficiency of intact attached leaves in the field. (•), WT; (Δ), GL. F_v/F_m value was measured as PSII efficiency with a portable chlorophyll a fluorometer (PAM-2000). Each data point is the mean of 10 separate measurements ± SD.

Photosynthetic Performance

We compared photosynthetic performance between GL and WT to confirm that the fundamental photosynthetic machinery of GL is normal. Shade-leaves of GL and WT, both of which are green, were used for this purpose. Figure 3 shows the light response curves of CO_2 assimilation rate (based on surface area) in detached shade-leaves. The chlorophyll content of shade-leaves in GL (green) was 0.58 g Chl m^{-2} and that in WT (green) was 0.70 g Chl m^{-2}. GL showed a very similar light-response curve to WT and there was no substantial difference in the maximum activity of CO_2 assimilation between them. These results clearly demonstrate that the photosynthetic capacity of GL is almost identical to that of WT, suggesting that photosynthetic machinery of GL is functionally not defective.

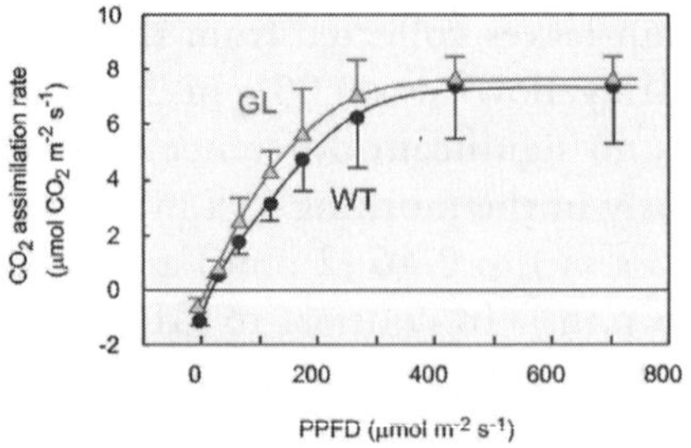

Figure 3. Light-response curves for the rate of photosynthetic CO_2 assimilation in WT and GL. Shade-leaves, which exhibited identical green coloration in both WT and GL, were used for measurements. Light intensities of irradiation were changed from high to low. (•), WT; (Δ), GL. Points are the means of three (•) or four (Δ) separate measurements ± SD.

Photoinhibition in Shade-Leaves of Golden Leaves

Figure 4 shows effects of high-light on several indicators of photoinhibition: D1 protein content, F_v/F_m value, O_2 evolution rate and CO_2 assimilation rate. Before high-light exposure, there was no substantial difference between WT and GL shade-leaves in D1 protein content on protein basis (Fig. 4A) and F_v/F_m value (Fig. 4C). GL showed a significant decrease in the D1 protein content and Fv/Fm value upon high-light illumination (Fig. 4A, 4B,4C). Similar to the indicators specific for PSII activity, those for net photosynthetic activity (i.e. O_2 evolution and CO_2 assimilation rates) also decreased upon high-light illumination (Fig. 4D, 4E). In contrast to these responses observed in GL, WT showed only a small decrease in D1 protein content and F_v/F_m value (Fig. 4A, 4B, 4C). The O_2 evolution and CO_2 assimilation rates in WT did not change even under high-light condition (Fig. 4D, 4E).

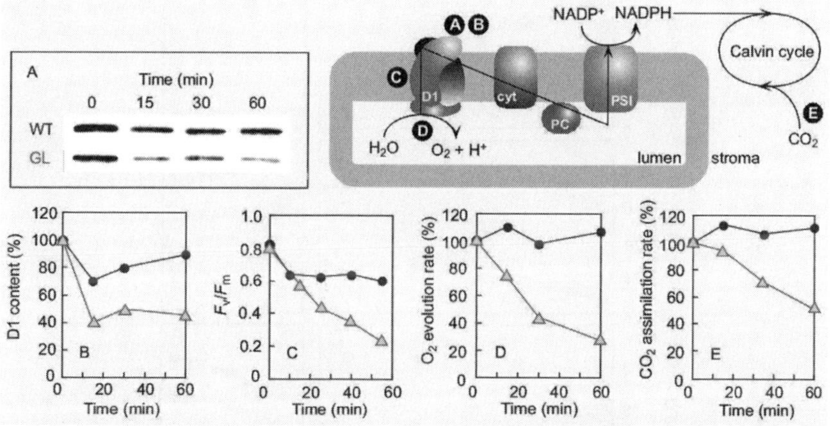

Figure 4. High-light induced photoinhibition of GL assessed by several parameters. Shade-leaves were exposed to high-light (2300 μmol m^{-2} s^{-1}) at zero time after preexposure to medium-light (1000 μmol m^{-2} s^{-1}) for 30 min. A schematic illustration shows the outline of photosynthesis including the electron transport process in thylakoid membranes. Letters (A-D) in filled circles, which correspond to those of the panels A-D, represent the sites where the photo synthetic parameters can be measured. A, Immuno blotting of the D1 protein contained in leaf-extract. B, Degradation of the in vivo D1 protein induced by high-light. Each point was plotted using a relative density of band on the gel as shown in A. C, Decline of F_v/F_m induced by high-light. Note that the time indicated represents total illumination time. D, High-light induced inhibition of the activity of O_2 evolution from intact detached leaves. E, High-light induced inhibition of the activity of CO_2 assimilation in intact detached leaves. Points are taken by a separate measurement in different leaves (A, B, D) or a leaf (C, E). (•), WT; (Δ), GL.

Reduced NPQ Capacity in Golden Leaves

When plants are exposed to high-light that exceeds the capacity of photosynthesis, excess absorbed light energy can be safely dissipated as heat [22]. This energy dissipation can be measured as the NPQ of PAM chlorophyll a fluorescence analysis. NPQ includes the quantum yield of PSII as an index of heat energy dissipation, photoinactivation of PSII and distribution of photon acceptance between PSII and PSI due to the state transition. It is known that the sate transition can be controlled by the redox state of Q_A [23], which can be measured as 1-qP. Figure 5 shows that there is no significant difference between GL and WT in yield and 1-qP values. Furthermore, measurements of the electron transport rate (ETR) showed essentially no difference in the sensitivity of PSII photoinactivation between GL and WT (data not shown). However, GL showed a much lower level of NPQ compared to WT (Fig. 5C, 5F). These results suggest that low NPQ in GL can be attributed to dysfunction of the heat dissipation mechanism.

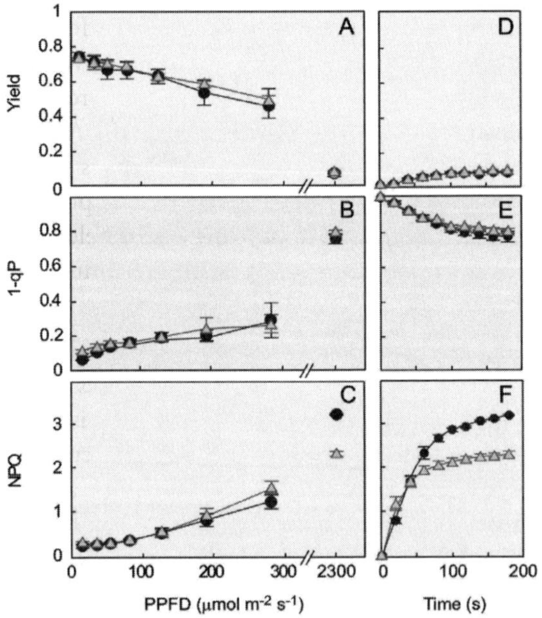

Figure 5. Yield of PSII (A, D), 1-qp reflecting the reduction state of the qa pool (B, E), NPQ (C, F) in shade-leaves of WT (•) and GL (Δ). Each measurement was made after the exposure of leaves to actinic light for 3 min at different intensities (A-C). In D, E and F, high light (2300 μmol m^{-2} s^{-1}) was used as actinic light to record time courses of Yield (D), 1-qP (E), and NPQ (F). Points are presented as the means of 5 (A-C) or 3 (D-F) separate measurements ± SEM.

Discussion

This study has demonstrated that photosynthetic activities of GL are highly susceptible to photoinhibition under high-light conditions (Figs. 2, 4). There was essentially no difference in photosynthetic efficiency between GL and WT at least up to 700 μmol m^{-2} s^{-1} for a short time (Fig. 3), but NPQ in GL was lower than that in WT at 2300 μmol m^{-2} s^{-1} (Fig. 5C, 5F). It has been reported that a npq1 mutant of Arabidopsis thaliana shows increased susceptibility to photoinhibition, but is unaffected in short-term photosynthetic O_2 evolution and CO_2 assimilation [24, 25]. For the operation of NPQ, special xanthophyll pigments in the light-harvesting complexes of PSII are known to have critical roles [26]. The extent of NPQ in plants is strongly correlated with the levels of zeaxanthin and antheraxanthin that are formed from violaxanthin by the enzyme violaxanthin deepoxidase (VDE) located on the luminal side of thylakoid membranes [22]. In excessive light conditions, VDE is activated and converts violaxanthin to zeaxanthin via antheraxanthin when acidification in thylakoid lumen reaches a critical threshold [27]. The activity of VDE is also regulated by the concentration of ascorbate in the lumen because VDE can utilize only ascorbate as a reductant [28]. It has been shown that decreases in ascorbate availability severely affect VDE activity in vivo [28]. Because thylakoid membranes do not have any active transport mechanisms for ascorbate, the concentration of ascorbate in the stroma determines that in the lumen by a passive diffusion mechanism [29]. In Arabidopsis, it has been clearly demonstrated that NPQ performance in vivo is reduced by a mutation that causes ascorbate deficiency [13]. We previously reported that the ascorbate content of GL is severally decreased in daytime under field conditions [30]. Thus, it is reasonable to assume that decrease of ascorbate content would be a cause of dysfunction of NPQ in GL.

Among antioxidant enzymes, we have shown that GL lacks heat-stable DHAR activity [18]. Plant cells possess several kinds of enzymes that exhibit DHAR activity: thioredoxin [31], glutaredoxin [32] and disulfide isomerase [33]. Recently, two distinct chloroplast proteins that exhibit DHAR activity have been isolated from spinach; one is Kunitz-type trypsin inhibitor [31] and the other is specific DHAR [34]. Because both chloroplastic DHARs show higher heat-stable activity than non-chloroplastic DHAR [31,34], GL may lack chloroplastic DHAR(s) though molecular information is not available. In fact, heat-stable activity cannot be detected in nonphotosynthetic organs such as fruits, that do not include chloroplasts [18]. Although chloroplastic DHAR(s) have been presumed to function in the ascorbate regeneration system for maintaining the ascorbate concentration in the chloroplasts [7], the physiological significance of DHAR in chloroplasts is still controversial [35-38]. Since the activity of stromal MDAR is high, the spontaneous disproportionation from MDA to DHA and ascorbate may not occur under in vivo situations. Therefore, there is an argument that DHAR might be dispensable for the ascorbate-glutathione cycle to protect chloroplasts from high-light stress [36]. However, there is another location in chloroplasts where the stromal enzymes cannot directly regenerate ascorbate, namely, the lumen space.

In the lumen, oxidation of ascorbate by xanthophylls, α-tochopherol and PSII result in the production of DHA and all of these reactions could be promoted under high-light conditions [13,35]. Therefore, stromal DHAR would be essential to reduce DHA produced in the lumen side. In this context, chloroplastic DHAR must be physiologically indispensable for photoprotection mechanisms particularly when leaves are exposed to high-light stress conditions. This is consistent with previous reports that DHAR activity in leaves of A. thaliana is increased by high-light conditions [39]. Thus, we consider that ascorbate availability of GL is decreased by high-light due to a lack of DHAR activity which eventually results in lowered performance of NPQ (Fig. 5C, 5F). Ascorbate is involved in photoprotection not only through NPQ but also primarily through ROS scavenging [7]. It should be noted that high susceptibility of GL to photoinhibition could be also explained by a low ROS scavenging capacity which may result in photodamage of target molecules in chloroplast [7] and the inhibition of D1 de novo synthesis [12].

The mechanism for yellow leaf production is still unknown. Sun-leaves of GL contain low amounts of chlorophyll and carotenoids but shade-leaves show comparable amount of those to WT, implying that exhibition of yellow leaves in GL is associated with a long-termed irradiation of high-light [21]. When leaves of GL were exposed to high-light, they showed significant symptoms of photoinhibition (Figs. 2, 4). It has been suggested that long-termed photoinhibition (or chronic photoinhibition) can lead photobleaching of photo synthetic pigments such as chlorophyll and carotenoids [3]. The npq1 Arabidopsis mutant which is partially defective in NPQ has been reported to show the photobleaching of chlorophyll in high-light conditions [40], a phenomenon apparently very similar to that observed in GL. It is plausible that photoinhibition is involved in an early stage of the yellow leaf-producing mechanism. Unlike herbaceous plants, the leaf-yellowing phenomenon does not cause senescence or cell death [21]. It has been shown that flavonoids accumulated in leaves could protect tissues from oxidative damage by complementing the ascorbate-glutathione cycle [15,41,42]. In addition to the ROS scavenging function, flavonoids have long been known to be effective sunscreen pigments particularly against UV radiation [43]. Increased flavonoid pigments in GL may play roles including ROS scavenging function, UV-protection and light-attenuation for shadeleaves. The mechanisms of photoinhibition susceptibility and, in particular of high-light tolerance in GL leaves remain to be fully clarified. Although it is difficult to apply molecular biology techniques to GL, we consider that the tropical tree could provide a unique opportunity for examining the important determinants of survival in high-light environments.

Materials and Methods

Plant Materials

Plant materials were Ficus microcarpa L. f. cv. Golden Leaves (GL) and Ficus microcarpa L. f. (WT) grown in the campus of the University of the Ryukyus on Okinawa island (26'15"N, 127'45"E) in Japan. The study period was from March to June in 1999–2000. The photo synthetic photon flux density (PPFD) on the plant canopy was c.a. 2500 μmol m^{-2} s^{-1} in full sunlight conditions during the period. There was no significant difference in growth conditions of PPFD between WT and GL. Green leaves, grown under light conditions of about 100 μmol m^{-2} s^{-1} as the maximum, were harvested and used for laboratory experiments as shade-leaves. To avoid the drought of leaves, leaf petioles were continuously kept in distilled water throughout the measurements. To activate the photosynthetic activity, detached leaves were exposed to an illumination of 1000 μmol m^{-2} s^{-1} for 30 min before experiments.

Illumination of Leaves

Leaves were illuminated with white light from three halogen lumps (400 W). Leaf temperature was maintained at 30°C with an electric fan or a thermostatically controlled water bath during illumination. Various light intensities were obtained by placing wire screens in front of the light source.

Measurements of Chlorophyll Fluorescence

Chlorophyll fluorescence was measured with a PAM-2000 chlorophyll fluorometer system under atmospheric conditions (Heinz Walz, Effeltrich, Germany). After a dark adaptation period of 15 min, minimum fluorescence (F_o) was determined by a weak red light. Maximum fluorescence of dark-adapted leaf (F_m) was measured during a subsequent saturating light pulse of white light (8000 μmol m^{-2} s^{-1} for 0.4 s). Maximum fluorescence (F_m') and steady-state fluorescence (Fs) of illuminated leaf were measured upon a subsequent saturating light pulse of white light (8000 μmol m^{-2} s^{-1}, for 0.4 s) to determine NPQ (F_m-F_m')/F_m'), yield of PSII [(Fm'-Fs)/Fm] and qP (Fm'-Fs)/Fm'-Fo) [44,45]. A 650 nm light-emitting-diode (LED) equipped with PAM-2000 was used for the illumination of leaf as actinic light.

Measurements of Gas Exchange

CO_2 assimilation was determined by the difference in CO_2 concentration between inlet and outlet of the assimilation chamber with a CO_2 gas analyzer (LI-6251; LiCor, Inc., Lincoln, Nebraska) in 0.036% CO_2 and 21% O_2[46]. Measurements of CO_2 assimilation rates were carried out under an illumination of 2300 µmol m^{-2} s^{-1}, except for light response curve measurements. Approximately 25 cm^2 of leaf were employed for an experiment. The air in the chamber was stirred rapidly to maintain a high boundary layer conductance for CO_2 diffusion. O_2 evolution was determined using a gas-phase Walker-type oxygen electrode system (Model LD-1; Hansatech, Norfolk, U.K.) in 5% CO_2 and 21% O_2[47]. Measurements of O_2 evolution were carried out on 4 cm^2 of leaf disc under an illumination of 600 µmol m^{-2} s^{-1}.

Measurements of D1 Protein Content

Leaf (0.5 g) was homogenized for 30 s at 0°C in a grinding medium containing 50 mM potassium phosphate (pH 7.0), 5 mM phenylmethylsulfonyl fluoride and 5% (w/v) polyvinylpolypyrrolidone. The homogenate (4 ml) was mixed with 4 ml of 125 mM Tris (pH 6.8) containing 5% (w/v) SDS, and then incubated at 80°C for 3 min to solubilize proteins. After cooling down the sample to room temperature, aggregates were removed by the centrifugation at 1,500 × g for 30 s. The supernatant (800 µl) was mixed with the same amount of a solution that contained 0.4 M sucrose, 8 M urea, 5 mM EDTA and 5% (w/v) 2-mercaptoethanol.

SDS-PAGE was carried out according to a previously reported method [48]. Polyacrylamide gels containing 6.0 M urea were used for the stacking gel (4.5 %) and separation gel (13 %). Samples that contained 8 µg protein were loaded in each lane. Immunoblot analysis was performed using specific polyclonal antibodies raised against the D1 protein [48]. The intensities of the protein bands were determined with NIH image system version 1.61 (NIH, USA).

List of Abbreviations Used

APX, ascorbate peroxidase; ^{1}Chl, singlet excited chlorophyll; ^{3}Chl, triplet excited chlorophyll; DHA, dehydroascorbate; DHAR, dehydroascorbate reductase; D1, the reaction center-forming protein of photosystem II; F_v/F_m, ratio of variable to maximum chlorophyll a fluorescence; GL, Golden Leaves; MDA, monodehydroascorbate; MDAR, monodehydroascorbate reductase; NPQ, nonphotochemical quenching; •OH, hydroxyl radical; $^1O_2$, singlet excited O_2; O_2-, superoxide radical; PPFD, photosynthetic photon flux density; ROS, reactive oxygen species; WT, wild-type.

Acknowledgements

We are grateful to Dr. M. F. Cohen of University of the Ryukyus for critical reading of the manuscript. This work was supported by a Grant-in-Aid for Scientific Research (B) and (C) from Japan Society for the Promotion of Science to HY.

References

1. Demming-Adams B, Adams WW: Photoprotection and other responses of plants to high light stress. Annu Rev Plant Physiol Plant Mol Biol 1992, 43:599–626.
2. Long SP, Humphries S, Falkowski PG: Photoinhibition of photosynthesis in nature. Annu Rev Plant Physiol Plant Mol Biol 1994, 45:633–662.
3. Powles SB: Photoinhibition of photosynthesis induced by visible light. Ann Rev Plant Physiol 1984, 35:15–44.
4. Osmond CB: What is photoinhibition? Some insights from comparisons of shade and sun plants. In: Photoinhibition of photosynthesis: from molecular mechanisms to the field (Edited by: Baker NR, Bowyer JR). Oxford, BIOS Scientific Publishers 1994, 1–24.
5. Ball MC: The role of photoinhibition during tree seedling establishment at low temperatures. In: Photoinhibition of photosynthesis: from molecular mechanisms to the field (Edited by: Baker NR, Bowyer JR). Oxford, BIOS Scientific Publishers 1994, 365–376.
6. Kitao M, Lei TT, Koike T, Tobita H, Maruyama Y, Matsumoto Y, Ang LH: Temperature response and photoinhibition investigated by chlorophyll fluorescence measurements for four distinct species of dipterocarp trees. Physiol Plant 2000, 109:284–290.
7. Asada K: The water-water cycle in chloroplasts: Scavenging of active oxygens and dissipation of excess photons. Annu Rev Plant Physiol Plant Mol Biol 1999, 50:601–639.
8. Macpherson AN, Telfer A, Barber J, Truscott TG: Direct detection of singlet oxygen from isolated photosystem II reaction centers. Biochim Biophys Acta 1993, 1143:301–309.
9. Yamamoto Y: Quality control of photosystem II. Plant Cell Physiol 2001, 42:121–128.
10. Kaiser WM: Reversible inhibition of the Calvin cycle and activation of oxidative pentose phosphate cycle in isolated intact chloroplasts by hydrogen peroxide. Planta 1979, 145:377–382.

11. Tanaka K, Otsubo T, Kondo N: Participation of hydrogen peroxide in the inactivation of Calvin-cycle SH enzymes in SO_2-fumgated spinach leaves. Plant Cell Physiol 1982, 23:1009–1018.
12. Nishiyama Y, Yamamoto H, Allakhverdiev SI, Inaba M, Yokota A, Murata N: Oxidative stress inhibits the repair of photodamage to the photosynthetic machinery. EMBO J 2001, 20:5587–5594.
13. Smirnoff N: Ascorbate biosynthesis and function in photoprotection. Phil Trans R Soc Lond B 2000, 355:1455–1464.
14. Asada K: Ascorbate peroxidase – a hydrogen peroxide-scavenging enzyme in plants. Physiol Plant 1992, 85:235–241.
15. Sakihama Y, Mano J, Sano S, Asada K, Yamasaki H: Reduction of phenoxyl radicals mediated by monodehydroascorbate reductase. Biochem Biophys Res Commun 2000, 279:949–954.
16. Miyake C, Asada K: Ferredoxin-dependent photoreduction of the monodehydroascorbate radical in spinach thylakoids. Plant Cell Physiol 1994, 35:539–549.
17. Hossain MA, Asada K: Purification of dehydroascorbate reductase from spinach and its characterization as a thiol enzyme. Plant Cell Physiol 1984, 25:85–92.
18. Yamasaki H, Takahashi S, Heshiki R: The tropical fig Ficus microcarpa L. f. cv. Golden Leaves lacks heat-stable dehydroascorbate reductase activity. Plant Cell Physiol 1999, 40:640–646.
19. Sicher RC: Yellowing and photosynthetic decline of barley primary leaves in response to atmospheric CO2 enrichment. Physiol Plant 1998, 103:193–200.
20. Sicher RC: Photosystem-II activity is decreased by yellowing of barley primary leaves during growth in elevated carbon dioxide. Int J Plant Sci 1999, 160:849–854.
21. Yamasaki H, Heshiki R, Ikehara N: Leaf-goldenning induced by high light in Ficus microcarpa L. f., a tropical fig. J Plant Res 1995, 108:171–180.
22. Gilmore AM: Mechanistic aspects of xanthophyll cycle-dependent photoprotection in higher plant chloroplasts and leaves. Physiol Plant 1997, 99:197–209.
23. Krause GH, Weis E: Chlorophyll fluorescence and photosynthesis: the basics. Annu Rev Plant Physiol Plant Mol 1991, 42:313–349.
24. Niyogi KK, Grossman AR, Bjorkman O: Arabidopsis mutants define a central role for the xanthophyll cycle in the regulation of photosynthetic energy converiion. Plant Cell 1998, 10:1121–1134.
25. Havaux M, Niyogi KK: The violaxanthin cycle protects plants from photooxidative damage by more than one mechanism. Proc Natl Acad Sci USA 1999, 96:8762–8767.

26. Gilmore AM: Xanthophyll cycle-dependent nonphotochemical quenching in Photosystem II: Mechanistic insights gained from Arabidopsis thaliana L. mutants that lack violaxanthin deepoxidase activity and/or lutein. Photosynth Res 2001, 67:89–101.

27. Gilmore AM, Yamasaki H: 9-aminoacridine and dibucaine exhibit competitive interactions and complicated inhibitory effects that interfere with measurements of Δ pH and xanthophyll cycle-dependent photosystem II energy dissipation. Photosynth Res 1998, 57:159–174.

28. Bratt CE, Arvidsson P-O, Carlsson M, Åkerlund H-E: Regulation of violaxanthin de-epoxidase activity by pH and ascorbate concentration. Photosynth Res 1995, 45:169–175.

29. Foyer CH, Leiandais M: A comparison of the relative rates of transport of ascorbate and glucose across the thylakoid, chloroplast and plasmalemma membranes of pea leaf mesophyll cell. J Plant Physiol 1996, 148:391–398.

30. Yamasaki H, Heshiki R, Yamasu T, Sakihama Y, Ikehara N: Physiological significance of the ascorbate regenerating system for the high-light tolerance of chloroplasts. In: Photosynthesis: from light to biosphere (Edited by: Mathis P). Dordrecht, Kluwer Academic 1995, 291–294.

31. Trümper S, Follmann H, Häberlein I: A novel dehydroascorbate reductase from spinach chloroplasts homologous to plant trypsin inhibitor. FEBS Lett 1994, 352:159–162.

32. Sha S, Minakuchi K, Higaki N, Sato K, Ohtsuki K, Kurata A, Yoshikawa H, Kotaru M, Masumura T, Ichihara K, Tanaka K: Purification and characterization of glutaredoxin (thioltransferase) from rice (Oryza sativa L.). J Biochem 1997, 121:842–848.

33. Wells WW, Xu DP: Dehydroascorbate reduction. J Bioenerg Biomembr 1994, 26:369–77.

34. Shimaoka T, Yokota A, Miyake C: Purification and characterization of chloroplast dehydroascorbate reductase from spinach leaves. Plant Cell Physiol 2000, 41:1110–1118.

35. Mano J, Ushimaru T, Asada K: Ascorbate in thylakoid lumen as an endogenous electron donor to Photosystem II: Protection of thylakoids from photoinhibition and regeneration of ascorbate in stroma by dehydroascorbate reductase. Photosynth Res 1997, 53:197–204.

36. Morell S, Follmann H, DeTullio M, Häberlein I: Dehydroascorbate and dehydroascorbate reductase are phantom indicators of oxidative stress in plants. FEBS Lett 1997, 414:567–570.

37. Morell S, Follmann H, De Tullio M, Häberlein I: Dehydroascorbate reduction: the phantom remaining. FEBS Lett 1998, 425:530–531.

38. Foyer CH, Mullineaux PM: The presence of dehydroascorbate and dehydroascorbate reductase in plant tissue. FEBS Lett 1998, 425:528–529.

39. Kubo A, Aono M, Nakajima N, Saji H, Tanaka K, Kondo N: Differential responses in activity of antioxidant enzymes to different environmental stresses in Arabidopsis thaliana. J Plant Res 1999, 112:279–290.

40. Havaux M, Bonfils J-P, Lütz C, Niyogi KK: Photodamage of the photosynthetic apparatus and its dependence on the leaf developmental stage in the npq1 Arabidopsis mutant deficient in the xanthophyll cycle enzyme violaxanthin deepoxidase. Plant Physiol 2000, 124:273–284.

41. Yamasaki H, Sakihama Y, Ikehara N: Flavonoid-peroxidase reaction as a detoxification mechanism of plant cells against H_2O_2. Plant Physiol 1997, 115:1405–1412.

42. Yamasaki H: A function of colour. Trends Plant Sci 1997, 2:7–8.

43. Cohen MF, Sakihama Y, Yamasaki H: Roles of plant flavonoids in interactions with microbes: from protection against pathogens to the mediation of mutualism. In:Recent research developments in plant physiology (Edited by: Pandalai SG). Trivandrum, Research signpost 2001, 157–173.

44. Schreiber U: Detection of rapid induction kinetics with a new type of high-frequency modulated chlorophyll fluorometer. Photosynth Res 1986, 9:261–272.

45. Genty B, Briantais J-M, Baker NR: The relationship between the quantum yield of photosynthetic electron transport and quenching of chlorophyll fluorescence. Biochim Biophys Acta 1989, 990:87–92.

46. Agata W, Kawamitsu Y, Hakoyama S, Shima S: A system for measuring leaf gas exchange based on regulating vapour pressure difference. Photosynth Res 1986, 9:345–357.

47. Kawamitsu Y, Boyer JS: Photosynthesis and carbon storage between tides in a brown alga, Fucus vesiculosus L. Marine Biol 1999, 133:361–369.

48. Yamamoto Y, Akasaka T: Degradation of antenna chlorophyll-binding protein CP43 during photoinhibition of photosystem II. Biochemistry 1995, 34:9038–9045.

The Role of Chlorophyll b in Photosynthesis: Hypothesis

Laura L. Eggink, Hyoungshin Park and J. Kenneth Hoober

ABSTRACT

Background

The physico-chemical properties of chlorophylls b and c have been known for decades. Yet the mechanisms by which these secondary chlorophylls support assembly and accumulation of light-harvesting complexes in vivo have not been resolved.

Presentation

Biosynthetic modifications that introduce electronegative groups on the periphery of the chlorophyll molecule withdraw electrons from the pyrrole nitrogens and thus reduce their basicity. Consequently, the tendency of the central Mg to form coordination bonds with electron pairs in exogenous ligands, a reflection of its Lewis acid properties, is increased. Our hypothesis states that the stronger coordination bonds between the Mg atom in chlorophyll b and chlorophyll c and amino acid sidechain ligands in chlorophyll a/b- and a/c-binding apoproteins, respectively, enhance their import into the chloroplast and assembly of light-harvesting complexes.

Testing

Several apoproteins of light-harvesting complexes, in particular, the major protein Lhcb1, are not detectable in leaves of chlorophyll b-less plants. A direct test of the hypothesis – with positive selection – is expression, in mutant plants that synthesize only chlorophyll a, of forms of Lhcb1 in which weak ligands are replaced with stronger Lewis bases.

Implications

The mechanistic explanation for the effects of deficiencies in chlorophyll b or c points to the need for further research on manipulation of coordination bonds between these chlorophylls and chlorophyll-binding proteins. Understanding these interactions will possibly lead to engineering plants to expand their light-harvesting antenna and ultimately their productivity.

Background

In plants and algae, the reaction centers of photosystem I and II are enclosed within core complexes that contain a precisely defined set of proteins – essentially all encoded in the chloroplast genome. The primary cofactor for the photochemical reactions in these complexes, chlorophyll (Chl) a, is also required for assembly of these complexes. The end-product of the Chl biosynthetic pathway in plants in the dark, protochlorophyllide (Pchlide), is unable to support the assembly processes, which suggests that the light-dependent reduction of the double bond between C17 and C18 of Pchlide (see legend to Fig. 1) has a profound effect on the properties of the molecule. Plants and green algae (Chlorophyta) contain in addition Chl b, an accessory Chl found only in peripheral light-harvesting complexes (LHCs). These complexes usually contain three xanthophyll molecules, two luteins and one neoxanthin, and nearly equal amounts of Chl a and Chl b (7 or 8 Chl a and 5 or 6 Chl b molecules for the major LHCII, with an a/b ratio of 1.4) bound to proteins (LHCPs) that are encoded in the nuclear genome and imported into the plastid after synthesis in the cytosol. Chl b-less mutant plants are deficient in Chl and, although containing fully functional reaction centers, have a relatively low photosynthetic capacity and greater sensitivity to high-intensity light because of a deficiency in LHCs [1]. Algal species in the family Chromophyta contain Chl c (Fig. 1) instead of Chl b, which is also restricted to LHCs and seems to serve the same function in these organisms that Chl b provides in the green plants [2]. A large volume of data exists in the literature on these Chl derivatives. In this article we propose a mechanism for the important auxiliary roles these Chls play in photosynthesis.

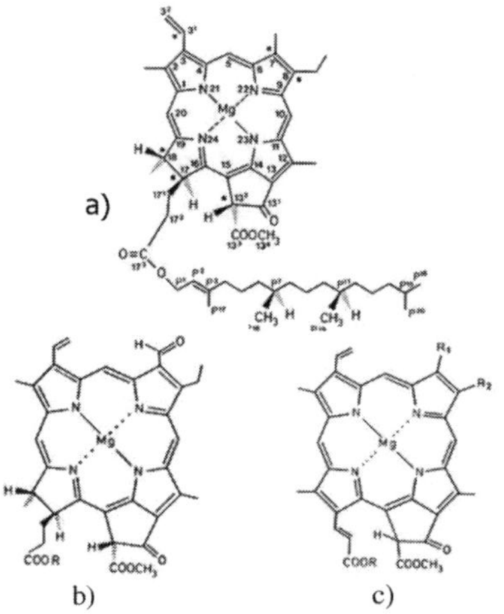

Figure 1. Structures of Chls. (a) Stereochemistry and numbering system in monovinyl-Chl a. Variations of Chl include (b) Chl b (7-formyl, R = phytyl); and (c) Chl c_1 (17^1-dehydro-Pchlide, R_1 = methyl; R_2 = ethyl or vinyl; R = H). Pchlide is similar to Chl but contains a saturated propionic rather than acrylic acid group on C17. (Structures as in [49]).

Presentation of the Hypothesis

Etioplasts, the form of the plastid that develops in dark-grown plants, were unable to insert LHCPs into membranes unless Chl was added [3]. In these experiments, in which the Zn derivatives were used because of their increased chemical stability over the Mg-containing molecules, Zn-pheophytin b was more effective in insertion than Zn-pheophytin a. An important role of Chl b was further revealed by experiments in which newly synthesized LHCPs were detected by pulse-labeling in Chl b-less mutant plants but the proteins were not recovered in chloroplasts isolated from these plants [4]. These Chl b-less plants did not accumulate several of the LHCPs, in particular Lhcb1, Lhcb6 and Lhca4 [5]. Chl b was not detected in plants exposed to intermittent light (cycles of 2 min of light and 98 min of darkness), which accumulated only small amounts of Chl a and thylakoid membranes [6,7]. Wild-type plants treated in this way accumulated only one LHCP (Lhcb5), while Chl b-less mutants exposed to intermittent light lacked all LHCPs [8]. In complementary fashion, Chl b did not accumulate when synthesis of LHCPs was inhibited [9]. When bean plants exposed to intermittent light were treated with chloramphenicol to inhibit synthesis of proteins on chloroplast ribosomes, Chl b and LHCPs accumulated in parallel with no increase in synthesis of total Chl [10]. These results indicate that photosystem I and II core complex proteins, which are synthesized in the chloroplast, compete effectively with

LHCP for small amounts of Chl a made under these conditions, and that Chl b does not accumulate until sufficient Chl a is made to satisfy core complexes.

Experiments with the model alga Chlamydomonas reinhardtii[11] showed that LHCPs were not detectably imported into the chloroplast in the absence of Chl synthesis and instead accumulated outside of the chloroplast in the cytosol and in vacuoles [12]. High concentrations of chloramphenicol caused strong suppression of total Chl synthesis when dark-grown algal cells were exposed to light, possibly by inhibition of Mg-chelatase [13] in addition to chloroplast protein synthesis. Synthesis of LHCPs on cytoplasmic ribosomes was not inhibited by chloramphenicol, and the proteins accumulated to the same level as in untreated cells [14,15]. However, because of the low rate of Chl synthesis, only a small fraction of the proteins were imported into the chloroplast and remained at the initial site of integration. As illustrated in Fig. 2b, immunoelectron microscopy detected LHCPs along the chloroplast envelope. LHCPs were not detected in the interior of the chloroplast, although cell fractionation recovered a substantial amount in a soluble form [14]. In control cells, the amount of Chl and thylakoid membranes increased rapidly when cells were illuminated, and LHCPs were detected in thylakoid membranes throughout the chloroplast (Fig. 2a). This result, obtained with cells incubated at 25°C, was consistent with localization of LHCPs on envelope membranes in cells immediately after initiation of thylakoid biogenesis at 38°C [16]. At the higher temperature, not all the newly synthesized LHCPs were incorporated into envelope membranes, and the excess accumulated in cytosolic vacuoles [16,17]. This evidence for the envelope as the site of initial interaction of LHCPs with Chl was also supported by proliferation of envelope-derived vesicles in dark-grown Chlamydomonas cells exposed to only a few minutes of light [18] and the lack of thylakoids in a mutant of Arabidopsis deficient in a protein apparently required for formation of vesicles from the inner membrane of the envelope [19].

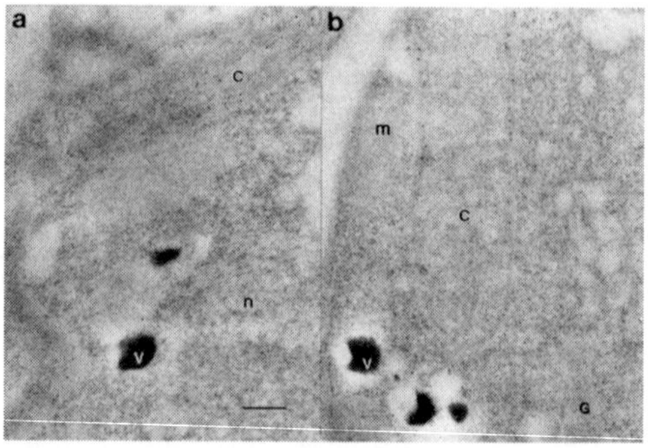

Figure 2. Immunoelectron microscopic localization of LHCPs in dark-grown cells of C. reinhardtii exposed to light for 6 h at 25°C (a) without or (b) in the presence of 200 μg chloramphenicol ml^{-1}. The experimental conditions were described previously [14]. Bound antibodies were detected with protein A conjugated to 10-nm gold particles [12]. c, chloroplast; G, Golgi; m, mitochondrion; n, nucleus; v, vacuoles. The bar = 0.5 μm.

These experiments with in vivo systems demonstrated that Chl b provides a function in LHC assembly that is not served by Chl a. Association of Chl with proteins occurs through coordination bonds between the Mg of Chl, as the Lewis acid, and amino acid sidechains as Lewis bases. The availability of an unshared pair of electrons in the Lewis base (the ligand) varies widely and is the primary factor in the strength of the coordination bond. The chemical properties of the central Mg in Chl also influence the strength of the resulting coordination bond. Biosynthetic modifications to the periphery of the tetrapyrrole ring progressively cause withdrawal of electrons from the pyrrole nitrogens, thereby decreasing their basicity [20,21]. For example, oxidation of the 7-methyl group in Chl a to the electronegative aldehyde of Chl b reduces the pK of the pyrrole nitrogens by 2 pH units. Similarly, oxidation of the propionyl sidechain on Pchlide to the acrylate group in Chl c brings its electronegative carboxyl group into conjugation with the π system of the macrocycle, with the same effect [22]. As a consequence, the central Mg atom of Chls b and c has a greater affinity for exogenous electrons, thus is a stronger Lewis acid. These considerations point to the possibility that proteins form stronger coordination bonds with Chls b and c than with Chl a, which may be particularly critical with ligands that are weak Lewis bases. The lack of an aldehyde group on the periphery of the macrocycle of Chl c, which replaces Chl b in homologous complexes in chromophytic algae, indicates that the primary interaction between Chls and the proteins does not involve such substituents. Whether phytylation of Chl is important for binding to proteins is not clear, because Chl c is incorporated into Chl a/c-protein complexes without esterification.

Testing the Hypothesis

Tamiaki et al. [23] demonstrated that introduction of an oxygen atom to the periphery of a Zn-tetrapyrrole macrocycle, as occurs in the conversion of Chl a to Chl b, increased about two-fold the equilibrium constant for formation of the coordination complex with pyridine in benzene. Consistent with this observation, studies of detergent-induced dissociation of LHCs suggested that Chl b is held by the proteins approximately two-times more tightly than Chl a[24]. Tighter binding of Chl b is apparently responsible for the well-known stability of light-harvesting complexes during mildly denaturing gel electrophoresis. The initial accumulation of LHCPs in the chloroplast envelope implies that Chl interacts with these proteins, likely by binding to the conserved motif in the first membrane-spanning region (helix-1) [25,26] when transit through the envelope is initiated. Molecular modeling suggested that this 'retention' motif – ExxHxR in the first and ExxNxR in the third membrane-spanning region – within all LHCPs and related proteins provides two ligands for Chl, an ion-pair between Glu (E) and Arg

(R) and the sidechain of either His (H) or Asn (N). Binding of Chl a to a 16-mer synthetic peptide was reduced by one-half when His within the motif sequence was replaced with Ala [27]. Replacement in addition of the Glu or Arg with Ala eliminated binding to the synthetic peptide. Import of a mutant LHCP into isolated chloroplasts was nearly abolished when His within the motif was substituted with Ala [28]. Association of Chl with this motif, therefore, appears essential for continuation of the proteins on the pathway of assembly of an LHC.

An illustration of the effect of binding two molecules of Chl with enhanced affinity to a retention motif is shown in Fig. 3. Assuming a relative equilibrium constant of 3.0 for Chl a and 5.0 for Chl b binding to a ligand in LHCP (numerals approximated from data obtained by Tamiaki et al. [23]), the increase in affinity of Chl b with the protein leads to a nearly three-fold increase in stability of the complex over that with Chl a when two molecules are bound. This conclusion is derived from the equations: R + Chl ↔ R·Chl; R·Chl + Chl↔ R·Chl2; R + 2Chl ↔ R·Chl2; Keq = [R·Chl2]/ [R] [Chl]2. The additional molecules of Chl b in LHCII would further enhance this effect by shifting the equilibria toward complex formation.

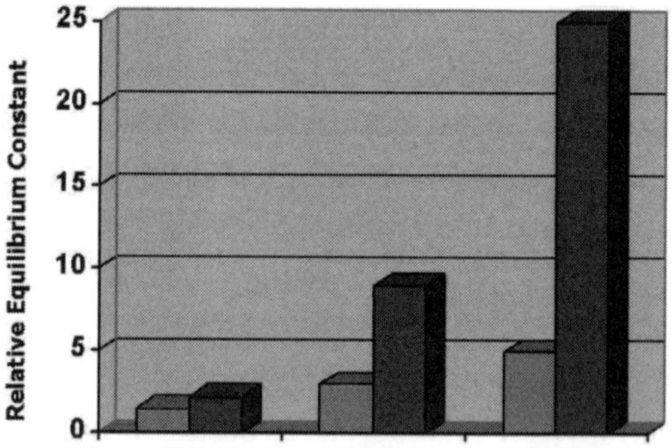

Figure 3. Graphical illustration of the relative equilibrium constants for complexes of Chl with retention motifs when one (blue) or two (magenta) molecules of Pchlide, Chl a or Chl b are bound.

The most electronegative ligand in LHCPs is the sidechain of His. Less strong Lewis bases are the charge-compensated Glu in an ion-pair with Arg, the amide group of Gln and Asn, and finally the carbonyl of the peptide backbone as the weakest [29]. The importance of the ligand was demonstrated by substitution of His with the weaker Lewis base Asn in the apoprotein of the bacterial light-harvesting complex LH1, which eliminated assembly of the complex in vivo and reconstitution in vitro [30]. Formation of a stable coordination bond with a weaker Lewis base is expected to require a stronger Lewis acid. Consistent with this prediction, a position in CP29, a minor LHCII, was preferentially filled during reconstitution by Chl a when the amino acid residue was the normal Glu, in an ion-pair with a bound Ca++ ion, but occupancy was shifted toward Chl b when the ligand was a weaker base, the amide group of Gln [31]. Although the on-rate for Chl b may be slower than that for Chl a, because binding may be impeded by a water molecule more strongly coordinated to the central Mg atom of Chl b, the greater Lewis acid strength of Chl b allows more stable bonds with the weaker ligands.

Our hypothesis on the biological role of Chl b should be reflected in the binding sites of Chl in LHCII. Resolution of the structure of native LHCII at 3.4 Å [29] revealed locations of individual Chls but did not provide identification of the Chl in each site or whether any site in the complex has mixed occupancy. The model developed from this work suggested that binding sites in the core of the complex, near the central lutein molecules, were occupied by Chl a, whereas Chl b was more peripheral. From measurements of ultrafast energy transfer kinetics within native LHCII, Gradinaru et al. [32] suggested that indeed lutein transferred excitation energy entirely to Chl a while neoxanthin, a xanthophyll bound near helix-2 (see Fig. 4), transferred energy to Chl b. With similar techniques, however, Croce et al. [33] presented evidence for detectable transfer of energy from lutein to Chl b, which suggested close contact of Chl b molecules with the central luteins. Several groups developed a more direct approach for determining occupancy by analyzing effects on the composition of the final complex, after in vitro reconstitution, when amino acid residues in LHCPs were replaced with substitutes that are unable to serve as a ligand. For example, steric hindrance caused by substitution of bulky Phe for Gly78 (residue numbers are given with reference to Lhcbl) in the position designated a6 [29] prevented this peptide carbonyl, non-H-bonded because of Pro82 one helical turn further, from serving as a ligand (see Fig. 4). This change resulted in loss of one Chl b after reconstitution [34]. Gln 131 (b6) and Glu 139 (in an ion-pair with Arg142) (b5) were also identified as ligands to Chl b [31,34-36]. Remelli et al. [35] found that substitution of ligand Gln197 (a3) or His212 (b3) with Leu or Val, respectively, led to sub-integral loss of Chl a and Chl b, which indicated mixed occupancy in each site. Studies by Rogl and Kühlbrandt [34], on the other hand, suggested that both sites were filled with Chl a. Assignments after in vitro reconstitution may have a degree of uncertainty, because the composition of the final complex varies as a function of the Chl a:b ratio in the reconstitution mixture [37]. Mixed sites, even when Chl a was present in excess in the mixture [35], probably reflected a preference for binding of Chl b to the protein. Based on ligand strength, and the likelihood that occupancy is unambiguous in vivo[34], His212 may serve as a ligand for Chl a and Gln197 for Chl b.

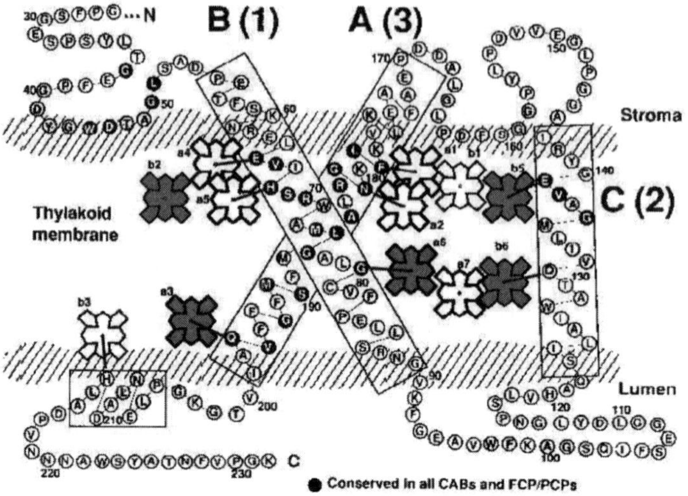

Figure 4. Model of the association of Chl with Lhcb1. The arrangement of the protein in thylakoid membranes is illustrated according to ref. 50. The "core" Chls (a1, a2, a4 and a5) are shown as Chl a according to ref. [35]. The green color marks positions of Chl b as proposed in the text. Sites a3 and b3, although mixed in occupancy after reconstitution [35], were assigned as shown based on ligand strength. At least four of the five Chl b molecules are coordinated directly to the protein. The biological requirement of Chl b for accumulation of Lhcb1 (see text) suggests an alternate assignment for a4, as also proposed in ref. [34].

Mutation of Glu65 (a4) or Asn183 (a2) each resulted in loss of one Chl a and one Chl b[35]. Rogl and Kühlbrandt [34] suggested that Glu65 (in an ion-pair with Arg185) may be a ligand for Chl b, with another site, occupied by Chl a, affected by loss of the protein-bound Chl. Chl b in site a4 would be consistent with the biological necessity of association of Chl b with helix-1 for retention of the protein in the chloroplast. However, based on similarity to results from reconstitution of the more simple CP29 (Lhcb4) [31], Remelli et al. [35] suggested that Glu65 (a4) and Asn183 (a2) are occupied by Chl a . Loss of the latter Chl apparently resulted in loss of 'out-lying' Chl b in site b2, which is near a2 in the 3-dimensional structure. These assignments thus account for the five Chl b molecules in the complex (Fig. 4). Site b1 must consequently be filled with a Chl a molecule [38]. The orientation of the transition moments of Chl b in sites b5 and b6 [38] suggest that an 'out-lying' Chl a molecule could coordinate with the formyl group of Chl b, a sterically more favorable arrangement than coordination to the 131-carbonyl oxygen because of the opposing orientations of the 132-carboxymethyl and 17-propionyl group (Fig. 1). Although coordination of an 'out-lying' Chl to a protein-bound Chl would enhance Lewis acid strength of the latter, the distances between Chls [29,35] suggest that interaction would require mediation by water molecules. Alternatively, these Chls may coordinate with peptide carbonyl groups.

Site a6, considered to be filled with Chl b[34,38], may play a role in retention of LHCPs in the chloroplast. Lhcb6, a minor LHCP, contains Gly instead of Pro at the position analogous to 82 in Lhcb1, thus eliminating the peptide bond carbonyl of Gly as a ligand, but Lhcb6 has a potential ligand for Chl b in Gln83 [39,40]. These two LHCP sub-species, along with Lhca4 are most affected by the lack of Chl b in vivo[5]. Lhcb4 (apoprotein of CP29) has Val instead of Pro at 'position 82', and the absence of site a6 in Lhcb4 may contribute to its drastic reduction in Chl b-less mutants [5,8]. However, Lhcb2, Lhcb3, Lhca1, Lhca2 and Lhca3 contain the Gly peptide carbonyl as a ligand (each has Pro at 'position 82' [40]) but are reduced only slightly, if any, in amount by the lack of Chl b. Site a6 may therefore not be essential to accumulation of the protein but serve in concert with initial involvement of Chl b, directly or indirectly, with the completely conserved retention motif. Because interactions that develop during import may be altered as the result of conformational changes as the complex assembles, in particular, as the retention motif loop [27] is stretched into a helical structure, the final occupancy in each site in the final complex may not reflect the initial associations. Understanding the constraints on assembly of the complex in vivo – including retraction into the cytosol when the amount of Chl is insufficient [12] – and the order in which Chls are bound, will require new experimental design. We expect that synthesis of Chl b by Chl(ide) a oxidase [41] will be determined by the local environment around specific Chl a molecules, created by the assembly process. It is interesting to note that the retention motif in all LHCs that contain Chl b is followed by a Trp residue, which may be involved in synthesis of Chl b.

A converse mutagenesis approach would provide a rigorous test of the hypothesis. A stable complex should be achieved with only Chl a, in a Chl b-less plant or by in vitro reconstitution, when weak ligands in LHCPs are replaced with stronger Lewis bases. Increased strength of the engineered coordination bonds with Chl a should compensate for the lack of Chl b. In particular, a stable complex should accumulate after Gln131, Glu139, Asn183 and Gln197 in Lhcb1 are replaced with His. A stronger ligand could also be introduced in the position of Gly78, which seems to be the weakest ligand in the complex. Substitution of these amino acids in the sequence of Lhcb1, a major LHCP that can not be detected in Chl b-less plants [5,8], would be expected to restore accumulation of the protein with only Chl a. This experiment provides a positive in vivo selection for validation of the hypothesis, in contrast to the dramatic decrease in accumulation of the proteins when ligands are removed by substitution with non-ligand amino acids [42]. Furthermore, whereas stable complexes can be achieved by reconstitution with wild-type Lhcb1 and only Chl b but not only Chl a[37,43], the hypothesis predicts that stable complexes can be reconstituted with the mutant protein containing these substitutions and Chl a.

Implications of the Hypothesis

An extensive amount of evidence in the literature supports the hypothesis presented in this article on the role of Chl b. It should be noted, however, that several LHCPs accumulate in chloroplasts in the absence of Chl b[5,8], perhaps because they integrate more easily into membranes, which implies that other features of the proteins are involved. The work already done has established that several LHCPs are imported into the chloroplast at a substantial rate only when sufficient Chl b is available and they accumulate initially in the envelope membrane. Results from in vivo experiments have shown that interaction of Chl b with the first membrane-spanning region, including the retention motif, is critical for progression of import of these proteins. The initial steps in assembly also require the abundant xanthophyll lutein [26], which has not been the focus of this article. The availability of Chl b thus strongly regulates import of LHCPs as well as assembly and eventual accumulation of light-harvesting complexes. The resulting dramatic enhancement in the efficiency of light capture for photosynthesis apparently provided a strong evolutionary pressure for development of the ability of photosynthetic organisms to synthesize Chl b or Chl c [44].

The structure of LHCs has been extensively studied and linkage of the complexes to reaction centers, physically and functionally, is well understood. Further understanding of LHC assembly requires a better knowledge of the characteristics of the reaction catalyzed by Chl(ide) a oxidase and whether Chl b is restricted to these complexes because LHCP serves as a specific effector of the oxidation of Chl(ide) a or whether the protein simply provides binding sites for Chl b and prevents its conversion back to Chl a[45]. The latter appears less likely as a specific effect, because similar ligands should occur in other proteins. In particular, the early-light induced proteins are homologous to LHCPs but bind little if any Chl b [46]. The mechanism of Chl b synthesis, an oxidation of the methyl group at position 7 [41], will be an area of active research in the future, now that the gene for Chl(ide) a oxidase has been identified [47,48]. Moreover, it is not known whether a pool of free Chl b exists in a local environment in chloroplast membranes that is mimicked by the amount of Chl b in reconstitution experiments. Attempts to understand assembly of the complex in vivo will provide ample opportunity for additional experimental work.

Abbreviations

Chl, chlorophyll; Chlide, chlorophyllide; Pchlide, protochlorophyllide; LHC, light-harvesting complex; LHCII, LHC associated primarily with photosystem II; LHCP, LHC apoprotein; Lhcb or Lhca, apoproteins of LHCs associated with photosystem II or I, respectively.

Acknowledgements

L.L.E was supported by Graduate Training Grant DGE9553456 from the National Science Foundation. This is publication number 498 from the Center for the Study of Early Events in Photosynthesis at Arizona State University.

References

1. Leverenz JW, Öquist G, Wingsle G: Photosynthesis and photoinhibition in leaves of chlorophyll b-less barley in relation to absorbed light. Physiol Plant 1992, 85:495–502.
2. Durnford DG, Deane JA, Tan S, McFadden GI, Gant E, Green BR: A plylogenetic assessment of the eukaryotic light-harvesting antenna proteins, with implications for plastid evolution. J Mol Evol 1999, 48:59–68.
3. Kuttkat A, Edhofer I, Eichacker LA, Paulsen H: Light-harvesting chlorophyll a/b-binding protein stably inserts into etioplast membranes supplemented with Zn-pheophytin a/b. J Biol Chem 1997, 272:20451–20455.
4. Preiss S, Thornber JP: Stability of the apoproteins of light-harvesting complex I and II during biogenesis of thylakoids in the chlorophyll b-less barley mutant chlorina f2. Plant Physiol 1995, 107:709–717.
5. Bossmann B, Grimme LH, Knoetzel J: Protease-stable integration of Lhcb1 into thylakoid membranes is dependent on chlorophyll b in allelic chlorina-f2 mutants of barley (Hordeum vulgare L.). Planta 1999, 207:551–558.
6. Akoyunoglou G, Argyroudi-Akoyunoglou JH: Effects of intermittent light and continuous light on the Chl formation in etiolated plants at various ages. Physiol Plant 1969, 22:228–295.
7. Akoyunoglou G: Development of the photosystem II unit in plastids of bean leaves greened in periodic light. Arch Biochem Biophys 1977, 183:571–580.
8. Król M, Spangfort MD, Huner NPA, Öquist G, Gustafsson P, Jansson S: Chlorophyll a/b-binding proteins, pigment conversions, and early light-induced proteins in a chlorophyll b-less barley mutant. Plant Physiol 1995, 107:873–883.
9. Maloney MA, Hoober JK, Marks DB: Kinetics of chlorophyll accumulation and formation of chlorophyll-protein complexes during greening of Chlamydomonas reinhardtii y-1 at 38°C. Plant Physiol 1989, 91:1100–1106.
10. Tzinas G, Argyroudi-Akoyunoglou JH: Chloramphenicol-induced stabilization of light-harvesting complexes in thylakoids during development. FEBS Lett 1988, 229:134–141.

11. Harris EH: Chlamydomonas as a model organism. Annu Rev Plant Physiol Plant Mol Biol 2001, 52:363–406.
12. Park H, Hoober JK: Chlorophyll synthesis modulates retention of apoproteins of light-harvesting complex II by the chloroplast in Chlamydomonas reinhardtii. Physiol Plant 1997, 101:135–142.
13. Kannangara CG, Vothknecht UC, Hansson M, von Wettstein D: Magnesium chelatase: association with ribosomes and mutant complementation studies identify barley subunit xantha-G as a functional counterpart of Rhodobacter subunit BchD. Mol Gen Genet 1997, 254:85–92.
14. Hoober JK: A major polypeptide of chloroplast membranes of Chlamydomonas reinhardi. Evidence for synthesis in the cytoplasm as a soluble component. J Cell Biol 1972, 52:84–96.
15. Hoober JK, Stegeman WJ: Kinetics and regulation of synthesis of the major polypeptides of thylakoid membranes in Chlamydomonas reinhardtii y-1 at elevated temperatures. J Cell Biol 1976, 70:326–337.
16. White RA, Wolfe GR, Komine Y, Hoober JK: Localization of light-harvesting complex apoproteins in the chloroplast and cytoplasm during greening of Chlamydomonas reinhardtii at 38°C. Photosynth Res 1996, 47:267–280.
17. Wolfe GR, Park H, Sharp WP, Hoober JK: Light-harvesting complex apoproteins in cytoplasmic vacuoles in Chlamydomonas reinhardtii (Chlorophyta). J Phycol 1997, 33:377–386.
18. Hoober JK, Boyd CO, Paavola LG: Origin of thylakoid membranes in Chlamydomonas reinhardtii y-1 at 38°C. Plant Physiol 1991, 96:1321–1328.
19. Kroll D, Meierhoff K, Bechtold N, Kinoshita M, Westphal S, Vothknecht U, Soll J, Westhoff P: VIPP1, a nuclear gene of Arabidopsis thaliana essential for thylakoid membrane formation. Proc Natl Acad Sci USA 2001, 98:4238–4242.
20. Phillips JN: Physico-chemical properties of porphyrins. In: Comprehensive Biochemistry, Vol 9 (edited by Florkin M, Stotz EH), Amsterdam, Elsevier 1963, 34–72.
21. Smith KM: General features of the structure and chemistry of porphyrin compounds. In: Porphyrins and Metalloporphyrins (Edited by Smith KM) Amsterdam, Elsevier Scientific 1975, 1–58.
22. Dougherty RC, Strain HH, Svec WA, Uphaus RA, Katz JJ: The structure, properties and distribution of chlorophyll c. J Am Chem Soc 1970, 92:2826–2833.

23. Tamiaki H, Yagai S, Miyatake T: Synthetic zinc tetrapyrroles complex with pyridine as a single axial ligand. Bioorg Med Chem 1998, 6:2171–2178.

24. Ruban AV, Lee PJ, Wentworth M, Young AJ, Horton P: Determination of the stoichiometry and strength of binding of xanthophylls to the photosystem II light harvesting complex. J Biol Chem 1999, 274:10458–10465.

25. Green BR, Pichersky E: Hypothesis for the evolution of three-helix Chl a/b and Chl a/c light-harvesting antenna proteins from two-helix and four-helix ancestors. Photosynth Res 1994, 39:149–162.

26. Hoober JK, Eggink LL: Assembly of light-harvesting complex II and biogenesis of thylakoid membranes in chloroplasts. Photosynth Res 1999, 61:197–215.

27. Eggink LL, Hoober JK: Chlorophyll binding to peptide maquettes containing a retention motif. J Biol Chem 2000, 275:9087–9090.

28. Kohorn BD: Replacement of histidines of light harvesting chlorophyll a/b binding protein II disrupts chlorophyll-protein complex assembly. Plant Physiol 1990, 93:330–342.

29. Kühlbrandt W, Wang DN, Fujiyoshi Y: Atomic model of plant light-harvesting complex by electron crystallography. Nature 1994, 367:614–621.

30. Davis CM, Bustamante PL, Todd JB, Parkes-Loach PS, McGlynn P, Olsen JD, McMaster L, Hunter CN, Loach PA: Evaluation of structure-function relationships in the core light-harvesting comlex of photosynthetic bacteria by reconstitution with mutant polypeptides. Biochemistry 1997, 36:3671–3679.

31. Bassi R, Croce R, Cugini D, Sandona D: Mutational analysis of a higher plant antenna protein provides identification of chromophores bound into multiple sites. Proc Natl Acad Sci USA 1999, 96:10056–10061.

32. Gradinaru CC, Stokkum van IHM, Pascal AA, van Grondelle R, van Amerongen H: Identifying the pathways of energy transfer between carotenoids and chlorophylls in LHCII and CP29. A multicolor, femtosecond pump-probe study. J Phys Chem B 2000, 104:9330–9342.

33. Croce R, Müller MG, Bassi R, Holzwarth AR: Carotenoid-to-chlorophyll energy transfer in recombinant major light-harvesting complex (LHCII) of higher plants. I. Femtosecond transient absorption measurements. Biophys J 2001, 80:901–915.

34. Rogl H, Külbrandt W: Mutant trimers of light-harvesting complex II exhibit altered pigment content and spectroscopic features. Biochemistry 1999, 38:16214–16222.

35. Remelli R, Varotto C, Sandoná D, Croce R, Bassi R: Chlorophyll binding to monomeric light-harvesting complex: a mutational analysis of chromophore-binding residues. J Biol Chem 1999, 274:33510–33521.

36. Yang C, Kosemund K, Comet C, Paulsen H: Exchange of pigment-binding amino acids in light-harvesting chlorophyll a/b protein. Biochemistry 1999, 38:16205–16213.

37. Kleima FJ, Hobe S, Calkoen F, Urbanus ML, Peterman EJG, van Grondelle R, Paulsen H, van Amerongen H: Decreasing the chlorophyll a/b ratio in reconstituted LHCII: structural and functional consequences. Biochemistry 1999, 38:6587–6596.

38. van Amerongen H, van Grondelle R: Understanding the energy transfer function of LHCII, the major light-harvesting complex of green plants. J Phys Chem B 2001, 105:604–617.

39. Jansson S: The light-harvesting chlorophyll a/b-binding proteins. Biochim Biophys Acta 1994, 1184:1–19.

40. Jansson S: A guide to the Lhc genes and their relatives in Arabidopsis. Trends Plant Sci 1999, 4:236–240.

41. Oster U, Tanaka R, Tanaka A, Rüdiger W: Cloning and functional expression of the gene encoding the key enzyme for chlorophyll b biosynthesis (CAO) from Arabidopsis thaliana. Plant J 2000, 21:305–310.

42. Flachmann R, Kühlbrandt W: Crystallization and identification of an assembly defect of recombinant antenna complexes produced in transgenic tobacco plants. Proc Natl Acad Sci USA 1996, 93:14966–14971.

43. Schmid VHR, Thomé P, Rühle W, Paulsen H, Kühlbrandt W, Rogl H: Chlorophyll b is involved in long-wavelength spectral properties of light-harvesting complexes LHC I and LHC II. FEBS Lett 2001, 499:27–31.

44. Tomitani A, Okada K, Miyashita H, Matthijs HCP, Ohno T, Tanaka A: Chlorophyll b and phycobilins in the common ancestor of cyanobacteria and chloroplasts. Nature 1999, 400:159–162.

45. Ohtsuka T, Ito H, Tanaka A: Conversion of chlorophyll b to chlorophyll a and the assembly of chlorophyll with apoproteins by isolated chloroplasts. Plant Physiol 1997, 113:137–147.

46. Adamska I, Kruse E, Kloppstech K: Stable insertion of the early light-induced proteins into etioplast membranes requires chlorophyll a. J Biol Chem 2001, 276:8582–8587.

47. Tanaka A, Ito H, Tanaka R, Tanaka NK, Yoshida K: Chlorophyll a oxygenase (CAO) is involved in chlorophyll b formation from chlorophyll a. Proc Natl Acad Sci USA 1998, 95:12719–12723.

48. Espineda CE, Linford AL, Devine D, Brusslan JA: The AtCAO gene, encoding chlorophyll a oxygenase, is required for chlorophyll b synthesis in Arabidopsis thaliana. Proc Natl Acad Sci USA 1999, 96:10507–10511.

49. Scheer H: Structure and occurrence of chlorophylls. In: Chlorophylls (Edited by Scheer H) Boca Raton, CRC Press 1991, 3–30.

50. Green BR, Durnford DG: The chlorophyll-carotenoid proteins of oxygenic photosynthesis. Annu Rev Plant Physiol Plant Mol Biol 1996, 47:685–714.

Exploring Photosynthesis Evolution by Comparative Analysis of Metabolic Networks between Chloroplasts and Photosynthetic Bacteria

Zhuo Wang, Xin-Guang Zhu, Yazhu Chen, Yuanyuan Li, Jing Hou and Yixue Li and Lei Liu

ABSTRACT

Background

Chloroplasts descended from cyanobacteria and have a drastically reduced genome following an endosymbiotic event. Many genes of the ancestral cyanobacterial genome have been transferred to the plant nuclear genome by horizontal gene transfer. However, a selective set of metabolism pathways is maintained in chloroplasts using both chloroplast genome encoded and nuclear genome encoded enzymes. As an organelle specialized for carrying out photosynthesis, does the chloroplast metabolic network have properties adapted for higher efficiency of photosynthesis? We compared metabolic network properties of chloroplasts and prokaryotic photosynthetic organisms, mostly cyanobacteria, based on metabolic maps derived from genome data to identify features of chloroplast network properties that are different from cyanobacteria and to analyze possible functional significance of those features.

Results

The properties of the entire metabolic network and the sub-network that consists of reactions directly connected to the Calvin Cycle have been analyzed using hypergraph representation. Results showed that the whole metabolic networks in chloroplast and cyanobacteria both possess small-world network properties. Although the number of compounds and reactions in chloroplasts is less than that in cyanobacteria, the chloroplast's metabolic network has longer average path length, a larger diameter, and is Calvin Cycle -centered, indicating an overall less-dense network structure with specific and local high density areas in chloroplasts. Moreover, chloroplast metabolic network exhibits a better modular organization than cyanobacterial ones. Enzymes involved in the same metabolic processes tend to cluster into the same module in chloroplasts.

Conclusion

In summary, the differences in metabolic network properties may reflect the evolutionary changes during endosymbiosis that led to the improvement of the photosynthesis efficiency in higher plants. Our findings are consistent with the notion that since the light energy absorption, transfer and conversion is highly efficient even in photosynthetic bacteria, the further improvements in photosynthetic efficiency in higher plants may rely on changes in metabolic network properties.

Background

Photosynthesis is one of the most important and fundamental metabolic processes in the biosphere. The appearance of photosynthesis in prokaryotic organisms early in the earth's history fundamentally changed the composition of the atmosphere and subsequently determined the evolution of organisms. According to the theory of endosymbiosis, chloroplasts descended from cyanobacteria [1,2]. During endosymbiosis, the ancestral cyanobacterial genome was drastically reduced, and many genes were transferred to the nuclear genome [1,3]. As a result, the majority of the enzymes in chloroplast metabolic networks are nucleus-encoded, translated in cytosol, and then imported into chloroplasts [4]. Such massive transportation of proteins requires a large amount of energy and sophisticated regulation from plant cells. Since the metabolic networks in chloroplasts are mostly constructed with proteins encoded in nuclear genome, do the networks exhibit some unique properties and characteristics that deviate from the ancestors' metabolic networks? To answer this question, we conducted a comparative study of the metabolic networks between chloroplasts and several photosynthetic bacteria.

Studies on the evolution of photosynthesis have mostly focused on individual proteins or protein complexes related to photosynthesis [1,5-7]. With the recent advancements in genomics and the development of metabolic pathway databases, we are now able to reconstruct metabolic networks from complete and annotated genomes and conduct system-level comparisons of the metabolic networks. Recently, there have been several such studies comparing system-wide network properties among many organisms [8,9]. In this study, we examined the similarity and differences of network properties between chloroplasts and the photosynthetic bacteria including connectivity, clustering coefficient, path length, network diameter [8,9], and modularity [10-13]. Comparisons of modular structures of the metabolic network provide insights about the modification of major metabolisms of chloroplasts, such as addition or loss of certain metabolisms and the changes in the organization of metabolism due to endosymbiosis.

Results

Chloroplast Metabolic Network Exhibits Different Characteristics Compared to Photosynthetic Bacteria

The basic statistics of reconstructed metabolic networks in chloroplasts and photosynthetic bacteria are shown in Table 1. The numbers of enzymes in all metabolic networks were similar. However, there were more cases of one enzyme catalyzing two or more reactions in the photosynthetic bacteria. For example, aminomethyltransferase (EC 2.1.2.10) catalyzes three reactions in synechorocus sp. WH8102 (syw):

Glycine + Tetrahydrofolate + NAD^+ <=> 5,10-Methylenetetrahydrofolate + NH_3 + CO_2 + NADH + H^+

5-Formyltetrahydrofolate <=> 5,10-Methenyltetrahydrofolate + H_2O

Tetrahydrofolate + S-Aminomethyldihydrolipoylprotein <=> 5,10-Methylenetetrahydrofolate + NH_3 + Dihydrolipoylprotein

Table 1. Structure and topological properties of whole network in chloroplasts and several photosynthetic bacteria.

Species	Enzyme number	Compound number	Reaction number	Average compound connectivity	Enzyme CC	Compound CC	Enzyme AL	Compound AL	Enzyme Diameter	Compound Diameter
Chloroplast	376	586	560	2.6185	0.534371	0.431872	5.07847	4.83902	19	19
syw	371	860	694	2.5615	0.59365	0.503954	4.07523	3.972854	11	12
ana	401	881	728	2.5182	0.590467	0.513945	4.15901	3.95608	11	12
cte	323	724	579	2.4627	0.577056	0.506881	4.12231	3.94473	12	12
gvi	377	830	683	2.4789	0.594211	0.518726	4.15974	3.95251	12	12
pma	338	718	578	2.5195	0.577878	0.487342	4.09658	3.92037	12	12
pmm	342	760	614	2.5447	0.590459	0.48967	4.06937	3.92196	10	11
pmt	352	791	626	2.4855	0.581159	0.495484	4.09455	3.98362	12	12
syn	387	823	681	2.5176	0.590339	0.501971	4.1349	3.91225	12	12
tel	350	653	585	2.6718	0.593009	0.488283	4.11994	3.87589	11	12

CC: clustering coefficient; AL: average path length.

In contrast, only the last reaction exists in the chloroplast network. When we compared enzymes in chloroplasts and photosynthetic bacteria, we found some differences among them. For example, there are 376 and 371 enzymes respectively in chloroplast and Synechococcus sp. WH8102 (syw) metabolic network, among which 210 enzymes are shared by them.

Even though the numbers of compounds and reactions in chloroplast network are fewer than those in photosynthetic bacteria, the average connectivity of compound nodes is very similar among them (Table 1). In addition, the distribution of compound connectivity in chloroplasts and cyanobacteria followed the Power law. The average clustering coefficients, the average path lengths and the diameters of both enzyme and compound nodes (Table 1) confirmed that the metabolic networks under study are scale-free and small-world networks using hypergraph model. It is evident from Table 1 that the topological properties are very similar among all photosynthetic bacteria, while chloroplasts exhibit some differences. Although the chloroplast network has fewer compound nodes and hyper-edges in its hypergraph representation, the average path lengths and diameters of both enzyme and compound nodes are longer than those in photosynthetic bacteria. The average clustering coefficient of both enzyme and compound nodes are lower in chloroplasts, suggesting an overall loose network structure in chloroplast. We also conducted an in-depth comparison of the densities of enzyme networks in chloroplasts and cyanobacteria by analyzing the cores using Pajek [14]. The k-core of a network is defined as a subnetwork of a given network where each vertex has at least k neighbors in the same core. For chloroplasts and Synechococcus sp. WH8102 (syw), the largest core includes 32 and 37 enzymes respectively, among which 24 enzymes are shared by the two cores.

The Network is Highly Clustered Around Calvin Cycle in Chloroplasts

For the SubNetwork, which includes reactions directly connected with the Calvin Cycle, the average clustering coefficient is higher and the average path length is shorter than the whole network, indicating tighter linkage between reactions in the SubNetwork, in both chloroplast and photosynthetic bacteria. Although the overall chloroplast network shows a lower average clustering coefficient and longer average path length compared to photosynthetic bacteria, the ratio of average clustering coefficient between the SubNetwork and the whole network is higher in chloroplasts than that in photosynthetic bacteria. The ratio of average path length between the SubNetwork and whole network is lower in chloroplasts than that in photosynthetic bacteria (Figure 1), suggesting that the chloroplast network is highly clustered around the Calvin Cycle.

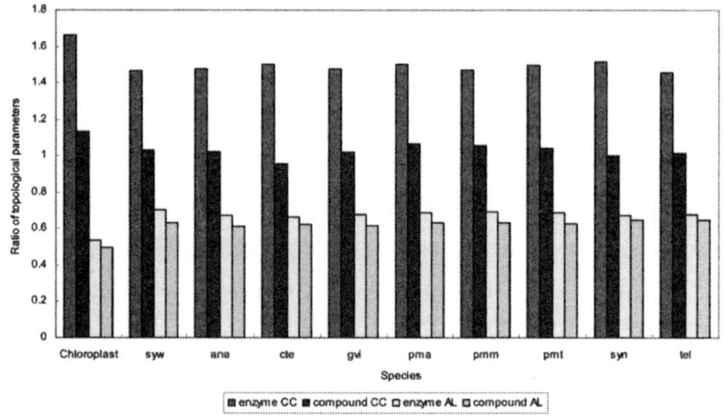

Figure 1. Ratio of topological properties in SubNetwork to whole network for chloroplasts and photosynthetic bacteria. CC: clustering coefficient; AL: average path length.

Furthermore, we made an interesting observation when we ranked the connectivity of different compounds in the network. We extracted the top ten connected (hub) compounds in the whole network and then checked their ranks in the SubNetwork. It is interesting to notice that glutamate, which is a crucial compound for nitrogen assimilation, is highly connected (hub) in the whole networks of both chloroplast and cyanobacteria. However, glutamate does not exist in the chloroplast SubNetwork but still exists in all cyanobacteria SubNetworks. The difference lies in the reaction L-Glutamate <=> 4-Aminobutanoate + CO_2 catalyzed by L-Glutamate 1-carboxy-lyase (EC 4.1.1.19), which is missing in chloroplast. This observation suggests that the nitrogen assimilation is not directly linked to carbon fixation in chloroplasts, but is linked in cyanobacteria.

Simulation of the Possible Impact of an Incomplete Dataset on the Topological Properties of Metabolic Network

Most data collected in this study were originated from genome annotations, which may be incomplete. In order to assess the effect of such incomplete data, we designed an experiment using the well-studied and most complete E. coli metabolic network. First, the topological properties of the entire network were calculated using the hypergraph model. Then, fractions of enzymes and reactions were randomly removed from the network and the network properties were again calculated. The results after random removal of nodes were used to simulate the impact of incomplete metabolic information on the full network. Table 2 demonstrates that the topological properties of the metabolic network remain nearly unaffected when 35% of the enzymes were randomly removed. Even after removal of 50% the topological parameters change by less than 5% from those of the complete network. The diameters increase by 8.33% over the original network, which represents the most significantly changed parameter, but this value is far lower than the differences of network parameters between chloroplasts and photosynthetic bacteria, indicating that the topological differences of the two networks are unlikely to be caused by an incomplete dataset. These results strongly validate the significance of our comparisons between chloroplasts and photosynthetic bacteria and support the conclusion that chloroplasts have an overall loose but strongly Calvin Cycle-centered network structure.

Table 2. Change of topological properties with randomly reducing size in E. coli metabolic network.

Topological properties	Whole network	5% reduced network	15% reduced network	25% reduced network	35% reduced network	50% reduced network
enzyme CC	0.584591	0.584634	0.593573	0.588487	0.599161	0.598578
compound CC	0.494804	0.498944	0.51094	0.515453	0.523472	0.530856
enzyme AL	4.1142	4.14179	4.20145	4.21134	4.24669	4.31382
compound AL	4.11805	4.14388	4.23782	4.28722	4.34033	4.30933
enzyme diameter	12	12	12	12	12	13
compound diameter	12	12	12	12	12	13

The Chloroplast Network Shows a Better Modular Structure than Photosynthetic Bacteria

A natural step after the study of overall properties of a complex network is to investigate the substructures within the network and possible functions of the substructures. One of the methods to decompose a complex network structure is to find modules within the network based on the connectivity among the nodes. In this study, we view modules as sub-networks where the nodes are highly connected within a module, but much less connected between modules.

Many approaches have been used to detect modules in metabolic network including elementary modes, extreme pathways, flux analysis [15-17], and graph clustering techniques such as Markov Clustering [MCL, 18], Iterative Conductance Cutting [ICC, 19], and Geometric Minimal Spanning Tree Clustering [GMC, 20]. After comparison, we adopted the method from Guimerà and Amaral [21,22] to identify modules in metabolic networks in chloroplasts and photosynthetic bacteria (see detailed description in the "Methods" section). This method is called the SA module-detection algorithm in the remainder of the text.

Modular structures differ among different organisms. The similarity of overall modular structure among chloroplasts, photosynthetic bacteria, E.coli, Arabidopsis thaliana and Cyanidioschyzon merolae has been calculated and is shown as a dendrogram in Figure 2 (see "Methods" section for detailed description of the similarity measurements of modules). Remarkably, all cyanobacteria exhibit very similar modular organization and are different from chloroplasts. Arabidopsis thaliana and Cyanidioschyzon merolae are clustered together with high similar modular structure. This result is consistent with the topological results (Table 1) that chloroplast metabolic network shows different characteristics.

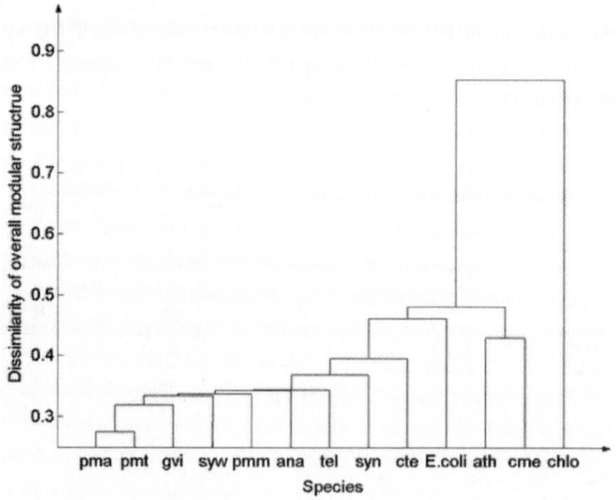

Figure 2. Similarity of overall modular structures among chloroplasts, photosynthetic bacteria, E.coli, Arabidopsis thaliana and Cyanidioschyzon merolae.

Matching modules to particular metabolisms reveals the possible biological significance of modularity [21,22]. The function of each enzyme module in chloroplast and photosynthetic bacteria was classified using the classification scheme proposed in KEGG which includes nine major pathways: carbohydrate metabolism, energy metabolism, lipid metabolism, nucleotide metabolism, amino-acid metabolism, glycan biosynthesis and metabolism, metabolism of cofactors and vitamins, biosynthesis of secondary metabolites, and biodegradation of xenobiotics. Based on Guimerà and Amaral [21,22], we mapped the modules to KEGG functional classifications; if more than 50% of the enzymes in a module belong to one major pathway, then the module is considered pathway specific. The match between modules and KEGG classifications for chloroplasts and Synechococcus sp. WH8102 (syw) are shown in Figure 3. Other cyanobacteria showed similar functional categories mapping to their corresponding modules. Interestingly, glycan biosynthesis and metabolism, and biodegradation of xenobiotics are absent in chloroplasts but present in cyanobacteria (Figure 3A,B). In addition, some metabolic processes related to gibberellins, abscisic acid, brassinolide, cytokinin, indole-3-acetic acid, ethylene, polyamine and jasmonic acid are specific to chloroplasts, which are mostly included in module 3. Most of these molecules are related to hormone synthesis or metabolism [23-25].

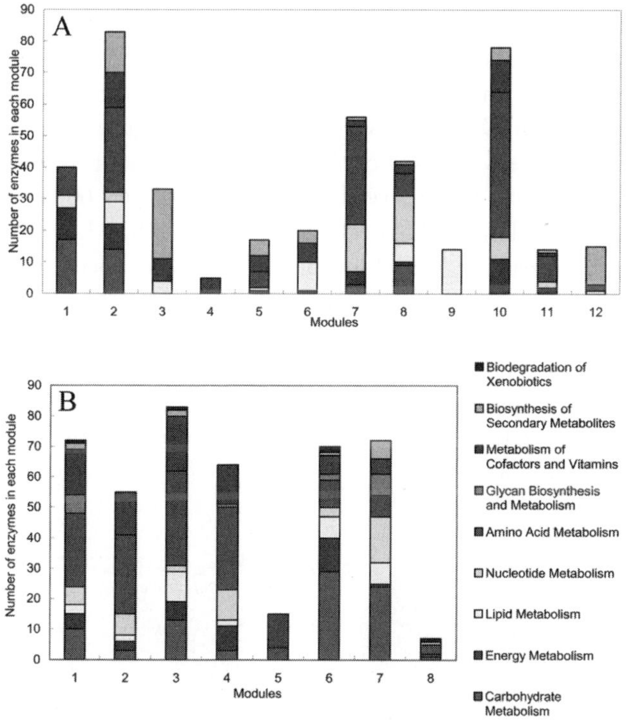

Figure 3. Comparison of functional modules in chloroplasts and cyanobacteria. (A) Chloroplast enzyme modules; (B) Synechococcus sp. WH8102 (syw) enzyme modules map according to KEGG classification.

Several modules were organized around amino-acid metabolic functions in both chloroplasts and Synechococcus sp. WH8102 networks, which are module 2, 7, 10, 11 in chloroplast and module 1, 2, 3, 4 in Synechococcus sp. WH8102, respectively. In chloroplasts, module 4 exclusively consists of enzymes in cofactor and vitamin metabolism, and all enzymes in module 9 belong to lipid metabolism (Figure 3A). However no module in Synechococcus sp. WH8102 completely corresponds to any one specific pathway (Figure 3B). Nearly 90% of the enzymes in module 3 in the chloroplast network are related to biosynthesis of secondary metabolites. Also 80% enzymes in module 12 relate to hormone metabolism in chloroplasts (Figure 3A). In contrast, only module 5 and module 8 in the cyanobacteria contain more than 50% enzymes belonging to cofactor and vitamin metabolism and to amino acid metabolism respectively (Figure 3B).

By comparing the similarity between any two modules in chloroplasts and each photosynthetic bacterium, we found for each bacterium 5 to 7 modules similar to corresponding modules in chloroplasts. Moreover five pairs of these modules are very conserved among chloroplasts and photosynthetic bacteria: three pairs correspond to amino-acid metabolism, two pairs belong to carbohydrate metabolism and nucleotide metabolism respectively, all of which are related to the core metabolism. It is evident that the core metabolic processes are conserved in evolution. As an example, the comparison of modules between chloroplast and Synechococcus sp. WH8102 was visualized in Figure 4. The five modules with the same color are composed of similar enzymes, mapped to the same functional pathways. These five conserved modules include 69.68% and 80.32% of all enzymes in chloroplasts and Synechococcus sp. WH8102, respectively. Of the common 210 enzymes between chloroplasts and Synechococcus sp. WH8102, approximately 60% of them exist in the conservative modules. The other modules in chloroplasts mainly correspond to metabolism of cofactors and vitamins, and biosynthesis of secondary metabolites. This result indicates that the core metabolisms of chloroplasts are similar to cynobacteria, including carbohydrate metabolisms, amino acid metabolisms and nucleotide metabolism. The difference lies on the specialized pathways.

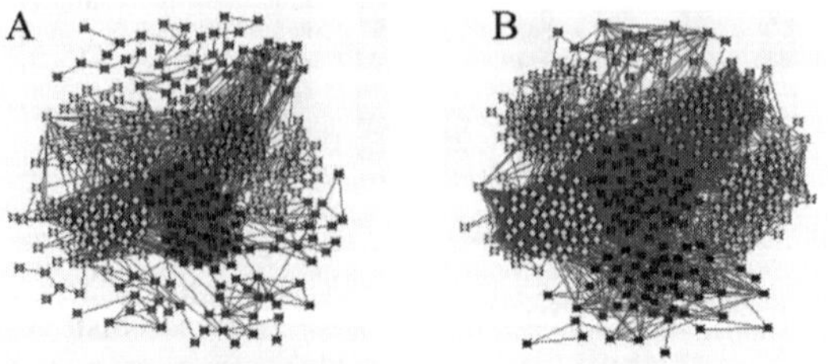

Figure 4. Conserved and different modules in metabolic network between chloroplasts and Synechococcus sp. WH8102 (syw). The modular structures of enzyme-centric networks for chloroplasts and syw are shown in (A) and (B) respectively. Each module is represented by a specific color. The five pairs of modules with same color are conserved modules between chloroplast and syw, among which the yellow, green and blue modules correspond to amino-acid metabolism, the light-orange and pink modules belong to carbohydrate metabolism and nucleotide metabolism respectively. The picture was drawn using the Pajek program.

Discussion

This study showed that the chloroplast metabolic network is less dense in comparison to photosynthetic bacteria as indicated by longer path length, larger diameter and fewer reactions. It has been suggested by Ma and Zeng [6] that the three domains of organisms exhibit quantitative differences in the metabolic network properties, i.e. eukaryotes and archaea seem to have a longer path length and a larger network diameter than bacteria. Our results suggest that global properties of chloroplast metabolic network are closer to eukaryotes than to bacteria, which may be a result of re-construction of metabolic networks by most of nucleus-coded proteins.

When comparing the SubNetwork properties, the chloroplast network is highly centered around the Calvin Cycle, indicating that the chloroplast network appears to be simplified on one hand but highly specialized on the other. This notion is further echoed by the subsequent investigation on modular structures (see below). The results could also support a view that the highly developed apparatus of light energy harvesting and its conversion to chemical energy has been optimized in cyanobacteria and that further metabolic advantages could be gained by improving the carbon fixation reactions in higher plants. Evolution of the different enzymes involved in photosynthesis has been studied extensively [26]. Our study suggests that overall network properties could be an addition to the phylogenetic analysis of individual enzymes, and might provide more information about the evolutionary history of chloroplasts.

In addition to being overall loose and Calvin Cycle-centered, chloroplast metabolic network shows a better modular structure than that of photosynthetic bacteria by SA module-detection algorithm. Our results showed that seven of the chloroplast modules are very pathway-specific in that more than 50% of the enzymes in the module belong to one pathway, such as amino acid synthesis, or carbohydrate metabolism (Figure 3A). In contrast, of the eight modules detected in Synechococcus sp. WH8102, only two modules show such pathway-specificity (Figure 3B). Moreover, two modules in chloroplasts are composed of enzymes of two pathways exclusively, lipid metabolism and the metabolism of cofactors and vitamins. Clearly, chloroplast metabolic network exhibits very different modular structure compared to cyanobacteria. Modules detected in this study represent the grouping of reactions based on their connections, which reflect in some degree the coordination of the whole metabolism. In chloroplasts, the overall complexity of the metabolic network seems reduced with fewer reactions and absence of some pathways, but the network becomes more organized with a highly modular structure.

All of the nine KEGG pathways exist in photosynthetic bacteria while two of them, glycan biosynthesis and biodegradation of xenobiotics, are absent in chloroplast. These two pathways are present in the cytosol of plant cells. Glycan biosynthesis, which underlines the synthesis of cellulose and glycol-protein on cell walls, is energetically favored to reside in cytosol instead of chloroplasts. If glycan synthesis resided in chloroplasts, the transfer of glycan from chloroplast to cell wall would need substantial energy input. Xenobiotic degradation is mostly carried out in peroxisomes in plant cells [27]. As the site of photosynthesis and O2 release, chloroplast stroma generate superoxide radicals [28], which could be a good place for xenobiotic degradation. However, these superoxides in chloroplast stroma would react with xenobiotics or xenobiotic degradation intermediates and form toxic radicals, which require a better control and subsequently reduce the efficiency of photosynthesis. Obviously, the compartmentalization of eukaryotic cells causes the specialization of functions and increase of efficiency in organelles. We also notice that metabolic processes related to hormones exist in chloroplasts, but not in any photosynthetic bacteria. It is quite intuitive that as multi-cellular organisms, plants need to communicate between cells. Hormones are the means of such communication. Those reactions related to hormones are probably a result of later addition from higher plants.

Despite the differences, some of the pathways are conserved between chloroplasts and photosynthetic bacteria. We noticed that five modules are common among all species in the study, which form a core of metabolism including carbohydrate metabolism, amino acid metabolism, and nucleotides metabolism. But the organization of these modules is different between chloroplasts and photosynthetic bacteria. The modules in chloroplasts show higher functional specificity than their counterparts in photosynthetic bacteria. The modules in photosynthetic bacteria appear to have a mixture of functions. For example, the Calvin Cycle is completely embedded in one module in chloroplasts, but split into two modules in Synechococcus sp. WH8102.

Recent studies have shown that cellular evolution might have been mainly driven by horizontal gene transfer (HGT) [29,30]. Since the metabolic network of chloroplasts exhibits a more highly modular organization, its evolution may be a result of multiple HGTs. In fact, multiple horizontal gene transfer events have been implied through the phylogenetic analysis of the key proteins involving photosynthetic light reactions [26]. Martin et al. found 1700 cyanobacteria genes in Arabidopsis nucleus including 166 genes with EC numbers, among which 92 enzymes are targeted to chloroplasts [3]. We mapped these 92 enzymes to modules in the chloroplast network and found 88% of the enzymes exist in the conserved modules corresponding to the core metabolism. The highly modular structure of chloroplast metabolism is possibly a prerequisite for a higher photosynthetic efficiency because a high modular structure can response to environmental or internal changes in a more coordinated and robust way. From another perspective, the light energy harvesting, transfer, and conversion to chemical energy in the form of ATP and NADPH has reached a high efficiency even in cyanobacteria [31,32]. As a result, changes in metabolic stoichiometry, in addition to changes in enzyme kinetics of certain key enzymes such as Rubisco [33] might represent the available options for higher photosynthetic efficiency. In this aspect, this is consistent with the results that chloroplast metabolism is centered on the Calvin Cycle.

Conclusion

In summary, by comparing the topological properties and features of metabolic networks between chloroplasts and photosynthetic bacteria, we showed that the chloroplast metabolic networks are reduced and simplified on one hand, but highly specialized and modular on the other. While overall density of the metabolic network in chloroplasts is reduced comparing to photosynthetic bacteria, the density of sub-networks directly linked to Calvin Cycle is increased. The chloroplast metabolic network also exhibits a highly modular structure compared to the metabolic network of photosynthetic bacteria. These special features of chloroplast metabolic network may reflect changes in the reconstruction of the network during endosymbiosis and the results of horizontal gene transfer. Functional mapping of the modules revealed that chloroplast metabolic network exhibited high functional specificity to the modules, indicating a better coordination of the overall metabolism and specialization of functions. Our findings are consistent with the notion that since the light energy absorption, transfer and conversion is highly efficient even in photosynthetic bacteria, the further improvements in photosynthetic efficiency in higher plants may rely on changes in metabolic network properties.

Methods

Dataset Preparation

The metabolic pathway data for chloroplasts were extracted from the Database of Chloroplast/Photosynthesis Related Genes collected by the Nagoya Plant Genome Group [34], which is a general dataset including all chloroplast enzymes in several plants, such as Arabidopsis thaliana, Oryza sativa and tobacco. For photosynthetic bacteria, we extracted the metabolic networks of nine species from KEGG: Anabaena sp. PCC7120 (ana), Chlorobium tepidum (cte), Gloeobacter violaceus (gvi), Prochlorococcus marinus SS120 (pma), Prochlorococcus marinus MED4 (pmm), Prochlorococcus marinus MIT9313 (pmt), Synechocystis sp. PCC6803 (syn), Synechococcus sp. WH8102 (syw), Thermosynechococcus elongates (tel). We also collected the metabolic pathways of E.coli, Arabidopsis thaliana and Cyanidioschyzon merolae (red algae) from KEGG. We coded enzymes and compounds by their corresponding EC number and compound ID number in the KEGG database, respectively. The direction of reactions was obtained based on the rules provided by Ma and Zeng [6]. A sub-network was constructed by including all reactions sharing metabolites with the Calvin Cycle. All enzymes and reactions in the Calvin Cycle are shown in Figure 5A.

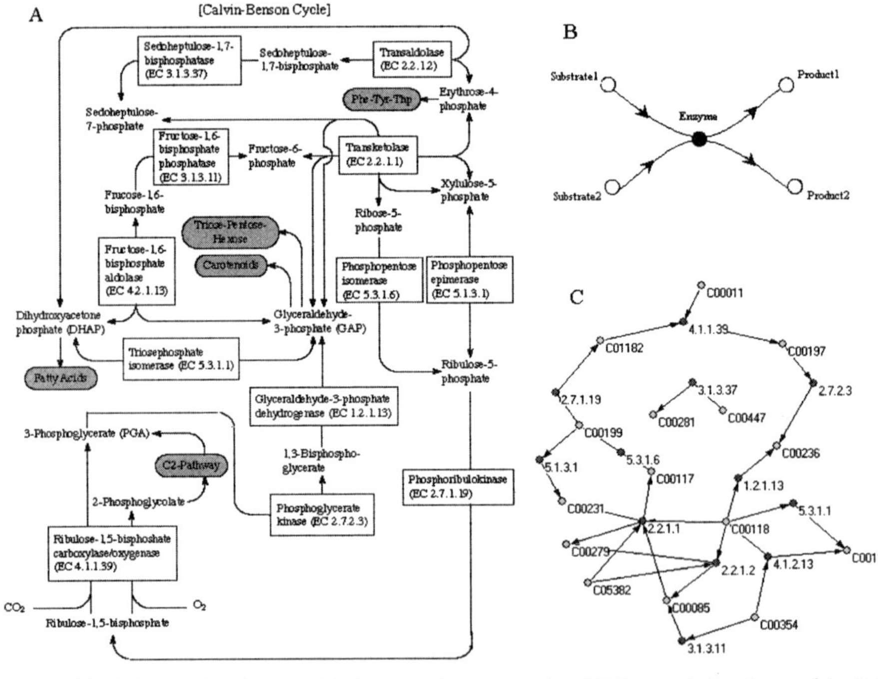

Figure 5. The Calvin Cycle pathway and its hypergraph representation.(A) The metabolic scheme of the Calvin Cycle, derived from the Database of Chloroplast/Photosynthesis Related Genes. (B) An example of hypergraph representation of biochemical reactions. (C) Graph visualization of the Calvin Cycle pathway in (A), where the red nodes and yellow nodes represent enzymes and compounds respectively. ATP, ADP, H2O, H^+, NAD^+, $NADP^+$, NADH, NADPH, Orthophosphate and Pyrophosphate have been omitted.

Network Reconstruction and Topological Properties of Networks

Most metabolic reactions have more than one substrate and/or more than one product, and therefore violate the condition of a one-to-one relationship between vertices and edges of a simple graph. Here we used a hypergraph model [35,36] to represent metabolic networks, where a hyper-edge represents a reaction and nodes represent different components involved in the reaction (i.e. enzymes and compounds). The hyper-edge relates a set of substrates to a set of products via enzymes. Figure 5B gives an example of a hypergraph, which offers an unambiguous representation of the enzymes and compounds in biochemical networks. The topological properties of both enzymes and compounds can be represented and analyzed simultaneously. The following topological properties were calculated:

Connectivity (Degree)

The connectivity of an enzyme node A is defined as the number of enzymes sharing compounds with the reaction catalyzed by A. For example, in Figure 5C, Fructose-1,6-bisphosphate phosphatase (3.1.3.11) catalyzes one reaction including two compounds C00354 and C00085. There are three enzymes catalyzing reactions sharing these two compounds, which are: fructose-1,6-bisphosphate aldolase (4.1.2.13), Transaldolase (2.2.1.2) and Transketolase (2.2.1.1). Therefore, the connectivity of Fructose-1,6-bisphosphate phosphatase (3.1.3.11) is three. The connectivity of a compound node is the number of hyper-edges containing the given compound. Average enzyme connectivity and compound connectivity are computed by averaging these two properties over all enzyme or compound nodes, respectively.

Path Length

Path length is the number of hyper-edges in the shortest path connecting two enzyme nodes or compound nodes. For example, in Figure 5C, the path length from C00354 to C00279 is two. The average path length (AL) of the entire hypergraph is the path length between each of two nodes, averaged over all pairs of nodes.

Diameter

The diameter of a hypergraph is the maximum path length between any pair of nodes.

Clustering Coefficient

This parameter measures the "cliquishness" of the neighborhood of a given node. Assuming k nodes are connected to a given node v and there are m hyper-edges between these k nodes (not including hyper-edges connecting them to v), the clustering coefficient of node v is: $C(v) = 2\,m/[k(k-1)]$. For example, Fructose-1-,6-bisphosphate phosphatase (3.1.3.11) has three enzymes connected to it, and every two of these three are connected, so m is 3 and C(v) is 1 for this enzyme.

The clustering coefficient (CC) of all enzyme or compound nodes in the hypergraph is defined as the average of C(v) over all enzyme or compound nodes.

Some small molecules, such as adenosine triphosphate (ATP), adenosine diphosphate (ADP), nicotinamide adenine dinucleotide (NAD) and H2O, are normally used as carriers for transferring electrons or energy and participate in many reactions, while typically not participating in product formation. The connections through these compounds should be treated differently when calculating the path length from one metabolite to another. The following small molecules were disregarded in the calculations as well as their connections when no product was formed: ATP, ADP, H2O, H+, NAD+, NADP+, NADH, NADPH, Orthophosphate, and Pyrophosphate. It should be noted that the omission is not determined by the compound, but by the reaction. For example, H2O is a small metabolite in many reactions, but in the following reaction:Putrescine + Oxygen + H2O <=> 4-Aminobutanal + NH3 + H2O2, H2O cannot be omitted because it participates in producing H2O2.

The Calvin Cycle is a key pathway in photosynthesis. We have defined the SubNetwork as a sub-network directly linked to the Calvin Cycle using the reactions that share all the compounds in the Calvin Cycle, with the exception of the small molecules listed before. We calculated the network properties of the SubNetwork and the ratios of each property between the SubNetwork and the total network.

Module Discovery of Enzyme-Centric Graphs

Module discovery methods based on metabolic flux are either intractable at the genome scale or have more overlap between modules [15-17]. The graph clustering techniques are regarded as appropriate for network modules detection; experimental study confirms MCL performs better than ICC and GMC in many cases [37]. In general, the MCL algorithm performs well for graph clustering except for graphs which are very homogeneous (such as weakly connected grids) and for graphs in which the natural cluster diameter (i.e. the diameter of a subgraph induced by a natural cluster) is large [38]. It has been successfully adapted to protein family classification, which has rather complete and definite data. However, MCL often gives a trivial clustering and is sensitive to signal noise, which may generate biologically insignificant modules. Guimerà and Amaral [21,22] identify modules in metabolic networks by maximizing the network's modularity using simulated annealing. By relating the metabolites in any given module to KEGG's nine major pathways, they validated that more than one-third of the metabolites in any module belong to a single pathway, which can provide a functional cartographic representation of the complex network.

We compared the modularity of metabolic networks by MCL and SA, and found MCL generated more small-size modules compared to SA, which were difficult to map to higher level functional categories. MCL decomposed the chloroplast enzyme network into 48 modules and the photosynthetic bacteria network into 30–40 modules. The size of the modules exhibits a power-law distribution, where one or two large modules include many enzymes from several unrelated biological pathways and many modules only consist of no more than four enzymes. SA, in contrast, gives a moderate number of modules. The SA algorithm detected 12 modules in the chloroplast enzyme network and 8 to 9 modules for the photosynthetic bacterial species. Each module consists of enzymes involved in one or several particular metabolic functions. This comparison indicates that SA might be more appropriate for the clustering analysis in this study. We selected modules detected by SA algorithm for similarity analysis and functional classification.

Deviating from Guimerà and Amaral [21,22], we used an enzyme-centric graph representation of the metabolic network where vertices were used to represent enzymes and edges were used to represent compounds. There will be a directed edge from enzyme E1 to enzyme E2, if E1 catalyzes a reaction generating a product A which is used as substrate of E2. Reversible reactions are considered as two separate reactions. Modularization of such enzyme-centric graph categorizes enzymes into different functional groups.

Similarity Measure of Modular Structures

To compare the modular structures among the networks from different species, we define a similarity measure based on Hamming distance [39]. For two modules a and b in two species, the number of enzymes in each module is N_a and N_b. First, we compute the similarity between any two enzyme members between module a and b. Any EC number is treated as a vector with 4 parts, which are given different weight 0.1, 0.2, 0.3, 0.4 according to EC hierarchy. For two EC numbers, one vector P emerges to describe their similarity. If they are same at the kth level, then P_k is 1, otherwise P_k is 0. Thus the similarity between any two enzymes i and j is defined as:

$$S_{ij}\sum_{k-1}^{4} w_k P_k \cdot$$

Note that the comparison of two EC numbers should be from high level to low level, if different at the kth level, then all Pt (t>=k) will be 0 regardless of whether they are the same at lower levels.

After collecting all similarities between any two enzymes, the most similar enzyme in module b for each enzyme i in module a is identified. This maximal similarity is represented as Sbesti. Then the global similarity between module a and module b should be defined as:

$$\mathrm{Simi}(a,b)=\frac{1}{N_a}\sum_{i=1}^{N_a}\mathrm{Sbest}_i\ .$$

Therefore, for any module in one species, its most similar module in another species can be identified. If two modules of two species are both most suited each other, they are regarded as conserved modules between these two species.

In order to investigate the overall modular structure among different species, we compared the modular similarity between two species based on the similarity between modules. Each module in each species is regarded as a sample, and the total of these samples as a large group. Thus the similarity between two species can be measured by the similarity between these two groups, which is defined according to the Hausdorff metric [40]. G1 and G2 are two groups representing two species, Sspecies(G1,G2) is the similarity between these two species, a and b are samples (modules above) belonging to G1 and G2, respectively. The similarity S(a, G2) between sample a belonging to group G1 and group G2 is defined as:

$$S(a,G_2)=\max\begin{bmatrix} Simi(a,b) \\ b\in G_2 \end{bmatrix}.$$

Then, the similarity between G1 and G2 is given by:

$$S(G_1,G_2)=\min_{a\in G_1}\left[S(a,G_2)\right].$$

It is important to note that this similarity is in general not symmetrical. Accordingly we introduce the similarity between G_2 and G_1:

$$S'(G_2,G_1)=\min_{b\in G_2}\left[\max\left\{\begin{matrix} Simi(b,a) \\ a\in G_1 \end{matrix}\right\}\right].$$

It is then convenient to introduce the similarity between two species as:

$S_{species}(G_1,G_2) = \min\{S(G_1,G_2),S'(G_2,G_1)\}$.

The Hausdorff metric provides a more accurate measurement of the structure similarity between two species, since the lower value of the forward and backward similarity is selected, which leads to a significantly underestimated assessment.

Author Contributions

ZW conducted the analysis of network properties and module discovery and implemented programs for the analysis. XGZ designed the experiments and analysis. YZC managed the project. YYL provided biological analysis. JH implemented software programs for the module comparison. YXL managed the project. LL designed and managed the project. All authors read and approved the final manuscript.

Acknowledgements

We would like to thank Dr. Roger Guimerà and Dr. Luís A. Nunes Amaral for kindly providing us the software Modul-w and the usage of the software for module discovery. We would like to thank Dr. Carl Woese, Dr. Hans Bohnert, and Dr. Peter Gogarten for their insightful discussion and comments. We would also like to acknowledge the contribution by Kilannin Krysiak, Kristen Aquino, and Tsai-Tien Tseng in data collection. This work was supported by grants from the National "973" Basic Research Program of China (2001CB510209; 2003CB715900; 2004CB518606), and the Fundamental Research Program of Shanghai Municipal Commission of Science and Technology (04DZ14003).

References

1. Martin W, Stoebe B, Goremykin V, Hansmann S, Hasegawa M, Kowallik KV: Gene transfer to the nucleus and the evolution of chloroplasts. Nature 1998, 393:162–165.
2. Chu KH, Qi J, Yu ZG, Anh V: Origin and phylogeny of chloroplasts revealed by a simple correlation analysis of complete genomes. Mol Biol Evol 2004, 21:200–206.
3. Martin W, Rujan T, Richly E, Penny D: Evolutionary analysis of Arabidopsis, cyanobacterial, and chloroplast genomes reveals plastid phylogeny and thousands of cyanobacterial genes in the nucleus. Proc Natl Acad Sci USA 2002, 99:12246–12251.

4. Leister D: Chloroplast research in the genomic age. Trends Genet 2003, 19:47–56.
5. Sugiura M, Hirose T, Sugita M: Evolution and mechanism of translation in chloroplasts. Annu Rev Genet 1998, 32:437–459.
6. Raven JA, Allen JF: Genomics and chloroplast evolution: what did cyanobacteria do for plants? Genome Biol 2003, 209:1–5.
7. Olson JM, Blankenship RE: Thinking about the evolution of photosynthesis. Photosynth Res 2004, 80:373–386.
8. Jeong H, Tombor B, Albert1 R, Oltvai ZN, Babarasi AL: The large-scale organization of metabolic networks. Nature 2000, 407:651–654.
9. Ma HW, Zeng AP: Reconstruction of metabolic networks from genome data and analysis of their global structure for various organisms. Bioinformatics 2003, 19:270–277.
10. Hartwell LH, Hopfield JJ, Leibler S, Murray AW: From molecular to modular cell biology. Nature 1999, 402:47–52.
11. Ravasz E, Somera AL, Mongru DA, Oltvai ZN, Barabási AL: Hierarchical organization of modularity in metabolic networks. Science 2002, 297:1551–1555.
12. Rives AW, Galitski T: Modular organization of cellular networks. Proc Natl Acad Sci USA 2003, 100:1128–1133.
13. Papin JA, Reed JL, Palsson BO: Hierarchical thinking in network biology: the unbiased modularization of biochemical networks. Trends Biochem Sci 2004, 29:641–647.
14. Batagelj V, Mrvar A: Pajek-program for large network analysis. Connections 1998, 21:47–57.
15. Schuster S, Fell DA, Dandekar T: A general definition of metabolic pathways useful for systematic organization and analysis of complex metabolic networks. Nature 2000, 18:326–332.
16. Burgard AP, Nikolaev EV, Schilling CH, Maranas CD: Flux coupling analysis of genome-scale metabolic network reconstructions. Genome Res 2004, 14:301–312.
17. Price ND, Reed JL, Palsson B: Genome-scale models of microbial cells: evaluating the consequences of constraints. Nat Rev Microbiol 2004, 2:886–897.
18. van Dongen S: Graph clustering by flow simulation. PhD thesis. University of Utrecht, Center of mathematics and computer science; 2000.
19. Kannan R, Vampala S, Vetta A: On clustering: good, bad and spectral. Proceedings of 41st Annual Symposium on Foundations of Computer Science 2000, 367–378.

20. Gaertler M: Clustering with spectral methods. In Master's thesis. University at Kon-stanz; 2002.

21. Guimerà R, Amaral LAN: Functional cartography of complex metabolic networks. Nature 2005, 433:895–900.

22. Guimerà R, Amaral LAN: Cartography of complex networks: modules and universal roles. J Stat Mech Theor Exp 2005, PO2001:1–13.

23. Bishop GJ, Yokota T: Plants steroid hormones, brassinosteroids: current highlights of molecular aspects on their synthesis/metabolism, transport, perception and response. Plant Cell Physiol 2001, 42:114–120.

24. Chae HS, Kieber JJ: Eto Brute? Role of ACS turnover in regulating ethylene biosynthesis. Trends In Plant Science 2005, 10:291–296.

25. Tanimoto E: Regulation of root growth by plant hormones-roles for auxin and gibberellin. Critical Reviews In Plant Sciences 2005, 24:249–265.

26. Xiong J, Bauer CE: Complex evolution of photosynthesis. Annu Rev Plant Biol 2002, 53:503–521.

27. Reddy JK: Peroxisome proliferators and peroxisome proliferation-activated receptor: biotic and xenobiotic sensing. American Journal of Pathology 2004, 164:2305–2321.

28. Asada K: The water-water cycle in chloropasts: scavenging of active oxygens and dissipation of excess photons. Annual Review of Plant Phyiology and Plant Molecular Biology 1999, 50:601–639.

29. Woese CR: Interpreting the universal phylogenetic tree. Proc Natl Acad Sci USA 2000, 15:8392–8396.

30. Woese CR: On the evolution of cells. Proc Natl Acad Sci USA 2002, 99:8742–8747.

31. Zhu XG, Govindjee NR, Baker NR, deSturler E, Ort DR, Long SP: Chlorophyll a fluorescence induction kinetics in leaves predicted from a model describing each discrete step of excitation energy and electron transfer associated with photosystem II. Planta 2005, 223:114–133.

32. van Grondelle R, Gobets B: Transfer and trapping of excitation in plant photosystems. In Chlorophyll a fluorescence a signature of photosynthesis. Edited by: Papageorgiou CC, Govindjee. Springer, Heidelberg, Germany; 2005:107–132.

33. Zhu XG, Portis AR Jr, Long SP: Would transformation of C3 crop plants with foreign Rubisco increases productivity? A computational analysis extrapolating form kinetic properties to canopy photosynthesis. Plant Cell and Environment 2004, 27:155–165.

34. Database of Chloroplast/Photosynthesis Related Genes [http://chloroplast.net/index.html].
35. Krishnamurth L, Nadeau J, Ozsoyoglu G, Ozsoyoglu M, Schaeffer G, Tasan M, Xu W: Pathways database system: an integrated system for biological pathways. Bioinformatics 2003, 19:930–937.
36. Klamt S, Stelling J, Ginkel M, Gilles ED: FluxAnalyzer: Exploring structure, pathways, and fluxes in balanced metabolic networks by interactive flux maps. Bioinformatics 2003, 19:261–269.
37. Brandes U, Gaertler M, Wagner D: Experiments on graph clustering algorithms. ESA LNCS 2003, 2832:568–579.
38. van Dongen S: Performance criteria for graph clustering and markov cluster experiments. In Report INS-R0012. National Research Institute for Mathematics and Computer Science; 2000.
39. Glazko GV, Mushegian AR: Detection of evolutionarily stable fragments of cellular pathways by hierarchical clustering of phyletic patterns. Genome Biol 2004, 5:1–13.
40. Nicolas A, Diego SC, Touradj E: MESH: measuring errors between surfaces using the hausdorff distance. Proceedings of the IEEE International Conference in Multimedia and Expo (ICME) 2002, 705–708.

Effects of Cu^{2+}, Ni^{2+}, Pb^{2+}, Zn^{2+} and Pentachlorophenol on Photosynthesis and Motility in Chlamydomonas Reinhardtii in Short-Term Exposure Experiments

Roman A. Danilov and Nils G. A. Ekelund

ABSTRACT

Background

Heavy metals, especially copper, nickel, lead and zinc, have adverse effects on terrestrial and in aquatic environments. However, their impact can vary depending on the nature of organisms. Taking into account the ability of heavy metals to accumulate in sediments, extended knowledge of their effects on aquatic biota is needed. In this context the use of model organisms (often unicellular), which allows for rapid assessment of pollutants in freshwater, can be of advantage. Pentachlorophenol has been extensively used for decades as a bleaching agent by pulp- and paper industry. Pentachlorophenol tends to accumulate in the nature. We aim to determine if photosynthesis and motility can be used as sensitive physiological parameters in toxicological studies of Chlamydomonas reinhardtii, a motile green unicellular alga. It is discussed if photosynthesis and motility can be used as sensitive physiological parameters in toxicological studies.

Results

The concentrations studied ranged from 0.1 to 2.0 mg l^{-1} for copper, nickel, lead and zinc, and from 0.1 to 10.0 mg l-1 for pentachlorophenol. Exposure time was set to 24 h. Copper and pentachlorophenol turned out to be especially toxic for photosynthetic efficiency (PE) in C. reinhardtii.

Conclusion

Copper and pentachlorophenol turned out to be especially toxic for PE in C. reinhardtii. Zinc has been concluded to be moderately toxic while nickel and lead had stimulatory effects on the PE. Because of high variance, motility was not considered a reliable physiological parameter when assessing toxicity of the substances using C. reinhardtii.

Background

Heavy metals, especially copper, nickel, lead and zinc, have adverse effects on terrestrial and in aquatic environments. However, their impact can vary depending of the nature of organisms [1, 2]. Taking into account the ability of heavy metals to accumulate in sediments, extended knowledge of their effects on aquatic biota is needed [3]. In this context the use of model organisms, which allow for rapid assessment of pollutants in freshwater, can be of advantage. Chlamydomonas reinhardtii has been shown to be one of those especially suited organisms for different kinds of studies [4, 5]. Previous investigations showed C. reinhardtii to be sensitive to copper, nickel and zinc [6,7,8,9]. Although, most studies concentrated on the impacts on growth rates and ultrastructure. The photosynthetic apparatus in C. reinhardtii was shown to be highly vulnerable to toxic substances thus making it a suitable parameter for toxicity estimation [10,11,12]. Motility has been shown to be one of the possible physiological markers for toxicity assessment using Euglena gracilis [13]. The aim of this study was to carry out comprehensive experiments in order to investigate effects of different concentrations of copper, nickel, lead and zinc on the photosynthetic efficiency and motility of C. reinhardtii. As an additional test substance pentachlorophenol was used to its previous use as a bleaching agent at pulp and paper factories in Sweden thus making it to a spread contaminant in some areas.

Results and Discussion

The photosynthetic response curve (PRC) of the control was characterised by an increase in oxygen evolution according to the increase in PFD up to 612 μmol m^{-2} s^{-1} (the highest PFD-value used in the experiments) and a decrease in oxygen evolution due to the inhibition of photosynthesis when PFD became constant (612 μmol m^{-2} s^{-1}, Fig. 1a). A peak of higher respiration immediately after the cessation of illumination can be explained by the light-enhanced dark respiration (LEDR, the rate of change of oxygen consumption - an acceleration), which contributed to the basal dark respiration. The behaviour of PRC is important evidence of how favourable the conditions are for photosynthesis [14]. The type of PRC described above was found in similar investigations to be typical in the green flagellate Euglena gracilis [15]. Similarly, this type of PRC can be considered as common in C. reinhardtii, too. At all treatments this basic type of PRC was observed. The stepwise drop in oxygen evolution at the maximum irradiance value in fig. 3a,b and fig. 5a,b should be considered as an artefact specific to the Light Pipette model used (when the value of oxygen saturation in the cuvette exceeds 200 %, unpublished results of a methodical study). Increasing concentrations of copper led to decrease in maximum values of oxygen evolution compared to the control, demonstrating especially severe impacts at the concentrations of 0.5 mg l^{-1} and higher (Fig. 1b,c,d,e,f). The treatments with nickel, lead or zinc did not caused such strong inhibitory effects as in the case of copper. Moreover, in the case of nickel (Fig. 2) maximum values of oxygen evolution were higher than those in the control and no inhibition of photosynthesis was observed when the PFD became constant (612 μmol m^{-2} s^{-1}). Increasing concentrations of nickel seemed to be stimulative based on the shape of PRCs. Only slight inhibition of photosynthesis at the constant maximal PFD values was detected in the cases of lead or zinc treatments (Figs. 3, 4, respectively) and maximum values of oxygen evolution were equal to or higher than in the control. The treatment with pentachlorophenol led to prolongated compensation points at lower concentrations (0.1 and 0.5 mg l^{-1}) and to severe impacts on photosynthesis at concentrations of 1.0 mg l^{-1} and higher (Fig. 5) comparable to those caused by copper treatments.

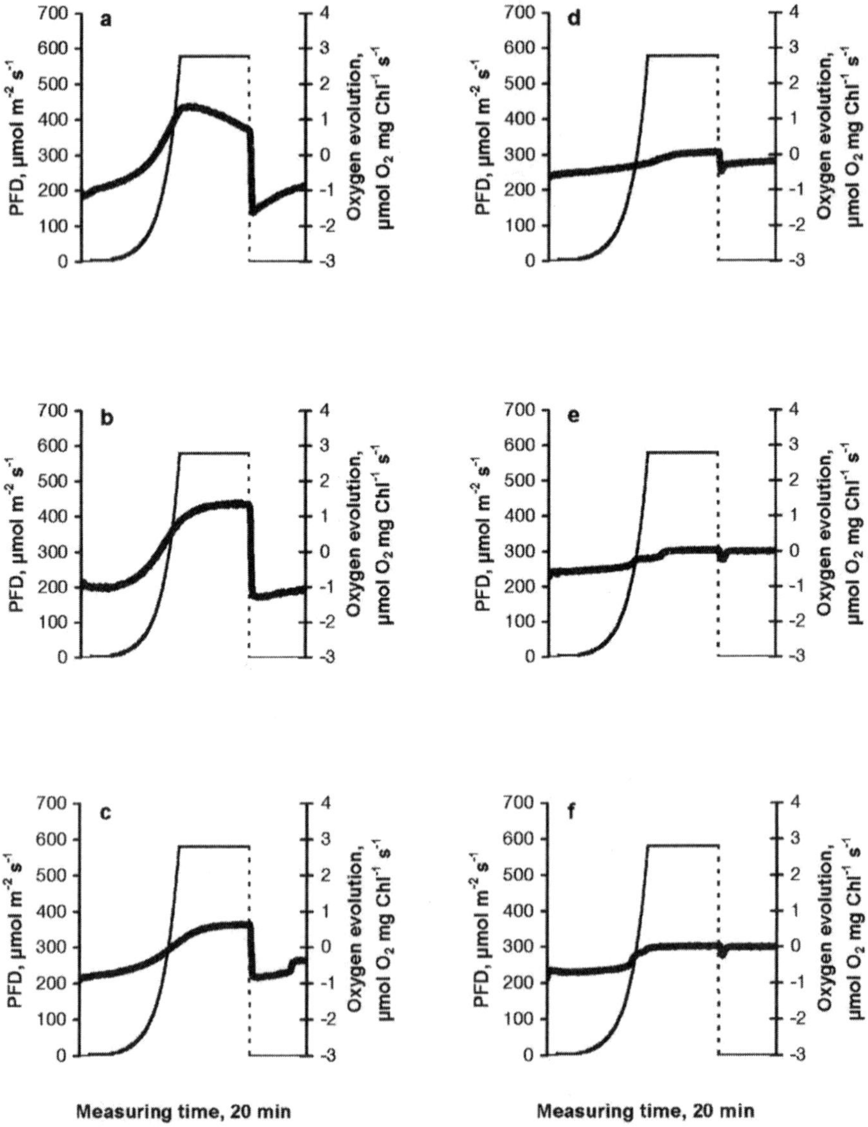

Figure 1. The dependence of photosynthetic response curves in C. reinhardtii to different concentrations of copper (mg l^{-1}) after 24 h exposure: a) control, b) 0.1, c) 0.5, d) 1.0, e) 1.5, f) 2.0. Thick solid line - oxygen evolution, thin dashed line - light evolution.

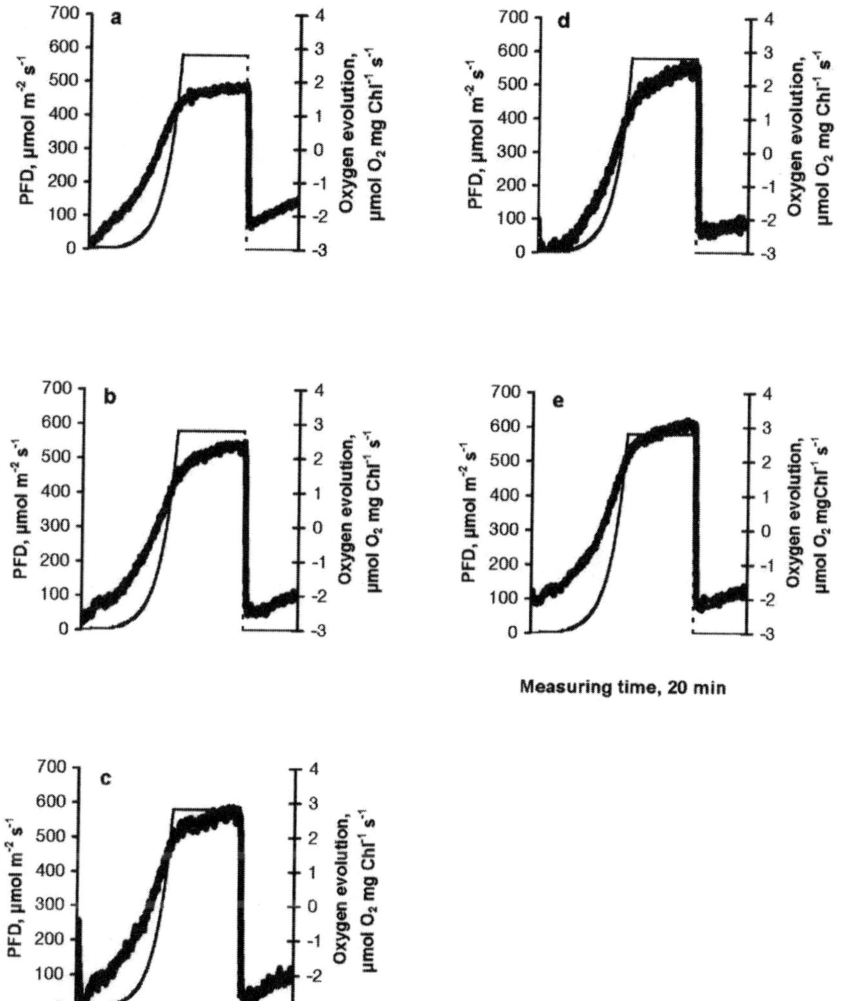

Figure 2. The dependence of photosynthetic response curves in C. reinhardtii to different concentrations of nickel (mg l^{-1}) after 24 h exposure: a) 0.1, b) 0.5, c) 1.0, d) 1.5, e) 2.0. Thick solid line - oxygen evolution, thin dashed line - light evolution.

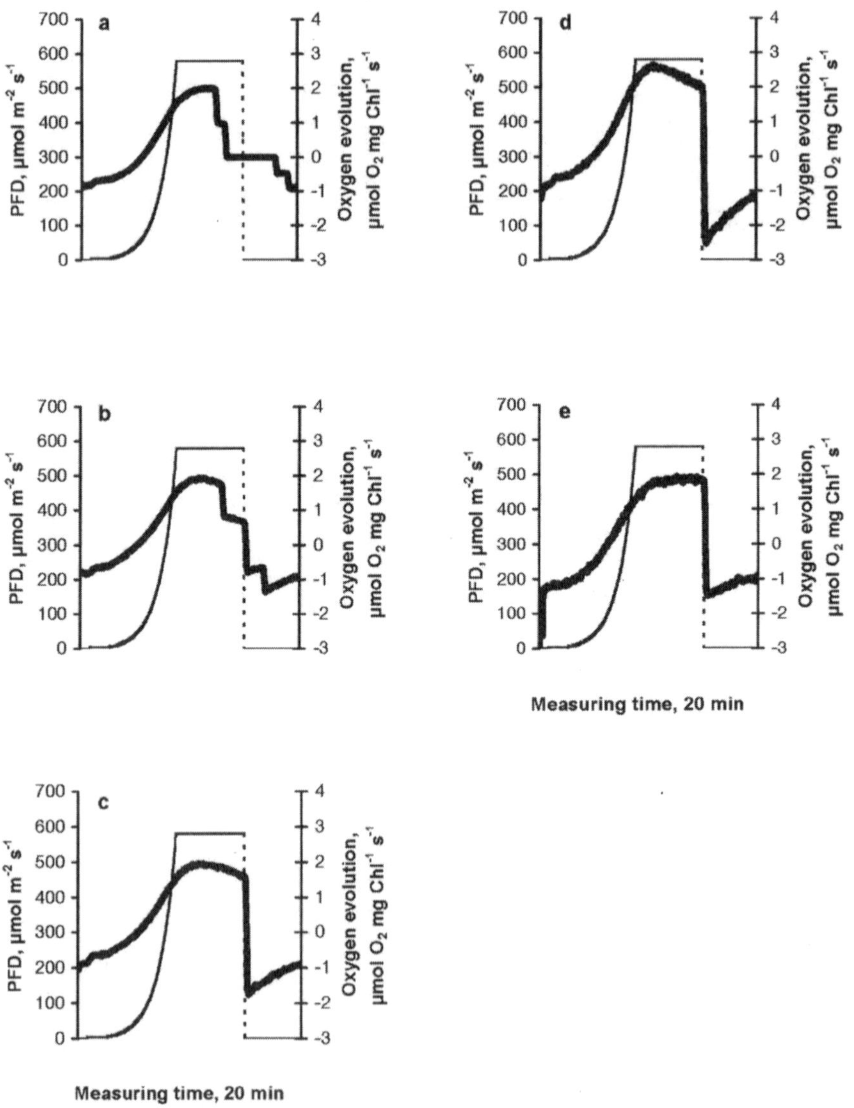

Figure 3. The dependence of photosynthetic response curves in C. reinhardtii to different concentrations of lead (mg l^{-1}) after 24 h exposure: a) 0.1, b) 0.5, c) 1.0, d) 1.5, e) 2.0. Thick solid line - oxygen evolution, thin dashed line - light evolution.

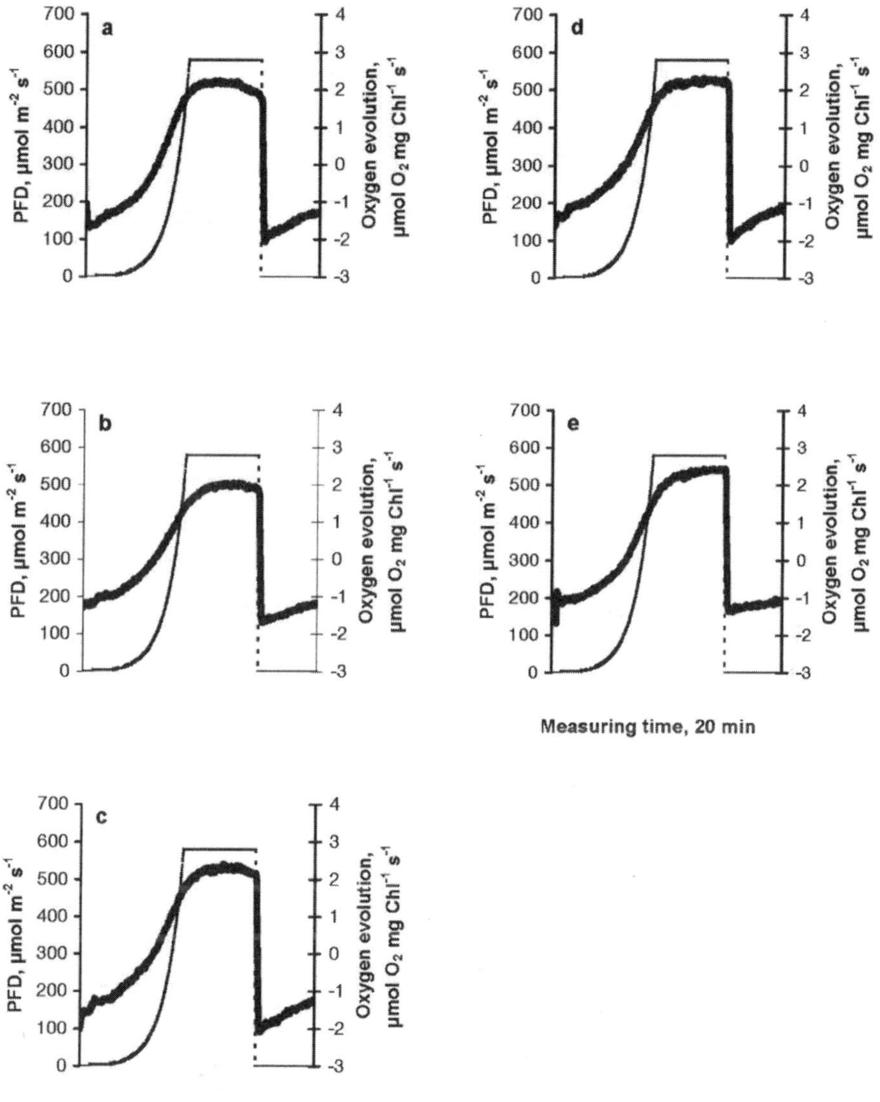

Figure 4. The dependence of photosynthetic response curves in C. reinhardtii to different concentrations of zinc (mg l^{-1}) after 24 h exposure: a) 0.1, b) 0.5, c) 1.0, d) 1.5, e) 2.0. Thick solid line - oxygen evolution, thin dashed line - light evolution.

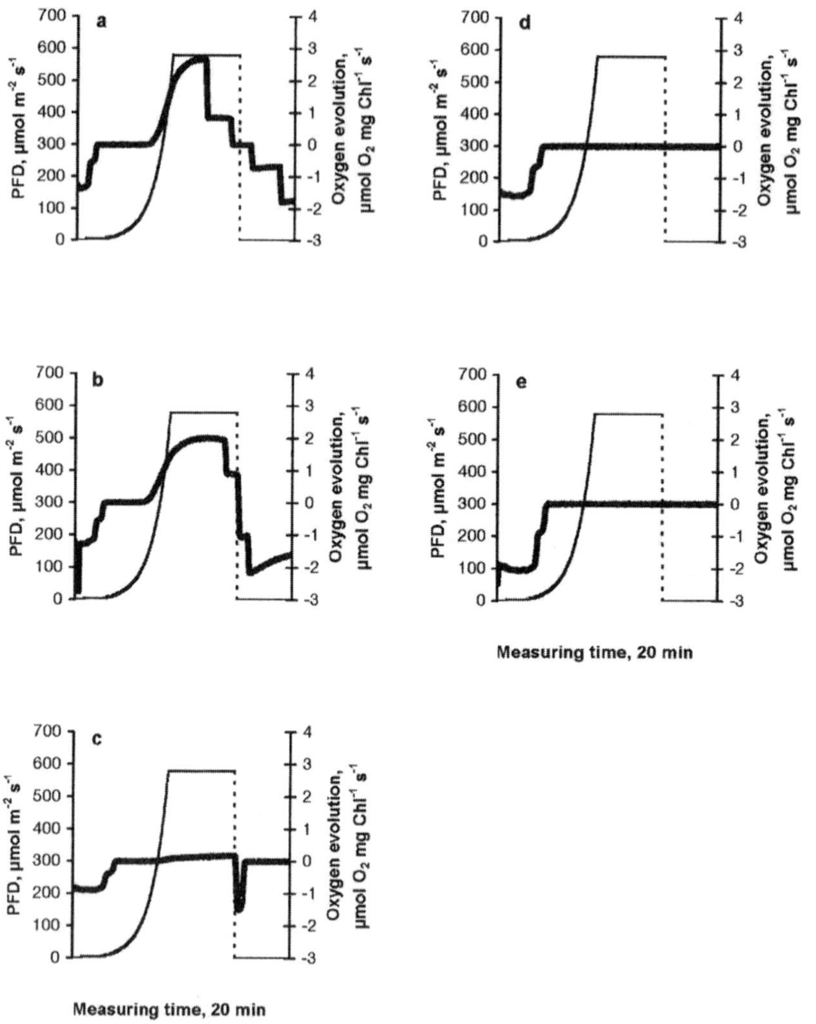

Figure 5. The dependence of photosynthetic response curves in C. reinhardtii to different concentrations of pentachlorophenol (mg l^{-1}) after 24 h exposure: a) 0.1, b) 0.5, c) 1.0, d) 5.0, e) 10.0. Thick solid line - oxygen evolution, thin dashed line - light evolution.

Because of often similar behaviour of PRCs, such parameters as photosynthetic efficiency (PE) and compensation point (CP) became important when comparing how favourable conditions for photosynthesis are after different treatments studied. PE measured as slope coefficients showed negative correlation with increasing concentrations of copper, zinc and pentachlorophenol (Tab. 1). While the decrease in PE could be predicted from the behaviour of PRCs by treatments with copper and pentachlorophenol, negative effects of increasing concentrations of zinc were not so obvious where all values of PE exceeded those of the control. Increasing concentrations of nickel and lead correlated positively with PE in C. reinhardtii. Thereby, in case of nickel all coefficients were higher than that in the control while in case of lead coefficients both below and above that in the control were observed.

Table 1. Effects of different concentrations of heavy metals and pentachlorophenol (Pent.) on the photosynthetic efficiency (slope coefficients) of C. reinhardtii after 24 h exposure.

Concentration, mg l^{-1}	Substances				
	Copper	Nickel	Lead	Zinc	Pent.
0.0	0.140	0.140	0.140	0.140	0.140
0.1	0.131	0.232	0.150	0.224	0.098
0.5	0.055	0.228	0.133	0.153	0.090
1.0	0.015	0.251	0.138	0.180	0.002
1.5	0.026	0.267	0.176	0.174	
2.0	0.047	0.244	0.160	0.165	
5.0					0.014
10.0					0.014
R	- 0.66	0.64	0.60	- 0.54	- 0.83

R - coefficient of correlation between increasing concentrations of the substances tested and the photosynthetic efficiency.

The capacity of copper to inhibit photosynthesis has been reported previously for some algae [16, 17]. Inhibition capacity of nickel, lead and zinc on different physiological parameters and photosynthesis in algae have been detected in earlier studies [9, 18,19,20,21,22,23]. Pentachlorophenol has been shown to be a severe environmental poison [24]. Based on the PE, our results let us conclude copper and pentachlorophenol to be especially toxic to the photosynthetic apparatus in C. reinhardtii. These results are in good agreement with the high inhibition capacity of copper to C. reinhardtii reported in other studies [7, 9, 11, 25]. Nickel and lead can be concluded to have stimulatory effects on the PE in C. reinhardtii. This fact can be explained by a possible short-term stimulatory effect of some usually toxic substances reported for some unicellular algae. This effect depends both on the nature of organism and of the exposure duration [26, 27].

Positive correlations were detected between CPs and increasing concentrations of copper, lead, zinc and pentachlorophenol (Tab. 2). In the cases of lead, zinc and pentachlorophenol, CPs were lower than in the control. When exposed to a stress, organism may need higher light intensity to reach the CP [14]. However, this trend is widely general and should be considered rather as a rule of thumb [15]. As pointed out by Ögren and Evans [14], both PE and CP are valuable indicators of how favourable conditions for photosynthesis are. On the other hand, it is not clear how efficient these parameters are when assessing the capacity to photosynthesise. Our previous results obtained on E. gracilis showed that discrepancies between trends in PE (measured as slope coefficients) and CPs could occur quite often [15]. Thus the question which parameter should be considered as more efficient still remains open. Our current opinion is that the PE provides more relevant information about the physiological state of the photosynthetic apparatus. However, more research is highly desirable to solve this question.

Table 2. Effects of different concentrations of heavy metals and pentachlorophenol (Pent.) on the compensation points (CP, $\mu mol\ m^{-2}\ s^{-1}$) of photosynthesis in C. reinhardtii after 24 h exposure.

Concentration,	Substances				
$mg\ l^{-1}$	Copper	Nickel	Lead	Zinc	Pent.
0.0	118	118	118	118	118
0.1	162	126	61	88	19
0.5	366	165	48	85	48
1.0	579	132	48	98	29
1.5	579	192	38	83	
2.0	579	105	98	108	
5.0					21
10.0					31
R	0.87		0.45	0.58	

R - coefficient of correlation between increasing concentrations of the substances tested and the CPs. Absolute values of R below 0.15 are not shown.

The calculated motility values showed high variance (Tab. 3). Only at the concentrations of pentachlorophenol of 5.0 and 10.0 mg l-1 the differences were significantly different (t-test). Hence, we do not consider motility in C. reinhardtii as a reliable parameter for toxicity testing. This is in disagreement with a general proposal of Häder et al. [13] to use motility as a valuable parameter in bioassays.

Table 3. Effects of different concentrations of heavy metals and pentachlorophenol (Pent.) on the cell motility (m s^{-1}) of C. reinhardtii after 24 h exposure.

Concentration, mg l^{-1}	Substances				
	Copper	Nickel	Lead	Zinc	Pent.
0.0	46.6 ± 34.9	46.6 ± 34.9	46.6 ± 34.9	46.6 ± 34.9	46.6 ± 34.9
0.1	26.6 ± 18.7	26.3 ± 17.8	33.1 ± 22.1	26.5 ± 16.6	52.1 ± 49.9
0.5	31.2 ± 26.3	31.2 ± 27.4	33.9 ± 21.5	27.8 ± 19.3	56.7 ± 41.9
1.0	25.4 ± 12.8	27.0 ± 18.6	40.0 ± 33.8	34.0 ± 24.5	53.1 ± 36.7
1.5	34.0 ± 25.5	32.2 ± 24.4	47.4 ± 38.7	29.3 ± 22.2	
2.0	28.9 ± 20.6	37.1 ± 28.3	46.4 ± 34.9	35.1 ± 26.1	
5.0					0
10.0					0

Values of standard deviation are shown.

Conclusions

Copper and pentachlorophenol turned out to be especially toxic for PE in C. reinhardtii. Zinc has been concluded to be moderately toxic while nickel and lead had stimulatory effects on the PE. Because of high variance, motility was not considered a reliable physiological parameter when assessing toxicity of the substances using C. reinhardtii.

Material and methods

One-week old cultures of C. reinhardtii (strain 137c mt+, Stefan Falk, Mid Sweden University, Östersund, Sweden) were used in the experiments. The strain has been grown for many years by the mentioned researcher and has been previously isolated from an intact site (no pollution). Algae were in the exponential growth phase when the experiments were started. Starting density of the cells was held at 10,000 cells ml-1. Freshwater medium described by Checcucci et al. [28] was used for cultures. The cells of C. reinhardtii were grown in 100 ml Erlenmeyer glass flasks at 20°C in a cultivation cabinet (Termax Klimatskåp, 6395 F/FL, Ninolab AB). The light/dark cycle was 16 h/8 h with an irradiance of 70 µmol m-2 s-1 (400-700 nm). The cells were exposed for 24 h to heavy metals as Cu2+,

Ni2+, Pb2+ and Zn2+ at the concentrations of 0.1, 0.5, 1.0, 1.5 and 2.0 mg l-1, respectively. Pentachlorophenol concentrations were 0.1, 0.5, 1.0, 5.0 and 10.0 mg l-1.

For each concentration studied the measurements were repeated three times and the mean values from the three replicates were calculated. Photosynthesis and respiration were measured as the rate of oxygen evolution/consumption per gram (g) chlorophyll and time (s-1) with a Light Pipette (Brammer, Illuminova, Uppsala, Sweden). The instrument consists of a light source that is connected to a cuvette with a micro-oxygen electrode (MI-730, Microelectrodes, Inc., 298 Rockingham Rd., Londonderry, NI), a quantum sensor, a temperature-controlled water bath and a computer. The light source provides precise photon flux density (PFD) of 0 to 3500 (µmol m-2 s-1). The rate of oxygen evolution was calculated on the basis of the ambient O2 concentration of 0.276 µmol ml-1, which in this investigation corresponds to 100%. Chlorophyll was estimated by extraction in 80% acetone solution and its absorbance was measured with a spectrophotometer (UV/VIS spectrometer Lambda Bio 20, PERKIN ELMER) at 652 nm. Chlorophyll content was measured according to the following equation:

$$\text{Chl (mg ml}^{-1}\text{)} = A_{652}/34.5,$$

where A_{652} corresponds to absorbance at 652 nm and 34.5 is the absorbance of 1 mg ml^{-1} Chl extracted in 80% acetone.

The cells of C. reinhardtii were placed in the cuvette and each measurement of oxygen evolution/consumption lasted for 20 min at 20°C. The photon flux density (PFD) provided by the light source continuously increased during 9 min from 0 up to 612 µmol m-2 s-1, remained unchanged for 6 min and was then turned off (darkness) for 5 min (Figure 1a). Measuring oxygen evolution for 20 min produced 600 measuring points at a rate of 5 Hz and every point represented the average of five subsamples. The photosynthetic efficiency (PE, the increase in photosynthetic rate vs. time immediately after illumination) was calculated as the slope coefficients of the photosynthetic curves between 20 and 500 µmol m-2s-1. The slope of the photosynthesis-irradiance curve is proportional to the maximum quantum yield of photosynthesis [29]. It means that a quicker increase in oxygen evolution would result in a higher coefficient.

A compound microscope Nikon Optiphot connected to a video CCD Camera Ikegami ICD-44L (Ikegami Tsushinki Co., Ltd., Japan) was used to study the motility of C. reinhardtii. The image was translated to a PC. All calculations were processed with Motile System V. 1.7 (motility) software developed by D.-P. Häder and K. Vogel, Erlangen, Germany [13]. The motility was measured in term of µm s-1. An average value from 2000 measurements was calculated in each measurement. All statistical analyses were performed in the computer package Minitab 11.0.

References

1. Clark RB: Marine pollution. 4th edition, Oxford: Clarendon Press 1997.
2. Seidl M, Huang V Mouchel JM: Toxicity of combined sewer overflows on river phytoplankton: the role of heavy metals. Environ Pollut 1998, 101:107–116.
3. Tikkanen M, Korhola A, Seppa H, Virkanen J: A long-term record of human impacts on an urban ecosystem in the sediments of Toolonlahti Bay in Helsinki, Finland. Environ Conserv 1997, 24:326–337.
4. Rochaix JD: Chlamydomonas reinhardtii as the photosynthetic yeast. Ann Review Genet 1995, 29:209–230.
5. Davies JP, Grossman AR: The use of Chlamydomonas (Chlorophyta: Volvocales) as a model algal system for genome studies and the elucidation of photosynthetic processes. J Phycol 1998, 34:907–917.
6. Garvey JE, Owen HA, Winner RW: Toxicity of copper to the green alga, Chlamydomonas reinhardtii (Chlorophyceae), as affected by humic substances of terrestrial and fresh water origin. Aquat Toxicol 1991, 19:89–96.
7. Winner RW, Owen HA: Toxicity of copper to Chlamydomonas reinhardtii (Chlorophyceae) and Ceriodaphnia dubia (Crustaceae) in relation to changes in water chemistry of a fresh water pond. Aquat Toxicol 1991, 21:157–170.
8. Macfie SM, Tarmohamed Y, Welbourn PM: Effects of cadmium, cobalt, copper, and nickel on growth of the green alga Chlamydomonas reinhardtii - the influences of the cell wall and pH. Arch Environ Contam Toxicol 1994, 27:454–458.
9. Sunda WG, Huntsman SA: Interactions among Cu2+, Zn2+, and Mg2+ in controlling cellular Mn, Zn, and growth rate in the coastal alga Chlamydomonas. Limnol Oceanogr 1998, 43:1055–1064.
10. Draber W, Hilp U, Likusa H, Schindler M, Trebst A: Inhibition of photosynthesis by 4-nitro-6-alkylphenols - structure activity studies in wild type and 5 mutants of Chlamydomonas reinhardtii thylakoids. Z Naturforsch C 1993, 48:213–223.
11. Hill KL, Merchant S: Coordinate expression of coproporphyrinogen oxidase and cytochrome C6 in the green alga Chlamydomonas reinhardtii in response to changes in copper availability. EMBO J 1995, 14:857–865.

12. Martin RE, Thomas DJ, Tucker DE, Herbert SK: The effects of photooxidative stress on photosystem I measured in vivo in Chlamydomonas. Plant Cell Develop 1997, 20:1451–1461.

13. Häder D-P, Lebert M, Tahedl H, Richter P: The Erlanger flagellate test (EFT): photosynthetic flagellates in biological dosimeters. J Photochem Photobiol 1997, 40:23–28.

14. Ögren E, Evans JR: Photosynthetic light-response curves. Planta 1993, 189:182–190.

15. Danilov RA, Ekelund NGA: Influence of waste water from the paper industry and UV-B radiation on the photosynthetic efficiency of Euglena gracilis. J App Phycol 1999, 11:157–163.

16. Cid A, Herrero C, Torres E, Abalde J: Copper toxicity on the marine microalga Phaeodactylum tricornutum - effects on photosynthesis and related parameters. Aquat Toxicol 1995, 31:165–174.

17. Nalewajko C, Olaveson MM: Differential responses of growth, photosynthesis, respiration, and phosphate-uptake to copper in copper tolerant and copper-intolerant strains of Scenedesmus acutus (Chlorophyceae). Can J Bot 1995, 73:1295–1303.

18. Navarro L, Torres-Marquez M, Gonzalez-Moreno S, Devars S, Hernandez R, Moreno-Sanchez R: Comparison of physiological changes in Euglena gracilis during exposure to heavy metals of heterotrophic and autotrophic cells. Comp Biochem Physiol, A: 1997, 116:265–272.

19. Abd-El-Monem HM, Corradi MG, Gorbi G: Toxicity of copper and zinc to two strains of Scenedesmus acutus having different sensitivity to chromium. Environ Exp Bot 1998, 40:59–66.

20. Angadi SB, Mathad P: Effect of copper, cadmium and mercury on the morphological, physiological and biochemical characteristics of Scenedesmus quadricauda (Turp.) de Breb.. J Environ Biol 1998, 19:119–124.

21. El-Naggar AH: Toxic effects of nickel on photosystem II of Chlamydomonasreinhardtii. Cytobios 1998, 93:93–101.

22. Fargasova A: Accumulation and toxin effects of Cu2+, Cu+, Mn2+, VO43-, Ni2+ and MoO42- and their associations: influence on respiratory rate and chlorophyll a content of the green alga Scenedesmus quadricauda. J Trace Microprobe Tech 1998, 16:481–490.

23. Zeisler R, Dekner R, Zeiller E, Doucha J, Mader P, Kucera J: Single cell green algae reference materials with managed levels of heavy metals. Fresenius J Anal Chem 1998, 360:429–432.

24. Tikoo V, Shales SW, Scragg AH: Effect of pentachlorophenol on the growth of microalgae. Environ Technol 1996, 17:1139–1144.
25. Visviki I, Rachlin JW: Acute and chronic exposure of Dunaliella salina and Chlamydomonas bullosa to copper and cadmium - effects on ultrastructure. Arch Environ Contam Toxicol 1994, 26:154–162.
26. Fargasova A, Kizlink J: Effect of organotin compounds on the growth of the freshwater alga Scenedesmus quadricauda. Ecotoxicol Environ Saf 1996, 34:156–159.
27. Vignoles P, Greyfuss G, Rondelaud D: Growth modification of Euglena gracilis Klebs after 2-benzamino-5-nitrothiazole derivates application. Ecotoxicol Environ Saf 1996, 34:118–124.
28. Checcucci A, Colombetti G, Ferrara R, Lenci F: Action spectra for photoaccumulation of green and colorless Euglena : evidence for identification of receptor pigments. Photochem Photobiol 1976, 23:51–54.
29. Platt T, Jassby A: The relationship between photosynthesis and light for natural assemblages of coastal marine phytoplankton. J Phycol 1976, 12:421–430.

Comparative Genomic Analysis of C_4 Photosynthetic Pathway Evolution in Grasses

Xiyin Wang, Udo Gowik, Haibao Tang, John E. Bowers, Peter Westhoff and Andrew H. Paterson

ABSTRACT

Background

Sorghum is the first C4 plant and the second grass with a full genome sequence available. This makes it possible to perform a whole-genome-level exploration of C4 pathway evolution by comparing key photosynthetic enzyme genes in sorghum, maize (C4) and rice (C3), and to investigate a long-standing hypothesis that a reservoir of duplicated genes is a prerequisite for the evolution of C4 photosynthesis from a C3 progenitor.

Results

We show that both whole-genome and individual gene duplication have contributed to the evolution of C4 photosynthesis. The C4 gene isoforms show differential duplicability, with some C4 genes being recruited from whole genome duplication duplicates by multiple modes of functional innovation. The sorghum and maize carbonic anhydrase genes display a novel mode of new gene formation, with recursive tandem duplication and gene fusion accompanied by adaptive evolution to produce C4 genes with one to three functional units. Other C4 enzymes in sorghum and maize also show evidence of adaptive evolution, though differing in level and mode. Intriguingly, a phosphoenolpyruvate carboxylase gene in the C3 plant rice has also been evolving rapidly and shows evidence of adaptive evolution, although lacking key mutations that are characteristic of C4 metabolism. We also found evidence that both gene redundancy and alternative splicing may have sheltered the evolution of new function.

Conclusions

Gene duplication followed by functional innovation is common to evolution of most but not all C4 genes. The apparently long time-lag between the availability of duplicates for recruitment into C4 and the appearance of C4 grasses, together with the heterogeneity of origins of C4 genes, suggests that there may have been a long transition process before the establishment of C4 photosynthesis.

Background

Many of the most productive crops in agriculture use the C4 photosynthetic pathway. Despite their multiple origins, they are all characterized by high rates of photosynthesis and efficient use of water and nitrogen. As a morphological and biochemical innovation [1], the C4 photosynthetic pathway is proposed to have been an adaptation to hot, dry environments or CO2 deficiency [2-5]. The C4 pathway independently appeared at least 50 times during angiosperm evolution [6,7]. Multiple origins of the C4 pathway within some angiosperm families [8,9] imply that its evolution may not be complex, perhaps suggesting that there may have been genetic pre-deposition in some C3 plants to C4 evolution [6].

The high photosynthetic capacity of C4 plants is due to their unique mode of CO2 assimilation, featuring strict compartmentation of photosynthetic enzymes into two distinct cell types, mesophyll and bundle-sheath (illustrated in Figure 1 for the NADP-malic enzyme (NADP-ME) type of C4 pathway). First, CO2 assimilation is carried out in mesophyll cells. The primary carboxylating enzyme, phosphoenolpyruvate carboxylase (PEPC), together with carbonic anhydrase (CA), which is crucial to facilitating rapid equilibrium between CO2 and HCO_3^-, is responsible for the hydration and fixation of CO2 to produce a C4 acid, oxaloacetate. In NADP-ME-type C4 species, oxaloacetate is then converted to another C4 acid, malate, catalyzed by malate dehydrogenase (MDH). Malate then diffuses into chloroplasts in the proximal bundle-sheath cells, where CO2 is released to yield pyruvate by the decarboxylating NADP-ME. The released CO2 concentrates around the secondary carboxylase, Rubisco, and is reassimilated by it through the Calvin cycle. Pyruvate is transferred back into mesophyll cells and catalyzed by pyruvate orthophosphate dikinase (PPDK) to regenerate the primary CO2 acceptor, phosphoenolpyruvate. Phosphorylation of a conserved serine residue close to the amino-terminal end of the PEPC polypeptide is essential to its activity by reducing sensitivity to the feedback inhibitor malate and a catalyst named PEPC kinase (PPCK). C4 photosynthesis results in more efficient carbon assimilation at high temperatures because its combination of morphological and biochemical features reduce photorespiration, a loss of CO2 that occurs during C3 photosynthesis at high temperatures [10]. PPDK regulatory protein (PPDK-RP), a bifunctional serine/threonine kinase-phosphatase, catalyzes both the ADP-dependent inactivation and the Pi-dependent activation of PPDK [11].

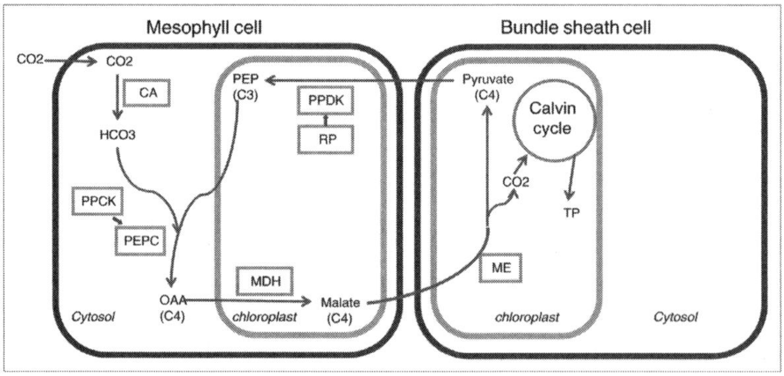

Figure 1. The NADP-ME type of C4 pathway in sorghum and maize. CA, carboxylating anhydrase; MDH, malate dehydrogenase; ME, malic enzyme; OAA, oxaloacetate; PEPC, phosphoenolpyruvate carboxylase; PPCK, PEPC kinase; PPDK, pyruvate orthophosphate dikinase; PPDK-RP, PPDK regulatory protein; TP, transit peptide.

The evolution of a novel biochemical pathway is based on the creation of new genes, or functional changes in existing genes. Gene duplication has been recognized as one of the principal mechanisms of the evolution of new genes. Genes encoding enzymes of the C4 cycle often belong to gene families having multiple copies. For example, in maize and sorghum, a single C4 PEPC gene and other non-C4 isoforms were discovered [12], whereas in Flaveria trinervia, a C4 eudicot, multiple copies of C4 PEPC genes were found [13]. These findings led to the proposition that gene duplication, followed by functional innovation, was the genetic foundation for photosynthetic pathway transformation [14].

All plant genomes, including grass genomes, have been enriched with duplicated genes derived from tandem duplications, single-gene duplications, and large-scale or whole-genome duplications [15-18]. A whole-genome duplication (WGD) occurred in a grass ancestor approximately 70 million years ago (mya), before the divergence of the panicoid, oryzoid, pooid, and other major cereal lineages [19,20]. A preliminary analysis of sorghum genome data suggested that duplicated genes from various sources have expanded the sizes of some families of C4 genes and their non-C4 isoforms [21]. However, different duplicated gene pairs often have divergent fates [22]. While most duplicated genes are lost, gene retention in some functional groups produces large gene families in plants [15,19,20]. Together with other lines of evidence, these have led to the interesting proposition of differential gene duplicability [23,24], or duplication-resistance [25], due to possible gene dosage imbalance, which can be deleterious [26]. Even when duplicated genes survive, there is rarely strong evidence supporting possible functional innovation [27].

Most C4 plants are grasses, and it has been inferred that C4 photosynthesis first arose in grasses during the Oligocene epoch (24 to 35 mya) [28,29]. Sorghum and maize, thought to have diverged from a common ancestor approximately 12 to 15 mya [21], are both in the Andropogoneae tribe, which is entirely composed of C4 plants [8]. Sorghum, a NADP-ME-type C4 plant grown for food, feed, fiber and fuel, is the second grass and the first C4 plant with its full genome sequence available [21]. The first grass genome sequenced was rice, a C3 plant. The availability of two grass genome sequences using different types of photosynthesis provides a valuable opportunity to explore C4 pathway evolution. In the present research, by using a comparative genomic approach and phylogenetic analysis, we compared C4 genes and their non-C4 isoforms in sorghum, maize and rice. The aims of this study are to investigate: the role of gene duplication in the evolution of C4 enzyme genes; the role of adaptive evolution in C4 pathway formation; the long-standing hypothesis that a reservoir of duplicated genes has been a prerequisite of C4 pathway evolution [14]; and whether codon usage bias has contributed to C4 gene evolution, as previously suggested [30]. Our results will help to clarify the evolution of the C4 pathway and may benefit efforts to transform C3 plants, such as rice, to C4 photosynthesis [31].

Results

PEPC Enzyme Genes

Grass PEPC enzyme genes form a small gene family. There are five plant-type and one bacteria-type PEPC (Sb03g008410 and Os01g0110700) [32] gene isoforms in sorghum and rice, respectively, excepting two likely pseudogenized rice isoforms (Os01g0208800, Os09g0315700) having only 217 and 70 codons. There is one sorghum C4 PEPC [33,34], Sb10g021330. Previous characterization indicated that its transcripts are more than 20 times more abundant in mesophyll than in bundle-sheath cells [35].

By analysis of gene colinearity, we investigated how genome duplication has affected the PEPC gene families in rice and sorghum. The PEPC gene in rice that is most similar to the sorghum C4 PEPC is Os01g0208700, sharing 73% amino acid identity. This similarity raised the possibility that the two genes are orthologous. Although the two genes under consideration are not in colinear locations, single-gene translocation is not rare in grasses [36]. The outparalogs, homologs produced by WGD in the common ancestor of sorghum and rice, of the sorghum C4 PEPC gene are located at the expected homoeologous locations in both sorghum and rice (Sb04g008720 and Os02g0244700). The rice gene Os01g0208700 and the C4 genes are grouped together, and outparalogs (Os02g0244700 and Sb04g008720) of the sorghum C4 gene form a sister group on the phylogenetic tree. The pattern can be explained if Os01g0208700 were orthologous to the sorghum C4 PEPC gene, implied by their high sequence similarity and shared high GC content (detailed below). In our view, the most parsimonious explanation of these data is that the oryzoid (rice) ortholog was translocated after the sorghum-rice (panicoid-oryzoid) divergence, then the panicoid (sorghum) ortholog was recruited into the C4 pathway. We cannot falsify a model invoking independent loss of alternative homeologs in sorghum (panicoids) and rice (oryzoids), respectively, although this model seems improbable in that such loss of alternative homoeologs has only occurred for approximately 1.8 to 3% of genome-wide gene duplicates in these taxa [21]. The other rice and sorghum PEPC genes form four orthologous pairs. Whether the genes from different orthologous groups are outparalogs could not be supported by colinearity inference associated with the pan-cereal genome duplication.

Grass PEPC genes show high GC content, like many other grass genes, apparently as a result of changes after the monocot-dicot split but before the radiation of the grasses [37]. The evolution of C4 PEPC genes in sorghum and maize was previously proposed to have been accompanied by GC elevation, resulting in codon usage bias [38]. We found that C4 PEPC genes do have higher GC content than other sorghum and maize PEPC genes, especially at the third codon sites (GC3). The sorghum and maize C4 PEPC genes have a GC3 content of approximately 84%, significantly higher than other genes in both species. The suspected rice ortholog Os01g0208700 has even higher GC3 content, approximately 92%. In contrast, the GC3 content of all Arabidopsis PEPC genes is <43%. This shows that the higher GC content in the C4 PEPC genes may not be related to the evolution of C4 function, as discussed below.

C4 PEPC genes show evidence of adaptive evolution. To characterize the evolution of C4 PEPC genes, we aligned the sequences and constructed gene trees without involving the possible pseudogenized rice gene. We found the genes to be in two groups, with one containing plant-type and the other bacteria-type PEPC genes. Careful inspection suggested problems with the tree, for orthologous genes were not grouped together as expected. After removing the bacteria-type genes and rooting the subtree containing the C4 genes with Arabidopsis PEPC genes, we obtained a tree in which orthologs are grouped together as expected (Figure 2a). The sorghum and maize C4 genes are on a remarkably long branch, suggesting that they are rapidly evolving compared to the other genes, and implying possible adaptive selection during the evolution of the C4 pathway, consistent with a previous proposal [39].

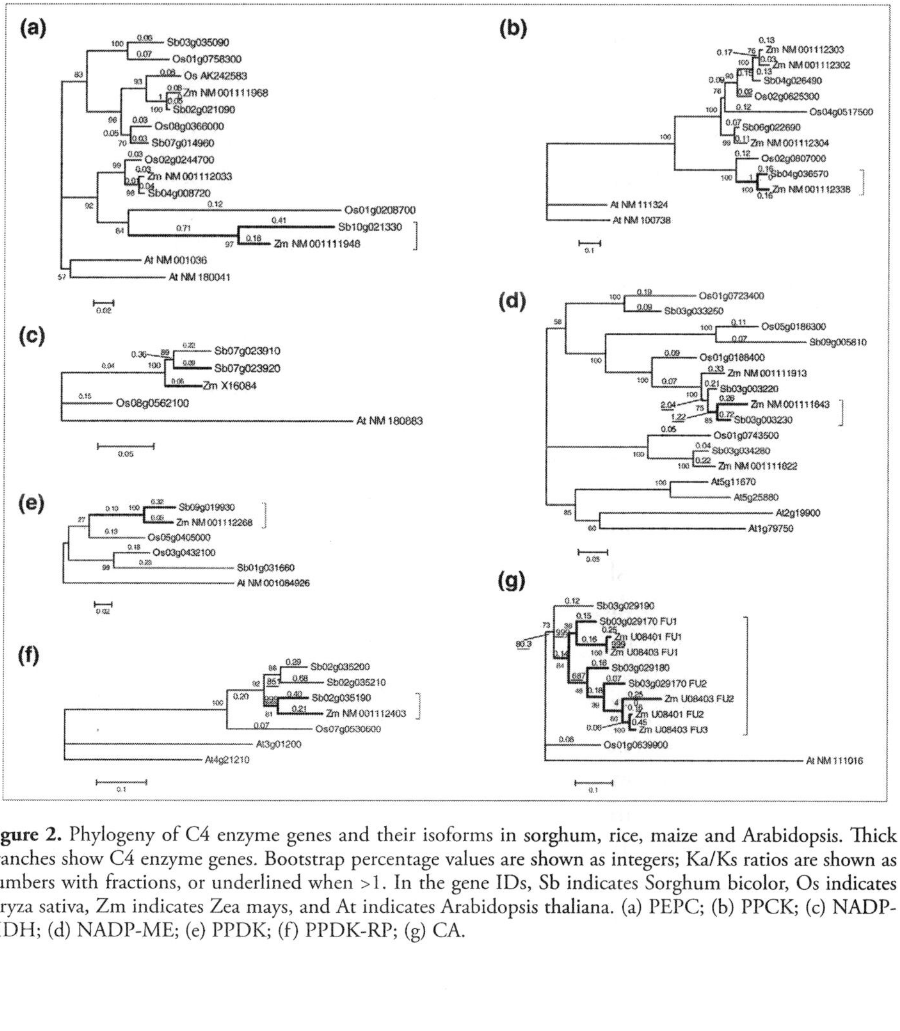

Figure 2. Phylogeny of C4 enzyme genes and their isoforms in sorghum, rice, maize and Arabidopsis. Thick branches show C4 enzyme genes. Bootstrap percentage values are shown as integers; Ka/Ks ratios are shown as numbers with fractions, or underlined when >1. In the gene IDs, Sb indicates Sorghum bicolor, Os indicates Oryza sativa, Zm indicates Zea mays, and At indicates Arabidopsis thaliana. (a) PEPC; (b) PPCK; (c) NADP-MDH; (d) NADP-ME; (e) PPDK; (f) PPDK-RP; (g) CA.

Maximum likelihood analysis supports possible adaptive evolution of C4 PEPC genes. First, characterization of nonsynonymous nucleotide substitution rates (Ka) supports rapid evolution of the C4 genes and their rice ortholog. Under a free-parameter model, Ka values are >0.048 on branches leading to C4 genes and their rice ortholog after the rice-sorghum split, as compared to ≤0.02 on branches leading to the non-C4 isoforms. Second, the C4 genes may have been positively selected. The Ka/Ks ratio is nearly tenfold higher (0.71) on the branch leading to the last common ancestor of the sorghum and maize C4 genes than on other branches after the rice-sorghum split (≤0.08). Though the ratio is <1, we propose that the striking difference in Ka/Ks between C4 and non-C4 genes may be evidence of positive selection in the C4 genes for the following reasons: the criterion Ka/Ks > 1 has been proposed to be unduly stringent to infer positive selection [40]; the maximum likelihood analysis is conservative, as reported previously [27]; and the similar slow evolutionary changes in all non-C4 genes in sorghum, maize and rice (Figure 1a) imply elevated rates in the C4 genes, rather than purifying selection in the non-C4 genes.

C4 PEPC genes show elevated and aggregated amino acid substitutions especially in function-specific regions, providing further evidence of adaptive evolution. Comparison to their outparalogs and their nearest outgroup sequence suggests that C4 PEPC genes have accumulated approximately 100 putative substitutions over their full length (Table 1), far more than non-C4 PEPC genes. The substitutions are referred to as putative since we cannot rule out the possibility of parallel and reverse mutations. However, the extremely significant difference strongly supports divergent evolution of C4 and non-C4 PEPC genes. The amino acid substitutions are not uniformly distributed along the lengths of the C4 genes, but concentrated in the carboxy-terminal half, including the critical mutation S780 (the serine at position 780 of the maize C4 PEPC protein that is essential to relieving feedback inhibition by malate [41]). This is consistent with previous findings [42].

Surprisingly, Os01g0208700 has also accumulated significantly more mutations than expected, and has a relatively larger selection pressure than other non-C4 PEPC genes, implying that it may also be under adaptive selection, as further discussed below.

Table 1. Aggregated amino acid substitution analysis results

Aggregated amino acid substitution analysis results

Gene 1	Gene 2	Outgroup	Alignment length	Alignment length without gaps	Average identity	Overall substitution number in gene 1	Overall substitution number in gene 2	*P*-value
PEPC								
Sb10g021330	Os02g0244700	Os01g0758300	972	958	0.76	110	26	5.89E-13
Zm_NM_001111968	Os02g0244700	Os01g0758300	971	968	0.78	92	33	1.31E-07
Sb10g021330	Os02g0244700	Sb03g035090	972	958	0.76	117	28	1.46E-13
Zm_NM_001111968	Os02g0244700	Sb03g035090	971	968	0.77	104	34	2.54E-09
PPCK								
Sb04g036570	Os02g0807000	Sb06g022690	309	284	0.65	15	14	8.53E-01
Zm_NM_001112338	Os02g0807000	Sb06g022690	309	281	0.63	18	11	1.94E-01
CA								
U08403_FU3	Os01g0639900	Sb03g029190.1	272	201	0.75	19	18	8.69E-01
U08403_FU2	Os01g0639900	Sb03g029190.1	273	200	0.73	20	18	7.46E-01
U08403_FU1	Os01g0639900	Sb03g029190.1	273	202	0.79	13	18	3.69E-01
U08401_FU2	Os01g0639900	Sb03g029190.1	272	201	0.75	18	18	1.00E+00
U08401_FU1	Os01g0639900	Sb03g029190.1	273	202	0.78	14	18	4.80E-01
Sb03g029170_FU2	Os01g0639900	Sb03g029190.1	272	201	0.78	14	16	7.15E-01
Sb03g029170_FU1	Os01g0639900	Sb03g029190.1	273	201	0.80	11	20	1.06E-01
Sb03g029180	Os01g0639900	Sb03g029190.1	274	202	0.80	11	19	1.44E-01
U08403_FU3	Os01g0639900	At_NM_111016	293	201	0.50	14	13	8.47E-01
U08403_FU2	Os01g0639900	At_NM_111016	293	200	0.49	16	14	7.15E-01
U08403_FU1	Os01g0639900	At_NM_111016	293	202	0.50	10	15	3.17E-01
U08401_FU2	Os01g0639900	At_NM_111016	293	201	0.50	12	13	8.41E-01
U08401_FU1	Os01g0639900	At_NM_111016	293	202	0.50	11	15	4.33E-01
Sb03g029170_FU2	Os01g0639900	At_NM_111016	293	201	0.50	10	10	1.00E+00
Sb03g029170_FU1	Os01g0639900	At_NM_111016	293	201	0.50	9	14	2.97E-01
Sb03g029180	Os01g0639900	At_NM_111016	293	202	0.50	8	11	4.91E-01
PPDK								
Sb09g019930	Os05g0405000	Os03g0432100	949	946	0.83	42	28	9.43E-02
Zm_NM_001112268	Os05g0405000	Os03g0432100	950	944	0.83	44	28	5.93E-02
Sb09g019930	Os05g0405000	Sb01g031660	958	946	0.76	37	15	2.28E-03
Zm_NM_001112268	Os05g0405000	Sb01g031660	961	942	0.78	32	18	4.77E-02
NADP-MDH								
Sb07g023920	Os08g0562100	At_NM_180883	443	427	0.77	22	19	6.39E-01
Sb07g023910	Os08g0562100	At_NM_180883	443	432	0.75	25	16	1.60E-01
ZM_X16084	Os08g0562100	At_NM_180883	443	430	0.75	25	13	5.16E-02
NADP-ME								
Sb03g003230	Os01g0188400	Os05g0186300	642	633	0.80	46	16	1.39E-04
Sb03g003230	Os01g0188400	Sb09g005810	642	633	0.80	41	20	7.17E-03
Sb03g003220	Os01g0188400	Os05g0186300	650	635	0.84	23	15	1.94E-01
ZM_NM_001111843	Os01g0188400	Os05g0186300	641	634	0.80	47	16	9.40E-05
ZM_NM_001111913	Os01g0188400	Os05g0186300	668	633	0.84	26	15	8.58E-02
PPDK-RP								
Sb02g035190	Os07g0530600	At4g21210	474	426	0.58	37	17	6.00E-03
Zm_NM_001112403	Os07g0530600	At4g21210	474	423	0.57	33	23	1.80E-01
Sb02g035190	Sb02g035200	Os07g0530600	476	408	0.69	19	22	6.40E-01
Sb02g035190	Sb02g035210	Os07g0530600	483	384	0.69	21	22	8.70E-01
Zm_NM_001112403	Sb02g035200	Os07g0530600	472	416	0.67	25	22	6.60E-01
Zm_NM_001112403	Sb02g035210	Os07g0530600	482	389	0.68	25	25	1.00E+00

PPCK Enzyme Genes

PPCK gene families have been enriched by duplication events, including the pan-cereal WGD and tandem duplication. We identified three PPCK gene isoforms in both sorghum and rice, respectively, which are in one-to-one correspondence in expected colinear locations between the two species (Figure 2b). These rice and sorghum isoforms correspond to four maize isoforms (ZmPPCK1 to ZmPPCK4; Figure 2b), with ZmPPCK2 and ZmPPCK3 likely produced in maize after its divergence from a lineage shared with sorghum. The sorghum C4 PPCK is encoded by Sb04g036570, and its maize ortholog is ZmPPCK1. Their C4 nature is supported by evidence that their expression is light-induced and their transcripts are more abundant in mesophyll than bundle-sheath cells [30]. In contrast, the expression of sorghum and maize non-C4 isoforms is not light- but cycloheximide-affected [30]. The outparalogs of the sorghum C4 gene and its rice ortholog were likely lost before the two species split, whereas the other four isoforms are outparalogs.

Maximum likelihood analysis and inference of aggregated amino acid substitutions found no evidence of adaptive selection during C4 PPCK gene evolution.

Consistent with a previous report [30], all studied grass PPCK genes have extremely high GC content, with a GC3 content from 88 to 97%. The grass C4 and non-C4 PPCK genes have similar GC content.

NADP-MDH Enzyme Genes

There are two NADP-MDH enzyme genes in sorghum, the non-C4 gene Sb07g023910 and the C4 gene Sb07g023920, tandemly located as previously reported [43]. They have only one homolog in both rice and maize [44], with the rice homolog (Os08g0562100) at the expected colinear location. This suggests that the NADP-MDH WGD outparalog was lost before the sorghum-rice split. Each of the sorghum tandem genes has an ortholog in Vetiveria and Saccharum, respectively [44], suggesting that the tandem duplication occurred before the divergence of sorghum and Vetiveria, but after the sorghum-maize split, an inference further supported by gene tree analysis in that they are more similar to one another than to the single maize homolog (Figure 2c).

The C4 NADP-MDH gene shows an interesting mode of adaptive evolution. Though the C4 NADP-MDH genes have accumulated more mutations than non-C4 genes, neither maximum likelihood analysis nor the inference of aggregated amino acid substitution suggest adaptive selection. However, the sorghum C3 and C4 genes were likely to have been produced by an ancestral C4 gene through duplication. One of the duplicates may have lost its C4 function as it is not light-induced and only constitutively expressed [43].

The NADP-MDH genes are chloroplastic. A chloroplast transit peptide (cTP) having approximately 40 amino acids is identified in all the genes from grasses and Arabidopsis. This indicates that the cTP was present in the common ancestor of angiosperms. Non-chloroplastic NADP-MDH genes identified in the sorghum genome share less than 40% protein sequence similarity with the chloroplastic ones.

All of the grass NADP-MDH enzyme genes studied have elevated GC content compared to the Arabidopsis ortholog, especially regarding GC3. The grass C4 genes have slightly higher GC content than the non-C4 genes.

NADP-ME Enzyme Genes

The NADP-ME gene family has been gradually expanding due to tandem duplication and the pan-cereal WGD. We identified five and four NADP-ME enzyme genes in sorghum and rice, respectively. The sorghum C4 gene is Sb03g003230, whose transcript is abundant in bundle-sheath but not mesophyll cells [35]. The C4 gene has a tandem duplicate that may have been produced before the sorghum-maize split based on gene similarity and tree topology (Figure 2d). The tandem genes share the same rice ortholog (Os01g0188400) at the expected colinear location, and their WGD duplicates can be found at the expected colinear location in both species. The other sorghum and rice NADP-ME genes form two orthologous pairs, having also remained at the colinear locations predicted based on the pan-cereal duplication.

Maximum likelihood analysis indicates that the sorghum and maize C4 NADP-ME genes are under positive selection. The branches leading to their two closest ancestral nodes have a Ka/Ks ratio > 1 (P-value = 8 x 10-10). Moreover, the C4 genes have a significant abundance of amino acid substitutions (Table 1). The most affected regions in sorghum and maize overlap with one another, from residue 141 to residue 230 in sorghum, and from residue 69 to residue 181 in maize.

The grass NADP-ME genes have higher GC content than their Arabidopsis homologs. The highest GC content (GC3 > 82%) is found not in the C4 genes but in their outparalogs, Sb09g005810 and Os05g0186300.

The C4 genes, their tandem paralogs in sorghum and maize, and their rice ortholog all share an approximately 39 amino acid cTP that is absent from their WGD paralogs in grasses, or homologs in Arabidopsis. This seems to suggest that the cTP was acquired by one member of a duplicated gene pair after the pan-grass WGD but before the sorghum-rice divergence.

PPDK Enzyme Genes

Sorghum and rice both have two PPDK enzyme genes. The sorghum C4 PPDK gene (Sb09g019930) is identified based on its approximately 90% amino acid identity with the maize C4 gene. Its transcript is abundant in mesophyll rather than bundle-sheath cells [35]. Its rice ortholog (Os05g0405000) can be inferred based on both gene trees (Figure 2e) and gene colinearity. The other rice and sorghum isoforms are orthologous to one another. Whether the four isoforms are outparalogs produced by the WGD could not be determined by gene colinearity inference due to possible gene translocations. However, synonymous nucleotide substitution rates and gene tree topologies support that the rice and sorghum paralogs were produced before the two species diverged, and approximately at the time of the pan-cereal WGD.

There are two PPDK genes in maize [10]. One of them encodes both a C4 transcript and a cytosolic transcript, controlled by distinct upstream regulatory elements [45]. The C4 copy has an extra exon encoding a cTP at a site upstream of the cytosolic gene [46]. We found that the sorghum C4 PPDK gene is highly similar to its maize counterpart along their respective full lengths, indicating their origin in a common maize-sorghum ancestor. The other maize PPDK gene has only a partial DNA sequence and, therefore, has been avoided in the present evolutionary analysis. A similarity search against the maize bacterial artificial chromosome (BAC) sequences indicates that it is on a different chromosome (chromosome 8) from the C4 gene (chromosome 6). The maize counterpart of the other sorghum PPDK isoform has not yet been identified in sequenced BACs.

The C4 PPDK genes may have experienced adaptive evolution. While maximum likelihood analysis did not find evidence of adaptive evolution of C4 PPDK genes (Figure 2e), the C4 genes have accumulated significantly or nearly significantly more amino acid substitutions than their rice orthologs, particularly in the region from approximately residue 207 to approximately residue 620 (Table 1).

All grass PPDK genes have higher GC content than their Arabidopsis homologs, with the C4 genes themselves being highest in GC content (GC3 content approximately 61 to 70%).

All of the characterized PPDK isoform sequences from grasses and Arabidopsis share an approximately 20 amino acid cTP, suggesting its origin before the monocot-dicot split.

PPDK-RP Enzyme Genes

Tandem duplication contributed to the expansion of PPDK-RP genes. Using the maize PPDK-RP gene sequence as a query, we determined its possible sorghum ortholog, Sb02g035190, which has two tandem paralogs. Their rice ortholog, Os07g0530600, was identified in the anticipated colinear region. However, we failed to find their WGD outparalogs in both sorghum and rice, suggesting possible gene loss in their common ancestor.

Gene trees indicate that the tandem duplication events may have occurred before the sorghum-maize divergence, but after the sorghum-rice divergence (Figure 2f). Maximum likelihood analysis suggests that both lineages leading to the maize PPDK-RP gene and its sorghum ortholog, and other isoforms, have been under significant positive selection (Ka/Ks >> 1, P-value = 2.5 x 10-8), implying possible functional changes in both lineages. Compared to their rice ortholog, sorghum and maize PPDK-RP genes have accumulated significantly more amino acid substitutions (Table 1), providing supporting evidence for functional innovation.

Both the C4 and non-C4 PPDK-RP genes in sorghum have similar GC content (GC3 content approximately 57 to 60%), while the maize PPDK-RP gene has higher GC content (GC3 content approximately 67%), especially in the third codon sites. All these grass PPDK-RP genes show higher GC content than their Arabidopsis homologs.

CA Enzyme Genes

Tandem duplication has profoundly affected the evolution of CA genes. There are two types of CA enzymes, the alpha and beta types in sorghum [21], and C4 CA genes are the beta type [47]. Here, we focus on beta-type CA genes. Our analysis indicates that there are four beta-type CA enzyme gene isoforms in sorghum, forming a tandem gene cluster with the same transcriptional orientation, on chromosome 3 (Figure 3a). Among them are two possible C4 genes (Sb03g029170 and Sb03g029180), which were shown by previous analysis of transcript abundance to be highly expressed in mesophyll but not bundle-sheath cells. The other two genes include one non-C4 gene (Sb03g029190) and one probable pseudogene (Sb03g029200) with only truncated coding sequence, a large DNA insertion in its second exon, and accumulated point mutations. These tandem genes have a common rice ortholog (Os01g0639900) at the expected colinear location, indicating that gene family expansion has occurred in sorghum (and maize; see below) since divergence from rice. The WGD outparalogs were not identified in either genome, implying possible gene loss after the WGD and before the rice-sorghum split.

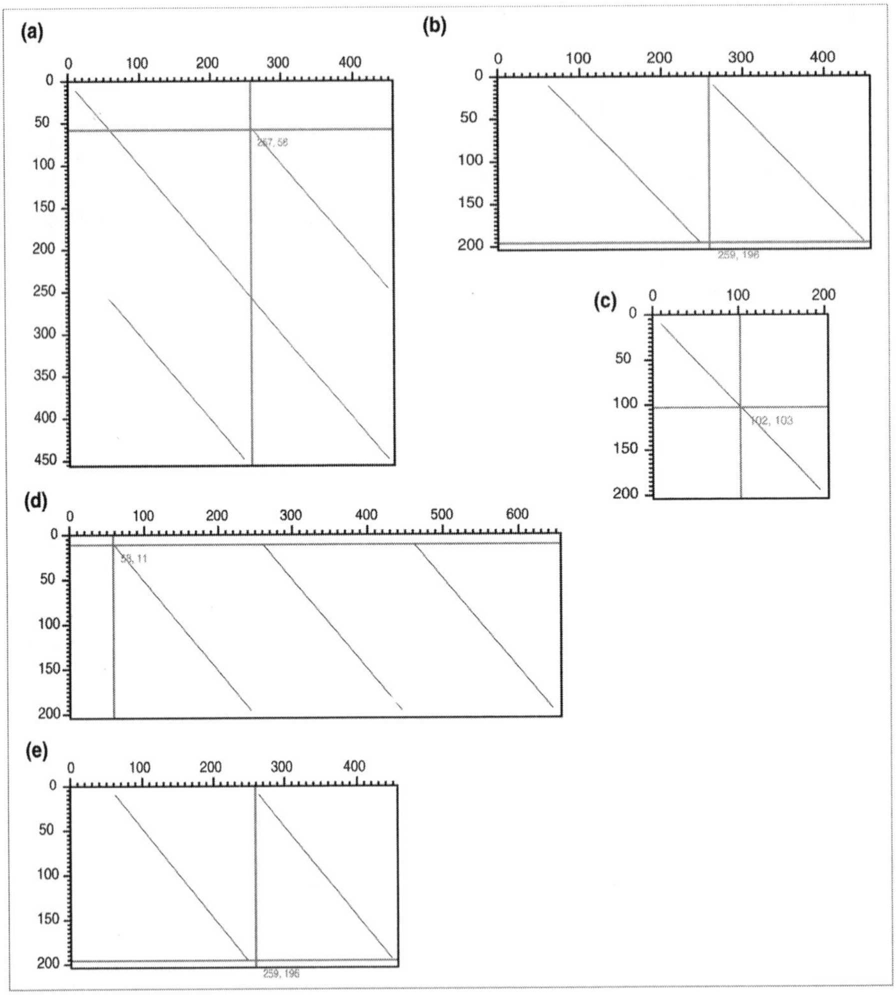

Figure 3. Dotplots between sorghum and maize CA enzyme protein sequences. (a) Self-comparison of protein sequence of Sb03g029170. (b) Sb03g029170 (horizontal) and Sb03g029180 (vertical); (c) Sb03g029190 (horizontal) and Sb03g029180 (vertical); (d) maize U08403 (horizontal) and Sb03g029180 (vertical); (e) maize U08401 (horizontal) and Sb03g029180 (vertical).

The two sorghum C4 CA genes differ in cDNA length [35]. We found that the larger C4 CA gene may have evolved by fusing two neighboring CA genes produced by tandem duplication. In spite of possible alternative splicing programs, Sb03g029170 has a gene length of approximately 10.4 kbp and includes 13 exons, as compared to 4.5 kbp in length and 6 exons for Sb03g029180. Pairwise dotplots between Sb03g029170 and Sb03g029180 show the former has an internal repeat structure absent from the latter (Figure 3ab). The duplication involves the last six of seven exons and intervening introns 1 to 6 of the ancestral gene (Figure 4a). Comparatively, the other sorghum genes have only exons 2 to 7, assumed to be a functional unit, both lacking the first exon in Sb03g029170, which encodes a cTP. This implies that several duplication events have recursively produced extra copies of the functional unit. Some functional units act as independent genes, while the other fused with the complete one to form an expanded gene including two functional units. We found that this fusion involved mutation of the stop codon in the leading gene. Each functional unit starts with an ATG codon, which we infer may increase the possibility of alternative splicing. This inference is supported by the finding that Sb03g029170 may have two distinct transcripts, identified by cDNA HHU69 and HHU22, respectively. The two transcripts have distinct lengths, 2,100 and 1,200 bp, respectively, with the expression of the longer one being light-inducible and C4-related but the shorter one not [35]. The non-C4 gene, Sb03g029190, has a normal structure (Figure 3c) and the pseudogene, Sb03g029200, has a truncated structure.

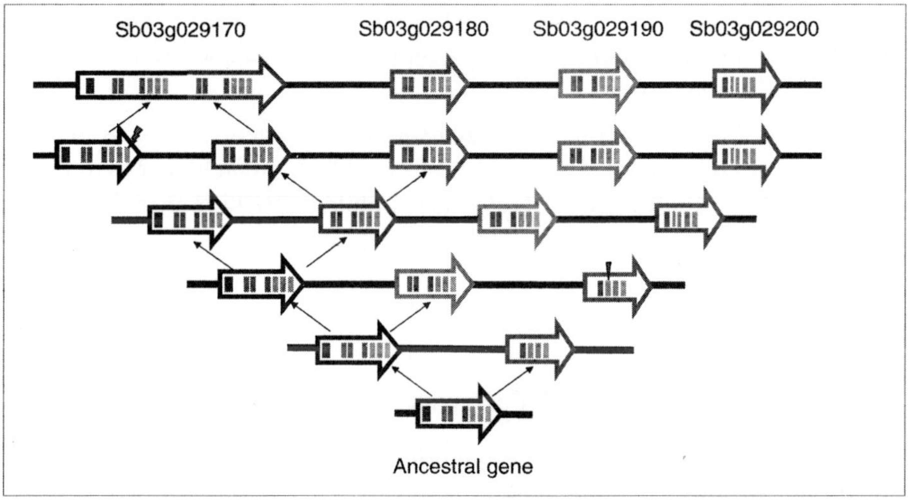

Figure 4. Tandem duplication and fusion of CA genes in sorghum. Postulated evolution of sorghum CA genes through four tandem duplication events and a gene fusion event is displayed. We show distribution and structures of CA genes, and their peptide-encoding exons, on sorghum chromosome 3. Genes are shown as the large arrows with differently colored outlines and exons are shown as colored blocks contained in the arrows. Homologous exons are in the same color. A chloroplast transit peptide is in dark red. A tandem duplication event is shown by two small black arrows pointing in divergent directions, and a gene fusion event is shown by two small black arrows pointing in convergent directions. A new gene produced by tandem duplication is shown with an arrow in a new color not used by the ancestral genes. A gene produced by fusion of two neighboring genes is shown as a bipartite structure, each part with the color of one of the fused genes. A stop codon mutation is shown by a lightning-bolt symbol, and an exon-splitting event by a narrow triangle.

The tandem duplication and gene fusion are shared by sorghum and maize, and maize furthermore has additional duplication. Interestingly, we found that the maize CA enzyme genes have two and three functional units, respectively (Figure 3de), implying further DNA sequence duplication and gene fusion in the maize lineage. Mutation of stop codons was also found in the leading gene sequences. Rice and Arabidopsis genes have only one functional unit preceded by a cTP.

To clarify the evolution of CA genes, we performed a phylogenetic analysis of the functional units (Figure 2f). The first functional units from sorghum and maize genes are grouped together, the second and third maize units and that of Sb03g029180 were in another group, and the rice gene and non-C4 sorghum gene Sb03g029190 were outgroups. This suggests the origin of the extra functional units to be after the Panicoideae-Ehrhartoideae divergence but before the sorghum-maize divergence, and continuing in the maize lineage. A possible evolutionary process in sorghum is illustrated in Figure 4b.

A gene tree of functional units suggested that C4 CA genes may have been affected by positive selection. According to the free-parameter model of the maximum likelihood approach, we found that the two functional unit groups revealed above may have experienced positive selection, in that Ka/Ks > 1 (Figure 2f), though this possibility is not significantly supported by statistical tests or by amino acid substitution analysis.

Excepting the possibly pseudogenized sorghum CA gene, the grass isoforms have very high GC content (GC3 content 82 to 92%), much higher than that of the Arabidopsis orthologs. The non-C4 gene, Sb03g029190, rather than any of the C4 genes, has the highest GC content in sorghum.

Discussion

Gene Duplication and C4 Pathway Evolution

In the case of the C4 pathway, the evolution of a novel biological pathway required the availability of gene families with multiple members, in which modification of both expression patterns and functional domains led to new adaptive phenotypes. An intuitive idea is that genetic novelty formation is simplified by exploiting available 'construction bricks', and the pathway genes that we are aware of were either 'subverted' from existing functions or were created through modification of existing genes. Three mechanisms of new gene formation have been proposed [48]: duplication of pre-existing genes followed by neofunctionalization; creation of mosaic genes from parts of other genes; and de novo invention of genes from DNA sequences.

Duplicated genes have long been suggested to contribute to the evolution of new biological functions. As early as 1932, Haldane suggested that gene duplication events might have contributed new genetic materials because they create initially identical copies of genes, which could be altered later to produce new genes without disadvantage to the organism [49]. Ohno proposed that gene duplication played an essential role in evolution [50], pointed out the importance that WGD might have had on speciation, and hypothesized that at least one WGD event facilitated the evolution of vertebrates [51]. This hypothesis has been supported by evidence from various gene families, and from the whole genome sequences of several metazoans [52,53]. Plant genomes have experienced recurring WGDs [15,54-57], and perhaps all angiosperms are ancient polyploids [54]. These polyploidy events contribute to the creation of important developmental and regulatory genes [58-61], and may have played an important role in the origin and diversification of the angiosperms [62]. About 20 million years before the divergence of the major grass clades [19,20], the ancestral grass genome was affected by a WGD, possibly preceded by still more ancient duplication events [17,63]. It is tempting to link this WGD to the evolutionary success of grasses, now including more than 10,000 species and covering about 20% of the Earth's land surface [64], though such a link has not yet been adequately justified.

Gene duplication has been related to the evolution of the C4 pathway, based on the finding that C4 enzyme genes are usually from families having multiple copies [14]. Consequently, an ability to create and maintain large numbers of duplicated genes has been supposed to be one precondition for certain taxa to develop C4 photosynthesis [6,14]. It was even suggested that evolution of the C4 pathway was largely a story of gene duplication while plants were still in the ancestral C3 state [14].

Different genes in the C4 pathway were affected in different ways and at different times by gene duplication. Firstly, the approximately 70-mya pan-cereal WGD enriched the reservoir of some genes but not others. For example, in sorghum, both duplicated copies were preserved for PEPC and NADP-ME genes, and one of the copies of each gene produced by WGD was later recruited into the C4 pathway. This finding highlights the contribution of WGD to the evolution of C4 photosynthesis. However, for NADP-MDH, CA, PPDK-RP and PPCK enzyme genes, one of the WGD duplicates was probably lost. For CA, NADP-ME, and PPDK-RP, tandem gains of new genes after the sorghum-rice divergence appears to have preceded C4 evolution. This seems to suggest that earlier availability of the pan-cereal duplicated copies was not by itself sufficient to initiate C4 evolution, although it is not clear whether what was lacking was genetic (a part of the machinery) or environmental (a sufficiently strong selective advantage to drive the transition).

Adaptive Evolution of C4 Genes

After duplication, there is evidence that some C4 genes experienced adaptive evolution; however, selection pressures and evolutionary modes have varied. Both maximum likelihood inference and patterns of aggregated amino acid differences indicate that the C4 NADP-ME and PPDK-RP enzyme genes have been under strong selective pressure. Maximum likelihood inference also implies that CA C4 enzymes have experienced positive selection, while aggregated amino acid differences indicate that C4 PEPC and PPDK genes may also have been under positive selection. The sorghum C4 genes of PPCK and NADP-MDH enzymes have also accumulated more substitutions than their rice orthologs, though the difference is not statistically significant. Compared to their rice orthologs, PEPC and NADP-MDH C4 genes evolve at a faster rate, providing further evidence of adaptation.

In many cases (NADP-ME, CA, PEPC) evidence from C4 plants supports adaptive evolution of the C4 gene family members only - the non-C4 homologs in C4 plants show no evidence of adaptive evolution, although the PEPC gene does show evidence in rice (C3). Further, the strongest evidence of adaptive evolution is in the period when the C4 pathway is thought to have evolved, after the divergence of sorghum and rice, but before the divergence of sorghum and maize.

Adaptive evolution is further supported by patterns of gene expression shown in previous reports. PEPC, PPDK, and CA C4 genes are expressed approximately 20 times more in sorghum mesophyll than bundle-sheath cells, while NADP-ME C4 genes are expressed much more in bundle-sheath than mesophyll cells [35]. The study of Flaveria intermediates shows that PEPC activity is increased approximately 40 times from C3 to full C4 species [65], and the NADP-ME activity is approximately 9 times higher in veins than mesophyll cells [66].

During the process of adaptive evolution, a duplicated gene may gradually acquire a new function (neofunctionalization) or subdivide the functions of its progenitor with the other duplicated copy (subfunctionalization). Laboratory evolution experiments indicated that an evolving new gene can initially acquire increased fitness for a new function without losing its original function [67]. This implies that a neofunctionalization process may begin with an initial subfunctionalization step, an implication that has been supported by theory [68]. It is unclear how long such a step may take. Here, we found that neofunctionalization of C3 genes to function in C4 photosynthesis could take a long time. Previous publications found that both C4 and non-C4 sorghum NADP-MDH genes were expressed in green leaves, though the C4 gene had higher transcript accumulation [43,44]. Together with maximum likelihood analysis involving more genes and different grasses, this finding indicated that C4 and non-C4 sorghum NADP-

MDH genes, produced before sorghum-Vetiveria divergence, have experienced subfunctionalization [44]. Sequence alignment here indicates that the sorghum non-C4 gene has been affected by three insertions and one deletion in its amino-terminal coding sequence, suggesting functional innovation. Regardless of whether the process is a division of functions of their progenitor gene, or evolution of a novel function in the non-C4 gene, co-expression, albeit at divergent levels, of the two genes in green leaves suggests that the process may not yet be finished.

In addition to the possible sheltering effect of a duplicated copy when evolving genetic novelty, alternative splicing may further shelter functional changes. The maize PPDK gene (and probably also its sorghum ortholog) encoding C4 transcripts also encodes cytosolic transcripts. Their rice homolog also has a dual promoter [69], implying that natural selection may have utilized this pre-existing functional duality to evolve C4 function. If C4 transcripts are products of a novel function, and non-C4 transcripts due to the original function, the genes may have retained the original function for millions of years while evolving a novel function. The state of bifunctionality may continue until possible genetic incompatibility, if any, accumulates to a point intolerable to fitness. Maize PPDK may not be the only case of such gene bifunctionality in the C4 pathway. As shown above, the sorghum CA gene, Sb03g029170, seems to have similar bifunctionality, encoding both C4 and non-C4 transcripts. Since its internally repeating structure may have been produced before sorghum-maize divergence, its maize homologs also share this bifunctionality, which may have existed for millions of years. These multiple cases in which alternative splicing may contribute a possible sheltering effect during evolution of new function by C4 genes imply that it (alternative splicing) may participate in other cases of evolution of genetic novelty.

We found that the sorghum and maize C4 PEPC genes are on a long branch of their gene tree, grouped together with their suspected rice ortholog, showing possible adaptive evolution based on both a high Ka/Ks ratio and elevated amino acid substitution. It is intriguing to ask whether possible adaptive evolution in the rice PEPC gene could be a foundation toward a new origin of the C4 pathway, or instead indicates non-C4 functional adaptation. Scrutiny of the rice PEPC sequence revealed only 2 of 12 amino acid substitutions that were previously inferred to be positively selected in C4 genes [42], and, in particular, it lacks the critical fixed mutation S780 that is shared by C4 PEPC genes in other angiosperms [41,65]. This rice gene was classified into the ppc-B1 group [42], found only in the C3 grasses, suggesting that its adaptive evolution is not leading to C4 photosynthesis, but possibly to other functional novelty.

Adaptive evolution of PEPC may have some valuable implications for the discovery of multiple groups of PEPC genes defined previously [42]. In some C4 grasses there are different groups of genes, such as ppc-B2 and ppc-C4; while another group of genes, ppc-B1, are found in only C3 grasses. These findings show that, in the C4 lineages after their divergence from the C3 lineages but perhaps prior to the evolution of the C4 pathway itself, further gene duplication(s) may have contributed to the establishment of C4 photosynthesis.

A Novel Mode of Gene Evolution

The CA enzyme genes display a novel mode of gene evolution and functional adaptation. As shown above, sorghum and maize C4 CA enzymes have one, two or three functional domains, produced through recursive duplications followed by a fusion process involving stop codon mutations in the leading domains. There have been at least four tandem duplication events in sorghum and its ancestral genomes. These tandem duplications started before sorghum-maize divergence, and appear to have continued in the maize lineage. The recurrence of tandem duplications together with the subsequent merger process may have acted as a mode of adaptive evolution. The present CA enzymes are beta-type, comprising a dimer having four zinc ions bound to the structure as active sites. Besides dimers, these enzymes can form tetramers, hexamers or octamers [47], suggesting that the dimer may be a building block. Recruiting extra domains through tandem duplication may contribute to the formation of more complex structures, with more functional binding sites making them work more efficiently to stabilize the balance between CO_2 and HCO_3^-. The expanded gene structure of these sorghum and maize CA genes are unusual, since the cDNAs of Urochloa paniculata and Flaveria bidentis, both C4 plants, are normal in size [70]. Nonetheless, there is precedent for internal repetition of CA gene structure in red algae, Porphyridium purpureum, resulting in two sets of functional binding sites [71]. This independent evolution of internally repeating structure in CA genes supports our hypothesis that such structure may confer functional advantages.

We found that the sorghum and maize C4 CA genes share a cTP, which had not been expected since the enzymes were not found to be chloroplastically localized in C4 plants. In C3 plants, the most abundant CA activity is in the chloroplast stroma, while in C4 plants, the exact location of CA is less clear [47], but the most abundant CA activity is localized in the cytosol of mesophyll cells [72]. The cTP of sorghum and maize C4 CA genes is similar to that of the Arabidopsis CA genes, suggesting its existence before monocot-dicot divergence. The preservation of a cTP in C4 genes for tens of millions of years cannot be explained as a mere relic but suggests possible multiple functionality. This inference is at least partially supported by the discovery of divergent functions implemented by two different transcripts produced by the single sorghum C4 CA gene, Sb03g029170. As shown above, the expression of the longer transcript is light-inducible, while that of the shorter one is not, indicating that the longer but not the shorter transcript may be involved in the C4 pathway.

A Long Transition Time From C3 to C4 Photosynthesis

Several evolutionary models have been proposed to explain the formation of the C4 pathway [73-75]. In summary, seven significant phases are recognized toward successful establishment of C4 photosynthesis: general preconditioning (for example, gene duplication); anatomical preconditioning (for example, close veins); enhancement of bundle-sheath organelles; establishment of photorespiratory CO_2 pump and transformation of glycine decarboxylase to bundle-sheath cells; enhancement of PEPC activity; integration; and optimization [6]. Although many biological and anatomical changes are needed, multiple origins in tens of angiosperm families suggest that it is not so difficult to evolve a novel C4 pathway. However, from an evolutionary viewpoint it is still interesting to ask whether a transition process of gene functional changes and/or enhancement is necessary before the final establishment, and how long such a transition might take. There has been a long time-lag between the initial decrease in CO_2 concentration and the appearance of C4 plants. The initial decrease in CO_2 concentration started at least 100 mya [6], while molecular clock analyses suggest that the earliest C4 plants (grasses) appeared about 24 to 35 mya [28,29] One proposed explanation for the time-lag was the lack of a sufficient reservoir of duplicated and neofunctionalized C3 genes to support C4 evolution [14]. Here, we found that the genes of key enzymes, such as PEPC and NADP-ME, were among the duplicated copies produced by the WGD approximately 70 mya [19,20]. Once again we note, however, that availability of the pan-cereal duplicated copies was not by itself sufficient to initiate C4 evolution, since some were lost from the common cereal ancestor and then, after divergence from rice, had to reduplicate in the sorghum-maize ancestor before C4 evolution could occur.

Differential duplicability of C4 genes and their non-C4 isoforms

The above characterization of gene duplication shows differential duplicability of C4 genes and their isoforms in grasses. Evidence from yeast indicates that gene redundancy tends to be preserved among some of the central proteins in the cellular interaction network [76]. Tens of plant genes were suggested to be duplication-resistant, and undergo convergent restoration to singleton status following several independent genome duplications [25]. The differential duplicability could be explained by gene dosage effects, organismal complexity, protein interaction centrality and protein domain preference [24-26,76]. Here, we have shown that some gene families, including PEPC, PPCK, CA, and NADP-ME genes, have been expanded by gene duplication, but not others such as PPDK genes. The families expanded by gene duplication tend to be multiply functional, such as PEPC and NADP-ME [14]. Different PEPC gene isoforms take on specific roles, including the regulation of ion balance, the production of amino-group acceptor molecules in symbiotic nitrogen fixation, and the initial fixation of C in C4 photosynthesis and Crassulacean acid metabolism [77]. NADP-ME catalyzes the oxidative breakdown of malate to form CO2 and pyruvate in the C4 pathway. Its non-C4 functions include the provision of carbon skeletons for ammonia assimilation [78] and reductant for wound-induced production of lignin and flavonoids [79,80]. CA genes are also prone to duplication, which may enhance their ability to form more complex structures, as discussed above. Though further duplication is not required when a former C3 gene is finally co-opted for C4 roles [14], we found that the sorghum NADP-MDH C4 gene did experience a tandem duplication event, with only one duplicated copy preserving the C4 function through possible subfunctionalization [44]. This implies that the sorghum NADP-MDH C4 gene itself may be duplication-resistant.

C4 Pathway and Codon Usage Bias

GC content elevation has resulted in codon usage bias [37], which has been hypothesized by some to have contributed to C4 adaptive evolution [30]. As shown above, though the grass C4 genes and their isoforms always have a higher GC content than their Arabidopsis counterparts, there is often a non-C4 grass gene having higher GC content than the C4 one(s). Thus, there is no clear evidence supporting co-variation between codon usage bias and C4 gene evolution. Base composition variation in grass genes has been a hot topic involving transcription, translation, modification and mutational bias [81-83].

Potential Contribution to Engineering New C4 Plants

A comprehensive characterization of the C4 pathway will help not only to understand how C4 photosynthesis evolves, but also may benefit crop improvement efforts. Of singular relevance are efforts to transform C3 plants into C4 plants. To perform such a transformation, one strategy is to incorporate the C4 pathway into C3 plants through recombinant DNA technology [84]. The strategy succeeds in transferring C4 genes into C3 plants and yielding high levels of C4 enzymes in desired locations [85,86]. It is of great interest to transform rice, a staple food for more than half of the world's population, to perform C4 function, as reviewed in a recent publication [31]. However, combined overproduction of C4 enzymes (PEPC, PPDK, NADP-ME, and NADP-MDH) resulted in only slightly higher levels of CO_2 assimilation than in wild-type rice [87]. This might indicate that not all components needed for C4 photosynthesis are known. There must be some transporters involved and there might also be some unknown regulatory factors. Knowledge of the complete sorghum genome might help to identify such components. As also shown above, though often not statistically significant, the sorghum and maize C4 genes appear to have been under adaptive evolution in different modes and levels, and show different duplicability. These findings may provide clues toward a successful transformation to C4 photosynthesis. Alternatively, perhaps adaptations such as we have suggested in the PEPC gene in C3 lineages have mitigated the perceived weaknesses of C3 photosynthesis.

Subtle differences in the C4 pathways used in different grasses are worthy of further investigation as well. For example, if our hypothesis is correct that internally repeating structure in CA genes may confer functional advantages, then engineering of the maize trimer into sorghum (for example) may be advantageous. Likewise, exploration of still more recent polyploids such as sugarcane might yield even more complex CA alleles. Tandem duplication of C4 NADP-MDH following the sorghum-maize divergence does not appear to have been essential to C4 evolution; indeed, one of the tandem genes appears to have lost C4 specificity. However, careful scrutiny of the physiological consequences of this change might suggest benefits that could be transferred to other crops.

Conclusions

Gene Duplication and C4 Pathway Evolution

Both WGD and single-gene duplication have contributed to C4 pathway evolution in sorghum and maize. Some C4 genes (PEPC, PPCK, and NADP-ME C4 genes) were recruited from duplicates produced by WGD. Sorghum NADP-MDH, NADP-ME and PPDK-RP C4 genes were affected by tandem duplication, with only one of the resulting copies involved in the C4 pathway. C4 genes show divergent duplicability. PEPC, NADP-ME, PPCK, and CA gene families were expanded by recursive duplication events, showing a duplication-philic nature, whereas NADP-MDH and PPDK are likely duplication-phobic. Further supporting evidence is that only one copy of NADP-MDH C4 gene duplicates preserves the C4 function.

Adaptive Evolution Divergent in Mode and Level

We found evidence of adaptive evolution of most C4 genes studied. However, the mode and level of adaptation is divergent among C4 genes. Adaptive evolution is achieved though rapid mutations in DNA sequences, aggregated amino acid substitutions, and/or considerable increases of expression levels in specific cells. Besides gene redundancy, we found that alternative splicing may have also sheltered the evolution of new function. Our analysis supports previous findings that maximum likelihood inference may be too conservative to find adaptive evolution. We found no evidence of co-variation between codon usage bias and C4 pathway development.

Special Evolutionary Mode of Grass CA Genes

Grass CA genes have evolved in a specific pattern featuring recursive tandem duplication and neighboring gene fusion, which produced distinct isoforms having one to three functional units. Two sorghum C4 CA genes have one and two functional units, while two characterized maize C4 CA genes have two and three functional units, respectively. The elongation of these genes by recruiting extra domains may contribute to the formation of more complex protein structures, as often observed in plants.

A Long Transition Time From C3 to C4 Photosynthesis

The hypothesis that a reservoir of duplicated genes in ancestral C3 plants was a prerequisite for C4 pathway development is only partially supported by present findings that some C4 genes were recruited from the duplicates. Availability of the pan-cereal duplicated copies was not sufficient to initiate C4 evolution, since some were lost from the common cereal ancestor, then had to reduplicate in the sorghum-maize ancestor before C4 evolution could occur. However, C4 gene isoforms show quite divergent duplicability, and there has been quite a long time-lag between the gene duplication events and the appearance of C4 grasses. These findings suggest a long transition process, including different modes of functional innovation, before the eventual establishment of C4 photosynthesis.

Materials and Methods

Known C4 enzyme genes and their non-C4 isoforms in sorghum, maize and Arabidopsis (Table 1) were downloaded from NCBI CoreNucleotide database [88]. Searching these known genes against sorghum [89] and rice [90] gene models by running BLAST [91] (E-value $< 1 \times 10^{-5}$), we identified other enzyme genes in these organisms. By characterizing sequence similarity and constructing gene trees, possible C4 genes were determined. The enzymes revealed here were linked to expression data reported previously [35] by comparing cDNA segments to gene sequences using BLAST.

Gene Colinearity Inference

The potential gene homology information defined by running BLAST was used as the input for MCscan [92] to find homologous gene pairs in colinearity. The built-in scoring scheme for MCscan is min($-\log_{10}$E_value, 50) for every matching gene pair and -1 for each 10 kb distance between anchors, and blocks that had scores >300 were kept. The resulting syntenic chains were evaluated using a procedure by ColinearScan [17] and an E-value $< 1 \times 10^{-10}$ was used as a significance cutoff.

Gene Phylogeny Construction

We constructed phylogenetic trees using several approaches, including the neighbor-joining, maximum likelihood, minimal evolution, and maximum parsimony methods, implemented in NADP-MEGA [93], PHYML [94], and PHYLIP [95], on both DNA and protein sequences. While running PHYML, parameters were set as adopted previously [42]. Bootstrap tests were performed with 100 repeats to produce percentage values, showing the stability of their topology. The trees mostly agree with one another. When there is inconsistency, the tree most strongly supported by bootstrap values was adopted for the subsequent adaptive evolution inference. For example, the trees of CA functional units were inconsistent among methods, and the best-supported neighbor-joining tree produced by protein sequences was adopted for further analysis.

Maximum Likelihood Inference of Adaptive Evolution

The tree constructed for the group of C4 enzyme genes and their non-C4 isoforms was used to perform further maximum likelihood analysis using the Codeml program in PAML [96]. To detect whether a specific C4 gene had been positively selected, we compared two types of competing models: a free-ratio model and a ratio-restriction model [97]. The free-ratio model assumes an independent Ka/Ks ratio for each branch, whereas the latter forces the Ka/Ks ratio to be 1 on the specific branch to be tested for positive selection, and for the other branches assumes independent ratios. Each model will produce a likelihood, and the twofold difference between them follows a Chi-squared distribution with 1 degree of freedom.

Aggregated Amino Acid Substitution Analysis

We adopted a comparative genomic approach initially proposed by Wagner [98] to detect genes potentially under positive selection. The Wagner approach inferred positive selection pressure by detecting possible aggregation of amino acid replacement. Here, we inferred possible amino acid replacements by comparing the homologous enzyme gene pair containing a C4 gene and non-C4 gene (often a rice gene) against the aligned outgroup sequence. A replacement site is identified in the C4 sequence that differs from the corresponding sites in both the homologous sequence and the outgroup sequence, which are identical. We found the number of all replacements, m, grouping the C4 and non-C4 protein sequences. If the occurrences of these replacement sites are assumed to be Poisson distributed with a parameter λ, we may evaluate the chance of observing a specific number of consecutive replacement sites along a sequence. For simplicity in description, for each sequence we first defined a replacement position array, $x = x_0, x_1, x_2,\ldots, x_t, x_{t+1}$, composed by all the positions x_i $(1 \leq i \leq t)$ of replacement sites and two ends of the sequence, that is, $x_0 = 0$ and $x_{t+1} = n - 1$, where n is the length of the alignment after purging gaps. Then we defined the replacement distance array, $d = (d_1,\ldots d_{t+1})$, where $d_1 = x_i - x_{i-1}$ $(1 \leq i \leq t + 1)$. The distance between two replacement sites $d_{i,k} = x_{i+k-1} - x_i$, where k is the number of the consecutive replacement sites in the corresponding sequence segment, follows a Pearson type III distribution following a probability density $\lambda(\lambda z)^{k-2} e^{-\lambda z}/\Gamma(k - 1)$ [99], where $\Gamma(k - 1) = (k - 1)!$. We can estimate the Poisson parameter λ with m/(2n). Supposing there are ti replacement sites along the i-th sequence, obviously, we get $\sum_{i=1}^{2} t_i = m$. Therefore, we could estimate the probability $P(d_{i,k})$ that k consecutive replacements in a distance between two replacement sites is smaller than the observed $d_{i,k}$ by the following integration:

$$P\left(d_{i,k}\right)=\frac{\lambda}{\Gamma(k-1)}\int_{0}^{d_{i,k}} {}^{k-2} e^{-\lambda z} dz .$$

We evaluated the occurrence probability of observed distance between any two replacement sites, and the smallest probability was used to locate a region with the most aggregated replacements, which was taken to be significant after a Bonferroni correction by considering the number of all combinations of replacement sites $\left|\left(\frac{t+2}{2}\right)\right|$. The occurrence probability was calculated using R [100]. If between two replacement sites there were gaps in the aligned sequences, they were omitted to check for possible selection. We composed Perl scripts to implement the described approach.

Maize Homolog Characterization

Maize BACs are from the MaizeSequence database [101]. The maize genes were searched against the BAC sequences to reveal their chromosomal locations, local DNA structures, and so on.

Chloroplastic Transit Peptide Inferrence

ChloroP1.1 [102] was used to predict the presence of cTPs in the enzyme protein sequences and the location of potential cTP cleavage sites.

Dotplotting

Dotplots between CA protein sequences were produced by running the public program DOTTER [103]. The Dotplots were produced by matched strings from two protein sequences in comparison. The expected score per residue of the matched strings was set to be 40.

Abbreviations

BAC: bacterial artificial chromosome; CA: carboxylating anhydrase; cTP: chloroplast transit peptide; MDH: malate dehydrogenase; mya: million years ago; NADP-ME: NADP-malic enzyme; PEPC: phosphoenolpyruvate carboxylase; PPCK: PEPC kinase; PPDK: pyruvate orthophosphate dikinase; PPDK-RP: PPDK regulatory protein; WGD: whole-genome duplication.

Author Contributions

XW designed and organized the present work. UG, PW and XW curated gene models. UG, HT, JEB and AHP contributed to this work through critical discussion. XW and AHP wrote the paper.

Acknowledgements

We appreciate financial support from the US National Science Foundation (MCB-0450260 to AHP). We thank Lifeng Lin for artwork.

References

1. Hatch MD, Slack CR: Photosynthesis by sugar-cane leaves. A new carboxylation reaction and the pathway of sugar formation. Biochem J 1966, 101(1):103–111.
2. Seemann JR, Sharkey TD, Wang J, Osmond CB: Environmental Effects on Photosynthesis, Nitrogen-Use Efficiency, and Metabolite Pools in Leaves of Sun and Shade Plants. Plant physiology 1987, 84(3):796–802.
3. Hattersley PG: The distribution of C3 and C4 grasses in Australia in relation to climate. Oecologia 1983, 57:113–128.
4. Ehleringer JR, Bjorkman O: A Comparison of Photosynthetic Characteristics of Encelia Species Possessing Glabrous and Pubescent Leaves. Plant physiology 1978, 62(2):185–190.
5. Cerling TE, Harris JM, MacFadden BJ, Leasey MG, Quade J, Eisenmann V, Ehleringer JR: Global vegetation change through the Miocene/Pliocene boundary. Nature 1997, 389:153–158.
6. Sage RF: The evolution of C4 photosynthesis. New Phytologist 2004, 161:341–370.
7. Mulhaidat R, Sage RF, Dengler NG: Diversity of kranz anatomy and biochemistry in C4 eudicots. American Journal of Botany 2007, 94(3):20.
8. Giussani LM, Cota-Sanchez JH, Zuloaga FO, Kellogg EA: A molecular phylogeny of the grass subfamily Panicoideae (Poaceae) shows multiple origins of C4 photosynthesis. American Journal of Botany 2001, 88(11):1993–2012.
9. Pyankov VI, Artyusheva EG, Edwards GE, Black CC Jr, Soltis PS: Phylogenetic analysis of tribe Salsoleae (Chenopodiaceae) based on ribosomal ITS sequences: implications for the evolution of photosynthesis types. Am J Bot 2001, 88(7):1189–1198.
10. Sheen J: C4 Gene Expression. Annu Rev Plant Physiol Plant Mol Biol 1999, 50:187–217.
11. Burnell JN, Chastain CJ: Cloning and expression of maize-leaf pyruvate, Pi dikinase regulatory protein gene. Biochem Biophys Res Commun 2006, 345(2):675–680.
12. Kawamura T, Shigesada K, Toh H, Okumura S, Yanagisawa S, Izui K: Molecular evolution of phosphoenolpyruvate carboxylase for C4 photosynthesis in maize: comparison of its cDNA sequence with a newly isolated cDNA encoding an isozyme involved in the anaplerotic function. J Biochem 1992, 112(1):147–154.

13. Poetsch W, Hermans J, Westhoff P: Multiple cDNAs of phosphoenolpyruvate carboxylase in the C4 dicot Flaveria trinervia. FEBS Letters 1991, 292:133–136.

14. Monson RK: Gene duplication, neofunctionalization, and the evolution of C4 photosynthesis. International Journal of Plant Science 2003, 164(6920):S43–S54.

15. Bowers JE, Chapman BA, Rong J, Paterson AH: Unravelling angiosperm genome evolution by phylogenetic analysis of chromosomal duplication events. Nature 2003, 422(6930):433–438.

16. Yu J, Wang J, Lin W, Li S, Li H, Zhou J, Ni P, Dong W, Hu S, Zeng C, Zhang J, Zhang Y, Li R, Xu Z, Li S, Li X, Zheng H, Cong L, Lin L, Yin J, Geng J, Li G, Shi J, Liu J, Lv H, Li J, Wang J, Deng Y, Ran L, Shi X, et al.: The Genomes of Oryza sativa: A History of Duplications. PLoS Biology 2005, 3(2):e38.

17. Wang X, Shi X, Li Z, Zhu Q, Kong L, Tang W, Ge S, Luo J: Statistical inference of chromosomal homology based on gene colinearity and applications to Arabidopsis and rice. BMC bioinformatics 2006, 7(1):447.

18. Blanc G, Wolfe KH: Widespread paleopolyploidy in model plant species inferred from age distributions of duplicate genes. The Plant cell 2004, 16(7):1667–1678.

19. Paterson AH, Bowers JE, Chapman BA: Ancient polyploidization predating divergence of the cereals, and its consequences for comparative genomics. Proc Natl Acad Sci USA 2004, 101(26):9903–9908.

20. Wang X, Shi X, Hao B, Ge S, Luo J: Duplication and DNA segmental loss in the rice genome: implications for diploidization. New Phytologist 2005, 165(3):937–946.

21. Paterson AH, Bowers JE, Bruggmann R, Dubchak I, Grimwood J, Gundlach H, Haberer G, Hellsten U, Mitros T, Poliakov A, Schmutz J, Spannagl M, Tang H, Wang X, Wicker T, Bharti AK, Chapman J, Feltus FA, Gowik U, Grigoriev IV, Lyons E, Maher CA, Martis M, Narechania A, Otillar RP, Penning BW, Salamov AA, Wang Y, Zhang L, Carpita NC, et al.: The Sorghum bicolor genome and the diversification of grasses. Nature 2009, 457(7229):551–556.

22. Lynch M, Conery JS: The evolutionary demography of duplicate genes. J Struct Funct Genomics 2003, 3(1-4):35–44.

23. He X, Zhang J: Gene complexity and gene duplicability. Curr Biol 2005, 15(11):1016–1021.

24. Liang H, Li WH: Gene essentiality, gene duplicability and protein connectivity in human and mouse. Trends Genet 2007, 23(8):375–378.

25. Paterson AH, Chapman BA, Kissinger JC, Bowers JE, Feltus FA, Estill JC: Many gene and domain families have convergent fates following independent whole-genome duplication events in Arabidopsis, Oryza, Saccharomyces and Tetraodon. Trends Genet 2006, 22(11):597–602.
26. Papp B, Pal C, Hurst LD: Dosage sensitivity and the evolution of gene families in yeast. Nature 2003, 424(6945):194–197.
27. Nielsen R, Bustamante C, Clark AG, Glanowski S, Sackton TB, Hubisz MJ, Fledel-Alon A, Tanenbaum DM, Civello D, White TJ, J JS, Adams MD, Cargill M: A scan for positively selected genes in the genomes of humans and chimpanzees. PLoS biology 2005, 3(6):e170.
28. Vicentini A, Barber JC, Aliscioni SS, Ciussani LM, Kellogg EA: The age of the grasses and clusters of origins of C4 photosynthesis. Global Change Biology 2008, 14:15.
29. Christin PA, Besnard G, Samaritani E, Duvall MR, Hodkinson TR, Savolainen V, Salamin N: Oligocene CO2 decline promoted C4 photosynthesis in grasses. Curr Biol 2008, 18(1):37–43.
30. Shenton M, Fontaine V, Hartwell J, Marsh JT, Jenkins GI, Nimmo HG: Distinct patterns of control and expression amongst members of the PEP carboxylase kinase gene family in C4 plants. Plant J 2006, 48(1): 45–53.
31. Sheehy JE, Mitchell PL, Hardy B: Charting New Pathways To C4 Rice. Los Banos (philippines): World Scientific Publishing Company; 2008.
32. Sanchez R, Cejudo FJ: Identification and expression analysis of a gene encoding a bacterial-type phosphoenolpyruvate carboxylase from Arabidopsis and rice. Plant physiology 2003, 132(2): 949–957.
33. Cretin C, Keryer E, Tagu D, Lepiniec L, Vidal J, Gadal P: Complete cDNA sequence of sorghum phosphoenolpyruvate carboxylase involved in C4 photosynthesis. Nucleic acids research 1990, 18(3): 658.
34. Cretin C, Santi S, Keryer E, Lepiniec L, Tagu D, Vidal J, Gadal P: The phosphoenolpyruvate carboxylase gene family of Sorghum: promoter structures, amino acid sequences and expression of genes. Gene 1991, 99(1):87–94.
35. Wyrich R, Dressen U, Brockmann S, Streubel M, Chang C, Qiang D, Paterson AH, Westhoff P: The molecular basis of C4 photosynthesis in sorghum: isolation, characterization and RFLP mapping of mesophyll- and bundle-sheath-specific cDNAs obtained by differential screening. Plant molecular biology 1998, 37(2):319–335.
36. Song R, Llaca V, Messing J: Mosaic organization of orthologous sequences in grass genomes. Genome Res 2002, 12(10): 1549–1555.

37. Carels N, Bernardi G: Two classes of genes in plants. Genetics 2000, 154(4):1819–1825.

38. Lepiniec L, Keryer E, Philippe H, Gadal P, Cretin C: Sorghum phosphoenolpyruvate carboxylase gene family: structure, function and molecular evolution. Plant molecular biology 1993, 21(3): 487–502.

39. Besnard G, Offmann B, Robert C, Rouch C, Cadet F: Assessment of the C(4) phosphoenolpyruvate carboxylase gene diversity in grasses (Poaceae). Theor Appl Genet 2002, 105(2-3): 404–412.

40. Roth C, Liberles DA: A systematic search for positive selection in higher plants (Embryophytes). BMC plant biology 2006, 6:12.

41. Wcsthoff P, Gowik U: Evolution of c4 phosphoenolpyruvate carboxylase. Genes and proteins: a case study with the genus Flaveria. Ann Bot (Lond) 2004, 93(1):13–23.

42. Christin PA, Salamin N, Savolainen V, Duvall MR, Besnard G: C4 Photosynthesis evolved in grasses via parallel adaptive genetic changes. Curr Biol 2007, 17(14):1241–1247.

43. Luchetta P, Cretin C, Gadal P: Organization and expression of the two homologous genes encoding the NADP-malate dehydrogenase in Sorghum vulgare leaves. Mol Gen Genet 1991, 228(3): 473–481.

44. Rondeau P, Rouch C, Besnard G: NADP-malate dehydrogenase gene evolution in Andropogoneae (Poaceae): gene duplication followed by sub-functionalization. Ann Bot (Lond) 2005, 96(7): 1307–1314.

45. Sheen J: Molecular mechanisms underlying the differential expression of maize pyruvate, orthophosphate dikinase genes. The Plant cell 1991, 3(3): 225–245.

46. Glackin CA, Grula JW: Organ-specific transcripts of different size and abundance derive from the same pyruvate, orthophosphate dikinase gene in maize. Proceedings of the National Academy of Sciences of the United States of America 1990, 87(8): 3004–3008.

47. Tiwari A, Kumar P, Singh S, Ansari S: Carbonic anhydrase in relation to higher plants. Photosynthetica 2005, 43(1): 1–11.

48. Wolfe KH, Li WH: Molecular evolution meets the genomics revolution. Nature genetics 2003, 33(Suppl): 255–265.

49. Haldane JBS: The causes of evolution. Ithaca: Cornell University Press; 1932.

50. Ohno S: Sex chromosomes and sex-linked genes. Berlin: Springler-Verlag; 1967.

51. Ohno S: Evolution by Gene Duplication. Berlin-Heidelberg_New York: Springer-Verlag; 1970.

52. Steinke D, Hoegg S, Brinkmann H, Meyer A: Three rounds (1R/2R/3R) of genome duplications and the evolution of the glycolytic pathway in vertebrates. BMC Biol 2006, 4:16.
53. Meyer A, Peer Y: From 2R to 3R: evidence for a fish-specific genome duplication (FSGD). Bioessays 2005, 27(9): 937–945.
54. Soltis PS: Ancient and recent polyploidy in angiosperms. The New phytologist 2005, 166(1):5–8.
55. Jaillon O, Aury JM, Noel B, Policriti A, Clepet C, Casagrande A, Choisne N, Aubourg S, Vitulo N, Jubin C, Vezzi A, Legeai F, Hugueney P, Dasilva C, Horner D, Mica E, Jublot D, Poulain J, Bruyere C, Billault A, Segurens B, Gouyvenoux M, Ugarte E, Cattonaro F, Anthouard V, Vico V, Del Fabbro C, Alaux M, Di Gaspero G, Dumas V, et al.: The grapevine genome sequence suggests ancestral hexaploidization in major angiosperm phyla. Nature 2007, 449(7161): 463–467.
56. Tang H, Bowers JE, Wang X, Ming R, Alam M, Paterson AH: Synteny and collinearity in plant genomes. Science 2008, 320(5875):486–488.
57. Chapman BA, Bowers JE, Feltus FA, Paterson AH: Buffering crucial functions by paleologous duplicated genes may impart cyclicality to angiosperm genome duplication. Proceedings of the National Academy of Sciences of the United States of America 2006, 103:2730–2735.
58. Maere S, De Bodt S, Raes J, Casneuf T, Van Montagu M, Kuiper M, Peer Y: Modeling gene and genome duplications in eukaryotes. Proceedings of the National Academy of Sciences of the United States of America 2005, 102(15): 5454–5459.
59. Blanc G, Wolfe KH: Functional divergence of duplicated genes formed by polyploidy during Arabidopsis evolution. The Plant cell 2004, 16(7): 1679–1691.
60. Seoighe C, Gehring C: Genome duplication led to highlyselective expansion of the Arabidopsis thaliana proteome. Trends Genet 2004, 20(10): 461–464.
61. Freeling M, Thomas BC: Gene-balanced duplications, like tetraploidy, provide predictable drive to increase morphological complexity. Genome research 2006, 16(7):805–814.
62. De Bodt S, Maere S, Peer Y: Genome duplication and the origin of angiosperms. Trends in ecology & evolution (Personal edition) 2005, 20(11): 591–597.
63. Salse J, Bolot S, Throude M, Jouffe V, Piegu B, Quraishi UM, Calcagno T, Cooke R, Delseny M, Feuillet C: Identification and characterization of shared duplications between rice and wheat provide new insight into grass genome evolution. The Plant cell 2008, 20:11–24.

64. Shantz HL: The place of grasslands in the earth's cover of vegetation. Ecology 1954, 35:143–145.

65. Svensson P, Blasing OE, Westhoff P: Evolution of C4 phosphoenolpyruvate carboxylase. Arch Biochem Biophys 2003, 414(2): 180–188.

66. Hibberd JM, Quick WP: Characteristics of C4 photosynthesis in stems and petioles of C3 flowering plants. Nature 2002, 415(6870): 451–454.

67. Aharoni A, Gaidukov L, Khersonsky O, McQ GS, Roodveldt C, Tawfik DS: The 'evolvability' of promiscuous protein functions. Nature genetics 2005, 37(1): 73–76.

68. He XL, Zhang JZ: Rapid subfunctionalization accompanied by prolonged and substantial neofunctionalization in duplicate gene evolution. Genetics 2005, 169(2): 1157–1164.

69. Matsuoka M: The gene for pyruvate, orthophosphate dikinase in C4 plants: structure, regulation and evolution. Plant & cell physiology 1995, 36(6):937–943.

70. Moroney JV, Bartlett SG, Samuelsson G: Carbonic anhydrases in plants and algae. Plant Cell and Environment 2001, 24:13.

71. Mitsuhashi S, Mizushima T, Yamashita E, Yamamoto M, Kumasaka T, Moriyama H, Ueki T, Miyachi S, Tsukihara T: X-ray structure of beta-carbonic anhydrase from the red alga, Porphyridium purpureum, reveals a novel catalytic site for CO(2) hydration. The Journal of biological chemistry 2000, 275(8): 5521–5526.

72. Ku MS, Kano-Murakami Y, Matsuoka M: Evolution and expression of C4 photosynthesis genes. Plant physiology 1996, 111(4): 949–957.

73. Edwards GE, Ku MSB: Biochemistry of C3-C4 intermediates. In The Biochemistry of Plants. Volume 10. Edited by: Hatch MD, Boardman NK. London: Academic Press; 1987:275–325.

74. Brown RH, Hattersley PW: Leaf Anatomy of C(3)-C(4) Species as Related to Evolution of C(4) Photosynthesis. Plant physiology 1989, 91:1543–1550.

75. Rawsthorne S: Towards an understanding of C3-C4 photosynthesis. Essays Biochem 1992, 27:135–146.

76. Kafri R, Dahan O, Levy J, Pilpel Y: Preferential protection of protein interaction network hubs in yeast: evolved functionality of genetic redundancy. Proceedings of the National Academy of Sciences of the United States of America 2008, 105(4):1243–1248.

77. Gehring HH, Heute V, Kluge M: Toward a better knowledge of the molecular evolution of phosphoenolpyruvate carboxylase by comparison of partial cDNA sequences. Journal of molecular evolution 1998, 46:107–114.

78. Chopra J, Kaur N, Gupta AK: A comparative developmental pattern of enzymes of carbon metabolism and pentose phosphate pathway in mungbean and lentil nodules. Acta Physiol Plant 2002, 24:67–72.

79. Casati P, Drincovich MF, Edwards GE, Andreo CS: Malate metabolism by NA-DP-malic enzyme in plant defense. Photosynth Res 1999, 61:99–105.

80. Maurino VG, Saigo M, Andreo CS, Drincovich MF: Non-photosynthetic 'malic enzyme' from maize: a constituvely expressed enzyme that responds to plant defence inducers. Plant molecular biology 2001, 45(4): 409–420.

81. Wong GK, Wang J, Tao L, Tan J, Zhang J, Passey DA, Yu J: Compositional gradients in Gramineae genes. Genome research 2002, 12(6): 851–856.

82. Wang HC, Singer GA, Hickey DA: Mutational bias affects protein evolution in flowering plants. Mol Biol Evol 2004, 21(1): 90–96.

83. Shi X, Wang X, Li Z, Zhu Q, Yang J, Ge S, Luo J: Evidence that natural selection is the primary cause of the GC content variation in rice genes. Journal of Integrative Plant Biology 2007.

84. Miyao M: Molecular evolution and genetic engineering of C4 photosynthetic enzymes. J Exp Bot 2003, 54(381): 179–189.

85. Ku MS, Agarie S, Nomura M, Fukayama H, Tsuchida H, Ono K, Hirose S, Toki S, Miyao M, Matsuoka M: High-level expression of maize phosphoenolpyruvate carboxylase in transgenic rice plants. Nat Biotechnol 1999, 17(1): 76–80.

86. Fukayama H, Tsuchida H, Agarie S, Nomura M, Onodera H, Ono K, Lee BH, Hirose S, Toki S, Ku MS, Makino A, Matsuoka M, Miyao M: Significant accumulation of C(4)-specific pyruvate, orthophosphate dikinase in a C(3) plant, rice. Plant physiology 2001, 127(3): 1136–1146.

87. Taniguchi Y, Ohkawa H, Masumoto C, Fukuda T, Tamai T, Lee K, Sudoh S, Tsuchida H, Sasaki H, Fukayama H, Miyao M: Overproduction of C4 photosynthetic enzymes in transgenic rice plants: an approach to introduce the C4-like photosynthetic pathway into rice. J Exp Bot 2008, 59(7): 1799–1809.

88. NCBI CoreNucleotide database [http://www.ncbi.nlm.nih.gov/].

89. Joint Genome Institute [http://www.jgi.doe.gov/].

90. Rice annotation project 2 [http://rgp.dna.affrc.go.jp/E/index.html/].

91. Altschul SF, Gish W, Miller W, Myers EW, Lipman DJ: Basic local alignment search tool. Journal of molecular biology 1990, 215:403–410.

92. Tang HB, Wang XY, Bowers JE, Ming R, Alam M, Paterson AH: Unreveling ancient hexaploidy throught multiply-aligned angiosperm gene maps. Genome research 2008.

93. Tamura K, Dudley J, Nei M, Kumar S: MEGA4: Molecular Evolutionary Genetics Analysis (MEGA) software version 4.0. Mol Biol Evol 2007, 24(8): 1596–1599.

94. Guindon S, Lethiec F, Duroux P, Gascuel O: PHYML Online--a web server for fast maximum likelihood-based phylogenetic inference. Nucleic acids research 2005, (33 Web Server): W557–559.

95. Felsenstein J: Phylogenies From Restriction Sites - a Maximum-Likelihood Approach. Evolution 1992, 46(1): 159–173.

96. Yang Z, Nielsen R: Synonymous and nonsynonymous rate variation in nuclear genes of mammals. Journal of molecular evolution 1998, 46(4): 409–418.

97. Yang Z: Likelihood ratio tests for detecting positive selection and application to primate lysozyme evolution. Mol Biol Evol 1998, 15:568–573.

98. Wagner A: Rapid detection of positive selection in genes and genomes through variation clusters. Genetics 2007, 176(4): 2451–2463.

99. Wagner A: A computational genomics approach to the identification of gene networks. Nucleic acids research 1997, 25(18): 3594–3604.

100. R language [http://www.r-project.org/].

101. MaizeSequence [http://www.maizesequence.org/].

102. Emanuelsson O, Nielsen H, von Heijne G: ChloroP, a neural network-based method for predicting chloroplast transit peptides and their cleavage sites. Protein Sci 1999, 8(5): 978–984.

103. Sonnhammer ELL, Durbin R: A dot-matrix program with dynamic threshold control suitable for genomic DNA and protein sequence analysis. Gene 1995, 167:1–10.

The Accent-Vocbas Field Campaign on Biosphere-Atmosphere Interactions in a Mediterranean Ecosystem of Castelporziano (Rome): Site Characteristics, Climatic and Meteorological Conditions, and Eco-Physiology of Vegetation

S. Fares, S. Mereu, G. Scarascia Mugnozza, M. Vitale, F. Manes, M. Frattoni, P. Ciccioli, G. Gerosa and F. Loreto

ABSTRACT

Biosphere-atmosphere interactions were investigated on a sandy dune Mediterranean ecosystem in a field campaign held in 2007 within the frame of the European Projects ACCENT and VOCBAS. The campaign was carried out in the Presidential estate of Castelporziano, a periurban park close to Rome. Former campaigns (e.g. BEMA) performed in Castelporziano investigated the emission of biogenic volatile organic compounds (BVOC). These campaigns focused on pseudosteppe and evergreen oak groves whereas the contribution of the largely biodiverse dune vegetation, a prominent component of the Mediterranean ecosystem, was overlooked. While specific aspects of the campaign will be discussed in companion papers, the general climatic and physiological aspects are presented here, together with information regarding BVOC emission from the most common plant species of the dune ecosystem. During the cam- paign regular air movements were observed, dominated by moderate nocturnal land breeze and diurnal sea breeze. A regular daily increase of ozone concentration in the air was also observed, but daily peaks of ozone were lower than those measured in summer on the same site. The site was ideal as a natural photochemical reactor to observe reaction, transport and deposition processes occurring in the Mediterranean basin, since the sea-land breeze circulation allowed a strong mixing between biogenic and anthropogenic emissions and secondary pollutants. Measurements were run in May, when plant physiological conditions were optimal, in absence of severe drought and heat stress. Foliar rates of photosynthesis and transpiration were as high as generally recorded in unstressed Mediterranean sclerophyllous plants. Most of the plant species emitted high level of monoterpenes, despite measurements being made in a period in which emissions of volatile isoprenoids could be restrained by development tal and environmental factors, such as leaf age and relatively low air temperature. Emission of isoprene was generally low. Accounting for the high monoterpene spring emission of the dune ecosystem may be important to correct algorithms at re gional and ecosystem levels, and to interpret measurements of fluxes of volatile isoprenoids and secondary pollutants.

Introduction

The Mediterranean ecosystems represent 1% of the Earth's land surface, and are concentrated mainly in the Mediterranean basin, with coastal ecosystems playing consequently an important role. Coastal sand dunes are natural structures which protect the coast by absorbing energy from wind, tide and wave action, and host ecosystems made by pioneering species which are adapted to life at the interface between see and land. The typical dune vegetation communities are part of the larger "macchia" or "maquis" ecosystem, which in its degraded state is referred to as "garrigue." In Italy, the macchia covers 2–3% of the territory (INFC, 2003) and is made primarily of evergreen shrubs. This vegetation stabilizes coastal sand dunes, and shelters inland vegetation from sea winds and consequent damages due to marine aerosol. Despite the large biodiversity characterizing the macchia vegetation, all plant species show ecological adaptations to the Mediterranean environmental conditions, which are characterized by hot and dry summers, mild and often rainy winters, recurrent exposure to salinity due to sea breeze and saline ground water table, and sandy soil with poor organic fraction and nutrients.

Plants establishing on coastal sand dunes are subjected to several environmental fluctuations which affect their growth, survival and community structure (Maun, 1994). Typically, the morphological and structural adaptations to drought, salinity and high temperatures that are observed in Mediterranean plants include small plant size, globular shape of the canopy, and sclerophytic leaves with thick cuticles and dense mesophyll (Thompson, 2005). To endure recurrent episodes of heat stress, drought, and salt stress, plants living in the dunes carry out two alternative ecological strategies. A few plants have a "water spending strategy" with stomata responding strongly to water availability and allowing excellent CO2 diffusion inside leaves and high rates of photosynthesis when water is available. Most plant species present a "water saving strategy," with stomatal opening permanently restricted by anatomical, morphological and physiological traits, and low rates of carbon assimilation even in the absence of environmental constraints. These plants show a more efficient water use under harsh conditions (Thompson, 2005).

Another line of defence against biotic and abiotic stresses is constituted by the synthesis and release of trace gases often referred to as biogenic volatile organic compounds (BVOC). The emission of isoprenoids and oxygenated volatiles is common in plants and is particularly widespread among the plant species of Mediterranean ecosystems, although reports about emissions from plants of the dune ecosystems are scarce. Isoprenoids are believed to strengthen membranes by protecting leaves against damage caused by heat and oxidative stresses (Vickers et al., 2009). Oxygenated compounds are indicators of damage occurring at cellular membranes and are released during membrane or cell wall degradation. Methanol and C-6 compounds are especially emitted at high rates under these conditions (Loreto et al., 2006).

Many of these trace gases are very reactive, and, once emitted in a polluted atmosphere that is enriched by anthropogenic emission of NOx, may fuel production of ozone (Chameides et al., 1988), secondary organic aerosols and particles (Claeys et al., 2004; Verheggen et al., 2007). These ancillary and unexpected effects of BVOC captured the attention of a multidisciplinary community of scientists that hopes to better understand the possible contribution of biogenic substances to pollution events. The impact of BVOC is expected to be more evident where biogenic and anthropogenic compounds are likely to react, for instance in urban areas, industrial parks and peri-urban green-belts and rural areas. When the sea breeze circulation is activated over the coast, the plants of the dune ecosystem are located at the entrance of the photochemical reactor regulating the ozone production downind. When the breeze switches into a land system at night, emitted BVOC can react with primary and secondary pollutants. Some of them, such as monoterpenes or sesquitepenes emitted with a temperature-dependent mechanism, can act as a night time sink of ozone leading to carbonyl and radical formation (Ciccioli and Mannozzi, 2007). Because of the complexity of factors influencing the behaviour of sandy dune plants and their critical location, coastal ecosystems of the Mediterranean region deserve specific studies, in which emission and deposition of photochemical oxidants and their precursors need to be determined, together with those of primary production and exchange of energy and matter with the atmosphere, namely water, CO2 and heat. In particular, fluxes of BVOC have been measured so far in coastal ecosystems (such as pseudosteppe and evergreen oak stands) which are different from the dune ecosystem both for the species-specific composition and for the environmental conditions (Ciccioli et al., 1997, 2003).

The Presidential Estate of Castelporziano is a large park at the southeastern edge of the large conurbation of Rome. Since 1951, Castelporziano has been an intact and preserved natural laboratory, where climate, ecological, geological and atmospheric researches are carried out by International research groups. Because of the large biodiversity preserved, the typical Mediterranean climate, and the proximity with urban pollution sources, the park has been a favoured site for the study of the interactions between biosphere and atmosphere. A large campaign was held in Castelporziano more than ten years ago within the framework of the European "Biogenic Emissions in the Mediterranean Area" (BEMA, 1997) project, to identify BVOC emitting plants and to estimate fluxes of isoprenoid emission in the atmosphere that could influence the chemistry of the troposphere (BEMA, 1997). Ten years later, two projects: the European Union Network of Excellence on Atmospheric Chemistry (ACCENT) and the European Science Foundation programme on Volatile Organic Compounds in the Biosphere-Atmosphere Interactions (VOCBAS), jointly organized a field campaign on the Castelporziano site. There were several reasons to expand BVOC measurements in Castelporziano. Since the pioneering BEMA campaign, outstanding progresses have been made in the inventory and understanding of BVOC emissions by plants. For instance, the advent of the Proton Transfer Reaction-Mass Spectrometry has allowed measurements of fluxes of trace gases with unprecedented speed and sensitivity (Lindinger et al., 1998), which makes flux measurements possible with eddy covariance techniques (Rinne et al., 2001; Davison et al., 2009). Moreover, scientific progresses have demonstrated that the emission of volatile isoprenoids is not only controlled by light and temperature (Guenther et al., 1995) but also change with leaf development (Fuentes and Wang, 1999), and in response to environmental constraints, especially salinity (Loreto and Delfine, 2000) and drought (Brilli et al., 2007). It is clear now that emission factors vary on different ecosystems, which makes essential knowledge of species-specific emission factors to correctly implement emission models, especially since the high resolution of the latest models at regional scale (Guenther et al., 2006) imposes a more refined assessment of emissions. Finally, the BEMA campaign characterized species composition and emission patterns of a retro-dunal macchia ecosystem (Owen et al., 1997; Ciccioli et al., 1997), during summer, that is, in a period in which vegetation already underwent strong physiological limitations due to drought. The ACCENT-VOCBAS campaign concentrated on the coastal dune, on a vegetation rich with BVOC-emitting species (Owen et al., 1997), and was performed on May–June 2007, when environmental constraints are still mild and fast vegetative growth typically occurs in the Mediterranean areas (Mooney and Dunn, 1970).

In summary, the field campaign in Castelporziano addressed the following main scientific objectives, as specifically reported in the referenced papers:

a) to assess with a new generation of instruments the species-specific BVOC emission by dune vegetation (this paper; Davison et al., 2009);

b) to determine, with joint field measurements, BVOC emission rates and physiological conditions from plants that were at the beginning of their growing season, and that were presumably yet unaffected by environmental stressors that strongly limit carbon fixation and allocation to secondary metabolites over dry and hot Mediterranean summers (this paper; Mereu et al., 2009);

c) to measure by disjunct eddy covariance (Karl et al., 2002) fluxes of BVOC at the ecosystem level on an ecosystem that has been never investigated previously (Davison et al., 2009);

d) to integrate concurrent leaf measurements of BVOC, CO2 and H2O fluxes driven by physiological processes with ecosystem measurements, supplying information for models of BVOC emissions and reactivity on the macchia vegetation (this paper; Vitale et al., 2009);

e) to assess, simultaneously to BVOC measurements and over the same area, the presence of air pollutants, focusing on ozone, NOx, and particles, in order to estimate whether the possibility exists for BVOC to react with anthropogenic compounds (Atkinson and Arey, 2003), initiating on site the complex chain of reactions that may lead to the formation of secondary pollutants, or protecting plants from pollution damage, as recently reported (Loreto and Fares, 2007) (Gerosa et al., 2009);

f) to assess, through concurrent measurements of meteorological and climatological parameters, whether the climatic factors may affect BVOC emissions and may explain air mass movements, and the consequent possibility that pollutants be formed or transported on peri-urban and rural areas (this paper; Davison et al., 2009).

Material and Methods

Site Information

The experimental site is located in the Presidential Estate of Castelporziano, 41° 41′54.56″ N, 12°21′ 9.50″ E, It covers and area of about 6000 ha located 25 km SW from the center of Rome, Italy. The Mediterranean ecosystems are well represented and preserved inside the Presidential Estate, which contains more than 1000 plant species. The part of the Estate facing the Tyrrhenian sea was chosen for the 2007 field campaign (Fig. 1). This area is characterized by sand dunes, 4–7 m high a.s.l., with mixed garrigue-type and maquis-type vegetation especially in the humid retro-dune area (Bernetti, 1997; Pignatti et al., 2001). The specific location of the experimental site was 100 m from the coast line, between a first and a second dune layer. An area of 1070 m^2 was used for the study on vegetation characteristics as described below.

According to phytoclimatic studies (Blasi, 1993), the experimental area belongs to a Thermo-Mediterranaean region, with long and prolonged stress aridity during summer periods, and a moderate cold stress during winter. The soil of the experimental site is a typical Regosol with a sandy texture and low water-holding capacity, which exacerbates early drought. This soil is not evolved, with an "A" horizon increasing in thickness with the distance from the sea. The chemical properties of the soil are dominated by the strong presence of carbonate elements (0.8% of CaCO3) that create an alkaline pH of 8.3 (Francaviglia et al., 2006). Organic matter is heterogeneously present in the first horizon of the soil which is also rich in fine roots, reflecting the patched presence of vegetation described in Table 1. The total content of soil organic carbon is in the range of 3.1 g(C) kg−1 soil, with high rates of microbial activity which also favoured a high mineralization rate (Trinchera et al., 1998; Pinzari et al., 1999). The retro-dune areas, very close to the experimental site, contain small water-pools, in which water temporarily accumulates, especially during winter, as a function of water-table level and rainfall rate.

Measurement of Environmental Parameters

Meteorological data were collected during the years 2005-2007 at two stations located inside the Castelporziano estate. The Tor Paterno station was located 6 km to the south of the experimental site, but at a similar distance from the coast.

Table 1. Physiognomic measurements of plant features and estimation of the plant cover of the dune ecosystem at Castelporziano experimental site, Rome, central Italy. H_{max} and H_{mean} (standard deviation in parenthesis related to all individuals in the study area) indicate the maximal and the mean height of plants, respectively. LAI is the Leaf Area Index. In parenthesis the standard deviations for H_{mean} and LAI (standard deviation in parenthesys related to all individuals in the study area) are shown. Leaf biomass, and the area covered by each species, in the two different ecosystems ("Garigue" and "Maquis"), is reported as a percentage and as total. Missing data are not available.

species	H_{max}	H_{mean}		LAI		Leaf biom.	Total cover		"Garigue" cover		"Maquis" cover	
	(cm)	(cm)		($m^2\,m^{-2}$)		($kg\,m^{-2}$)	(m^2)	(%)	(m^2)	(%)	(m^2)	(%)
Arbutus unedo	230	137	(55.2)	2.9	(0.31)	0.17	224.2	21.9	55.5	5.4	168.7	16.5
Rosmarinum officinalis	150	77	(28.6)	2.5	(0.38)	0.13	175.7	17.2	175.7	17.2		
Quercus ilex	300	187	(60.2)	2.9	(0.27)	0.20	149.1	14.6	12.1	1.2	137.0	13.4
Phillyrea latifolia	220	116	(45.4)	1.3	(0.21)	0.23	144.2	14.1	41.0	4.0	103.2	10.1
Erica multiflora	160	93	(27.3)	1.9	(0.17)	0.18	126.7	12.4	121.9	11.9	4.8	0.5
Cistus spp.	150	68	(28.6)	0.6	(0.23)	0.16	46.9	4.6	46.9	4.6		
Erica arborea	220	212	(10.0)	4.5	(1.12)	0.19	15.7	1.5			15.7	1.5
Pistacia lentiscus	143	90	(35.6)	2.2	(0.41)	0.29	12.2	1.2	12.2	1.2		
Smilax aspera	150	80	(49.7)	2.4	(0.63)	0.11	10.7	1.0	10.7	1.0		
Daphne gnidium	150	98	(30.0)	0.9	(0.30)	0.09	5.6	0.5	5.6	0.5		
Helicrisum litoreum	40	26	n.d.	1.5	(0.11)	0.16	4.3	0.4	4.3	0.4		
Juneperus phoenicea	170	120	n.d.	3.1	(0.21)	0.27	5.4	0.5	0.9	0.1	4.6	0.4
Juniperus oxycedrus	40	32	(13.7)	3.1	(0.34)	0.27	0.9	0.1	0.9	0.1		
Mean	163	103	(37.0)	2.3	(0.27)							

The Carboeurope station (serving the Carboeurope-IP European project, and managed by the University of Tuscia) was located at 500 m NE from the sandy dune site. In Tor Paterno, hourly values of temperature were recorded with a MP100A sensor (Rototronic, Huntington, NY, USA). A sonic anemometer (W200P, Vector instruments, Rhyl, UK) was used to instantaneously measure wind speed and directions, and a pluviometer (ARG 100, Environmental measurements, Sunderland, UK) was used to measure daily precipitation. All instruments were connected to a data logger (CR23X, Campbell scientifics, Shepshed, UK). Collected data were downloaded monthly and stored in the data-base of the Estate. The Carboeurope station was equipped with similar sensors, but meteorological data were complemented with sensors for the determination of atmospheric pressure and the flux of water and CO2, by Eddy Covariance.

The Bagnouls-Gaussen's diagrams were used to report the monthly averages of air temperatures and precipitation. The intersection (grey area) of the precipitation curve with the average temperatures indicates a period of aridity (Bagnouls and Gaussen, 1957).

Minimum and maximum monthly temperatures and precipitation were used to generate the Mitrakos' diagram, which identifies the Monthly Cold Stress (MCS) and the Monthly Drought Stress (MDS) indices, expressed in stress units and calculated according to the intensity and duration of cold and drought stress periods, as described by Mitrakos (1980). In brief, the MCS index was calculated according to Eq. (1):

$$MCS = 8* (10 - T) \quad (1)$$

where T is the average of the monthly minimal temperatures. For T =10°C, MCS=0 (considering 10°C as a threshold for the vegetative activity for Mediterranean plants); for T =–2.5°C, MCS is assumed to be 100 (assuming that minimal values of 2.5 correspond to the maximal cold stress).

The MDS index was calculated according the hypothesis that when monthly precipitation is below 50 mm drought stress occurs (Eq. 2):

$$MDS = 2* (50 - P) \quad (2)$$

where P is the sum of the monthly precipitation (mm). For P =0, MDS=100; for P =50, MDS=0.

Continuous monitoring of ozone was performed during the campaign from above canopy using a scaffold (6 m) built in the middle of our experimental area. Air was sampled at a rate of 3Lmin –1 with a vacuum pump through teflon tubing and passed into a photometric O3 analyzer (1008 Dasibi Environmental Corp., Glendale, CA, USA) mounted into a cabin 10 m distant from the scaffold. Ozone concentrations in Castelporziano were compared to those recorded at the EMEP station of Montelibretti located downwind the city of Rome, along the main direction followed by air masses in high pressure summer conditions (sea-land breeze circulation regime) (Ciccioli et al., 1999).

Measurement of Vegetation Composition and Distribution

An inventory of vascular species composition and their distribution in the 1070 m^2-wide experimental area around the scaffold where the monitoring devices were positioned was carried out during May. The projected area (PA – m^2) of the crown of each individual plant (n=12–17 per species) was obtained by measuring the crown radius along the four cardinal axes and averaging the four resulting areas. The sapwood area (SA – m^2) was assessed by measuring the stem diameter at 10 cm from the ground on each species, after the cork and phloem were removed, in order to obtain a species-specific sapwood area to projected area ratio (SA/PA). These samples were collected outside the plot in order to avoid wound-induced isoprenoids emissions inside the footprint of the experimental area.

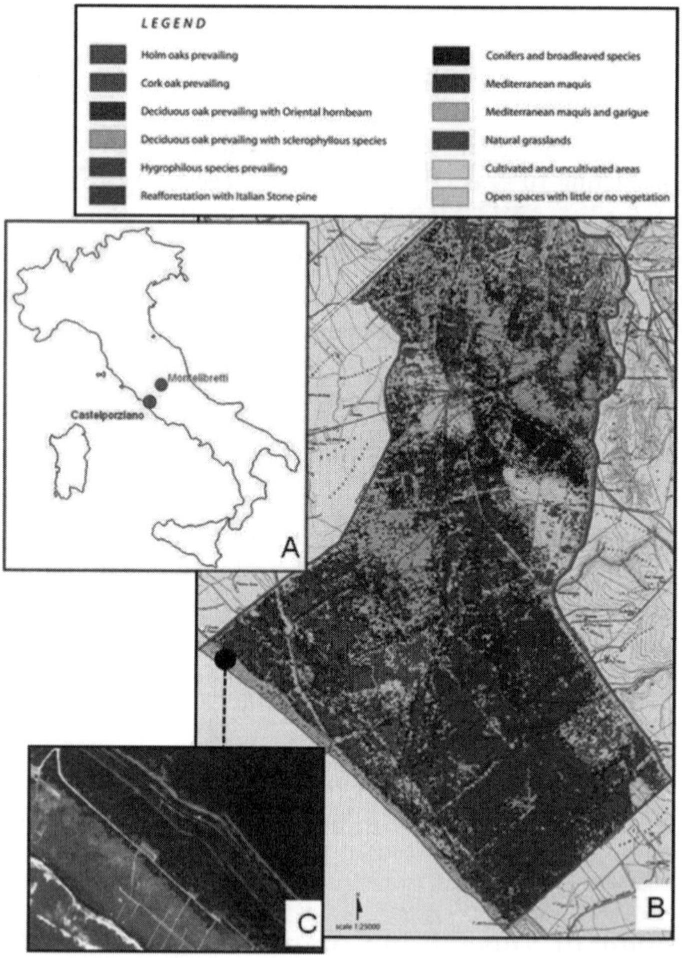

Figure 1. Overview of the experimental site located in Castelporziano. In the sketch showing the Italian peninsula (A), the red circle identifies the Castelporziano Estate (41° 41′54.56″ N, 12° 21′ 9.50″ E), an area of about 6000 ha located in the Lazio region, 25 km SW from the center of Rome; the blue circle identifies the CNR Research station in Montelibretti (42°6′ 26.82″ N, 12° 38′ 15.79″ E) The Castelporziano area is enlarged in (B). A legend explains the different vegetational types co-occurring in the estate. The black circle at the bottom left of panel (B) shows the specific location of the experimental site. Finally, panel (C) is an aerial picture of the experimental site, showing the coast (bottom left), the sand dune strip, the strip of Mediterranean maquis and garigue, surrounded by communication roads where the campaign took place (center), and the dense Meditterranean maquis (top right).

When it was not possible to distinguish to which stem the crowns of plants belonged, height and projected area were measured for the whole patch belonging to the same species. The percentage of soil covered (cover) by each species was obtained dividing the sum of the PA of each species by the area of the experimental plot.

The basal area of each species was calculated by:

$$BA=\frac{SA}{PA}\times cover\ . \qquad (3)$$

The leaf area of twenty leaves (Lal) per woody species collected from different plants was measured by computer software (Image Tool Software Roswell, GA). These leaves were dried at 80°C for a week to obtain the leaf mass (LM – g). The leaf mass to area ratio (LMA – g m2) of each leaf was calculated by:

$$LMA=\frac{La_l}{LM}\ . \qquad (4)$$

The leaf area (LA) to sapwood area ratio (LA/SA) of each species was estimated as the slope of the linear fit of the LAb vs. SA regression using thirty branches for each species with diameters ranging between 0.3 and 5 cm. The leaves of each branch were dried to obtain the total leaf material (LMtot – g) of the branch. The leaf area of each branch (LAb) was estimated by multiplying LMtot by the LMA:

$$LAb = LMtot \times LMA. \qquad (5)$$

The leaf area index (LAI) of Quercus ilex, Phillyrea latifolia, Arbutus unedo, and Erica arborea plants was derived by dividing the LA/SA ratio by the Projected Area (PA)/SA ratio:

$$LAI=\frac{LA}{PA}=\frac{LA}{SA}\times\frac{SA}{PA}\ . \qquad (6)$$

The LAI of each species was estimated as the average LAI calculated on at least ten different plants per species. The LAI of Cistus incanus, Rosmarinus officinalis and Erica multiflora was assessed in a different way because of the elevated number of stems. For these species the LAI was obtained by multiplying the LMA by the dry weight of the leaves present over 0.25 m2 of soil (three samples for each species).

Measurements of Plant Physiological Properties and Isoprenoids Emission

The measurements were performed during the month of May and in the central hours of the day (11:00 a.m.–03:00 p.m.). Photosynthesis, stomatal conductance and transpiration of all species were measured with a Li-6400-40 gas exchange open portable system (LI-COR, Lincoln, Neb., USA). A leaf was enclosed in a 6 cm^2 cuvette and an air flow was pumped to the cuvette after being filtered with an active carbon cartridge to scavenge pollutants or BVOC present in the air. The instrument allowed to control light intensity and leaf temperature, which were set, respectively, at 1000 µmol $m^{-2}s^{-1}$ of PAR (Photosynthetic Active Radiation) and 30oC. These are the basal conditions at which isoprenoid emission is commonly measured (Guenther et al., 1995). The outlet from the cuvette was diverted to the Li-6400-40 infrared gas analyzer to measure CO_2 and H_2O exchange. These exchange rates were recorded after an adaptation period of around 1 h after leaf enclosure, and in any case after stabilization of the photosynthetic parameters. The outlet of the cuvette was then diverted to a glass traps (16 cm in length, 4 mm I.D.) filled with 130 g Tenax GC,20–35 mesh, (Alldrich, USA) and 115 g Carbo-graph 1, 20–40 mesh, (LARA S.p.A., Rome, Italy) to collect BVOC. Air was passed through the trap at a flow rate of 200 ml min^{-1}, controlled by a pump (Pocket pump, SKC, PA, USA), until a total volume of 10 L of air was collected. Before measuring each individual leaf, emissions from the empty cuvette were also measured, to make sure that no enclosure contaminants were recorded.

Traps were stored in a refrigerated container until analyzed at CNR laboratory with a gas chromatograph (GC 5890, Hewelett Packard, Palo Alto, CA, USA) connected to a quadrupole mass spectrometer (MSD 5970, Hewelett Packard, Palo Alto, CA, USA). Traps were desorbed by keeping them at a temperature of 250°C for 5 min using helium as carrier gas at a flow rate of 20 ml min–1. Desorbed isoprenoids were concentrated in an empty liner at –190°C under liquid nitrogen and then injected into the GC column by rising the temperature of the liner to 200°C in 5s. A Chrompack, fused silica capillary column (50 m in length with 0.4 mm I.D.) coated with a 0.32 µm of CPsil5 (Middelburg, The Netherlands) was used for the chromatographic separation. The elution was carried out by rising the column temperature from 50 to 250°C ata rate of 5°C min–1. Compounds were identified and quantified following the procedure described in Ciccioli et al. (2002).

Artificial Irrigation on Cistus Incanus Plants

A $20m^{-2}$ area with Cistus incanus plants was located in the same experimental area. It was 500 m distant from the main experimental site, but with identical climatic and vegetation characteristics. This area was used to artificially irrigate plants, in order to better assess whether a drought stress was causing physiological effects during the experimental period. This area was far enough not to interfere with the micrometeorological measurements of the main site (see Davison et al., 2009). Plants were irrigated daily in the early morning. Irrigation started two weeks before the exprerimental measurements at the main site, and continued throughout the campaign. The measurements of physiological traits and isoprenoid emission, as detailed above, were also performed on the plants of this plot.

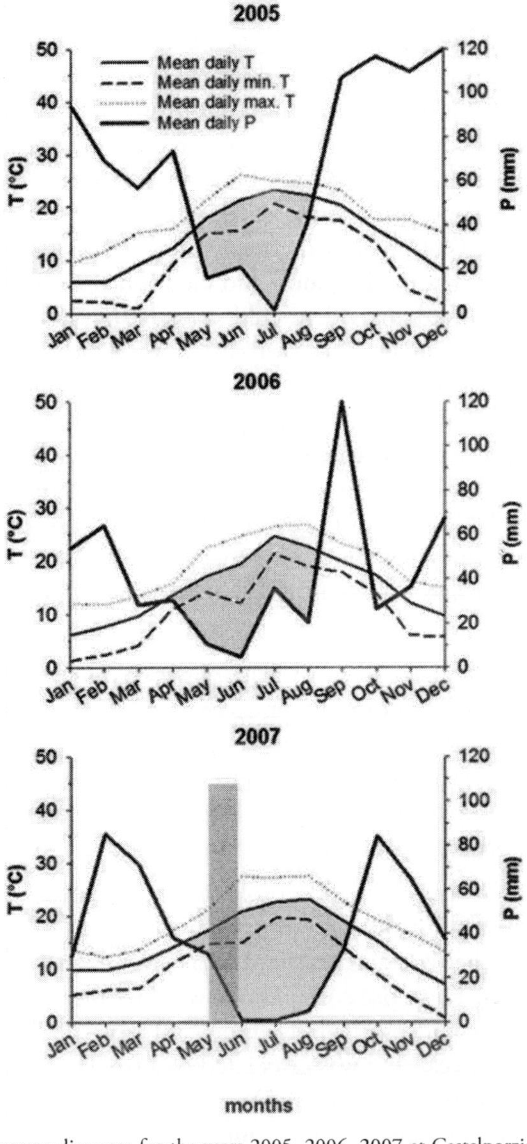

Figure 2. Bagnouls Gaussens diagrams for the years 2005, 2006, 2007 at Castelporziano Estate. 14 Monthly temperatures (T , left axis, oC) -are shown as daily mean (solid line), daily maximum 15 (dotted pattern) and daily minimum (dashed pattern). The sum of monthly rainfall (P, right axis, 16 mm) is also reported for the same periods. The yellow zone represents a drought period. The green 17 bar in the 2007 graph indicates the period of the campaign.

Results and Discussion

Climate and Meteorological Data

Meteorological data for the years 2005-2007 showed that the climate of the site is typically Mediterranean with mean monthly temperatures ranging between a minimum of 4°C and a maximum of 27°C, extreme summer temperatures occasionally exceeding 30°C (maximal values of 32°C in August 2007), absence of freezing events, and a pronounced summer drought, with rainfall events concentrated in autumn and spring. In particular, the mean annual precipitation over the years 2005–2007 was 713 mm, but only 480 mm were recorded during 2007. The summer dry period was recorded on each one of the three years, with low monthly rainfalls (<60 mm in the May–August period) as clearly shown in Bagnouls-Gaussen's diagrams (Bagnouls and Gaussen, 1957) (Fig. 2). The Mitrakos' MCS diagram showed an evident cold stress period in our experimental site from November to March (Fig. 3). More interestingly, the MDS index showed that drought stress was widespread over the entire dry seasons, but was particularly strong during the summer periods of years 2006 and 2007 (Fig. 3). Drought causes a severe stress in warm Mediterranean seasons, when high temperatures induce high evapotranspiration, with fast depletion of water reservoirs in the soil and in plant tissues. This explains the evolution of the "water saving" strategy, also observed in our measurements with dune species, which allows to slow down the main metabolic functions, and reduce water loss through stomata during summers.

The year 2007 was in general very warm, with prolonged drought stress also in winter, and monthly precipitation never exceeding 100 mm (Fig. 2). All climatic indices confirmed that the period of the campaign (May 2007) was favourable to vegetative growth, at least when compared to the hot and dry summer. Moreover, the proximity of the site to the sea led to a high humidity regime (with RH rarely below 60%) with formation of dew in the night-time.

Daily variations of the main meteorological parameters recorded in Castelporziano during the field campaign are shown in Fig. 4a to d, and consistently indicate that the weather was characterized by a certain variability, more typical of earlier spring periods. High pressure periods were frequently alternated to short low pressure episodes, and particularly important was the low pressure episode occurring on 28 May because it was associated to a front of humid air that generated strong winds blowing from the SW-W sector. Precipitations occurred also at the beginning of the campaign (4 May), but in this case the front was less intense and characterized by light winds blowing from the same sector as during the rest of the campaign. During low pressure episodes, a synoptic scale circulation drove the movement of the air masses. During high pressure periods, however, the circulation was mostly determined by a local sea-land breeze wind regime, with moderate to strong S-SW winds blowing during the day, and light N-NE winds after 02:00 a.m. (Figs. 4c and 5a). As already observed in previous campaigns (Kalabokas et al., 1997), the sea-land breeze regime generates a transport of biogenic emission toward the city of Rome during the day, but allows also primary (such as VOC, NOx) and secondary pollutants (such as ozone and PAN), respectively emitted and produced in the urban area of Rome, to reach Casteporziano at night, between 02:00–03:00 and 09:00 a.m. For most of the time the maximum air daily temperature ranged between 25 and 26°C and only in one day values of 30°C were reached. Because of these meteorological conditions, the vegetation of the sandy dunes never really reached stress conditions typically observed during summer, and gas-exchanges of CO2, H2O and BVOC were not limited.

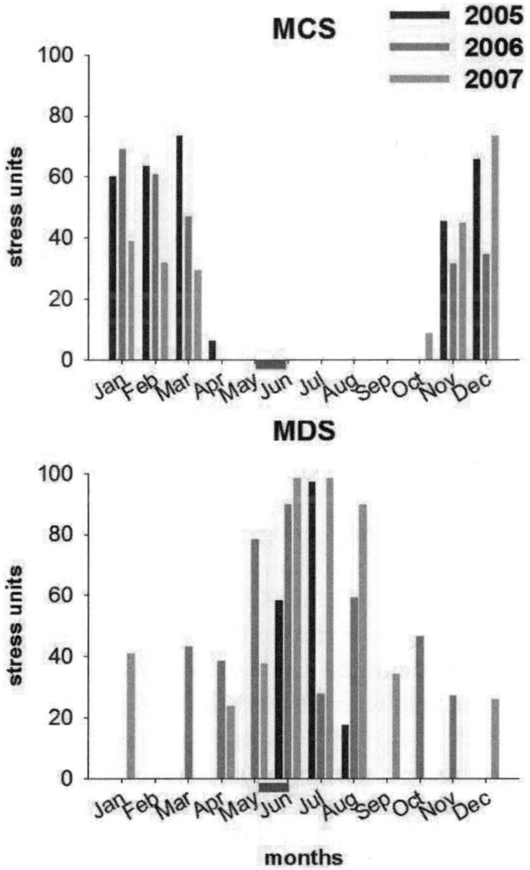

Figure 3. Mitrakos diagrams for monthly cold stress (MCS) and monthly drought stress (MDS) in the years 2005 (brown bars), 2006 (orange bars), 2007 (green bars) at Castelporziano Estate. The red line in correspondence of May and June indicates in the x-axes the the period of the campaign.

Ozone Levels During the Feld Campaign

The duration of high pressure periods (3–5 days, Fig. 4a) did not allow to generate the high levels of ozone typically observed in Castelporziano during summer (Kalabokas et al., 1997). As several studies performed in the Mediterranean coasts (Millan-Millan et al., 1998) have shown, high ozone levels occur during high pressure conditions, characterized by a local sea-land breeze circulation which last long enough to transport over the sea secondary pollutants generated in densely populated areas. Ozone formation is enhanced when air masses advected by the sea breeze penetrate deeply into the valleys and become progressively enriched in NOx and VOC emitted from urban areas. When moving out from these polluted areas, air masses become progressively depleted in NOx. Under these NOx-limited conditions ozone production progressively decreases, although its content in the polluted plume remains high because little removal occurs by secondary reactions (Finnlayson-Pitts and Pitts, 1999). After reaching a mountain range, the air masses are conveyed back over the sea by a return flow, located above the advection flow and often separated from it from a shear layer (Millan-Millan et al., 1998). Ozone present in the return flow can be transported up to 40 km over the sea, where it can stratify at night. After sunrise, the polluted layer containing the ozone transported from the city is mixed with the one resident below, resulting in an increase of ozone in the whole convective layer. When the sea breeze is activated (10:00–11:00 a.m.), ozone is driven back to the coast by advection (Millan-Millan et al., 1998). These effects have been also observed by Georgiadis et al. (1994) over the Adriatic sea. Such a continuous recirculation of ozone between the sea and inland sites where large emission of precursors take place causes a progressive increase in the ozone levels through all the areas interested by the air mass circulation, ultimately leading to photochemical smog episodes.

In the Tiber valley, the maximum ozone production usually occurs between the city limits of Rome and the suburban areas located 15 km from the city centre (Ciccioli et al., 1999). Under heavy smog conditions, the influx of the city plume of Rome gives rise to sharp peaks of ozone, peroxyacylnitrates (PAN), sulfates and nitrates in the suburban station of Montelibretti with maximum values that can be twice as high as those advected by the sea (Ciccioli et al., 1993, 1999; Cantuti et al., 1993). Ozone contained in these air masses is conveyed back to the Tyrrhenian sea by a return flow generated when they reach the Apennine range located 40–60 km from the coast (Baldi et al., 1993; Mastrantonio et al., 1994). The importance of the sea-land breeze regime in generating high levels of ozone in the area of Rome has been confirmed after application of the FARM model, whose results have been validated by field studies performed in different seasons (Gariazzo et al., 2007). These results show that in summer, when high pressure conditions last for more than 2 weeks, pollution generated in the area of Rome progressivelly increase the ozone levels present in the Tyrrhenian coast where Castelporziano is located.

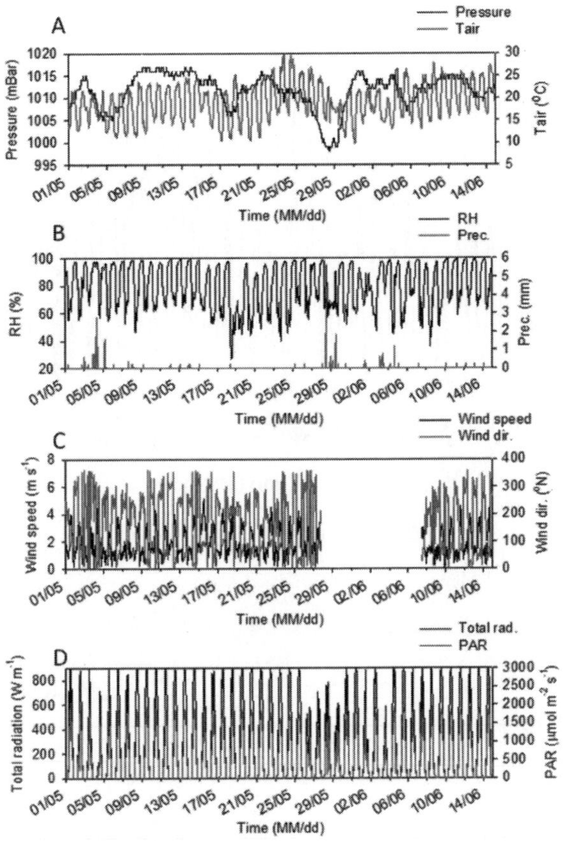

Figure 4. Daily profiles of the meteorological parameters collected at two stations inside the Castelporziano estate. The Tor Paterno station was located 6 km to the south of the experimental site, but at a similar distance from the coast. The Carboeurope station (serving the Carboeurope-IP European project, and managed by the University of Tuscia) was located 500 m NE from the sandy dune site. Panel (A) shows pressure and air temperature at Tor Paterno site, panel (B) shows relative humidity and precipitation at Tor Paterno, panel (C) shows wind speed and direction at the Carboeurope station, panel (D) shows the total radiation and PAR (photosynthetically active radiation) at the Carboeurope station.

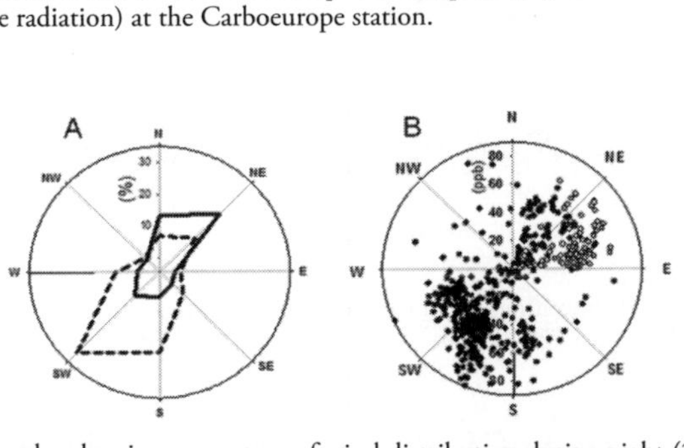

Figure 5. Panel (A): Polar plot showing percentage of wind distribution during night (21:00–06:00, solid line) and day hours (06:00– 21:00, dashed line) in Castelporziano Estate, central Italy. Panel (B): Polar plot showing ozone concentration (ppbv) during day (black circles) and night hours (white circles) in the same experimental area.

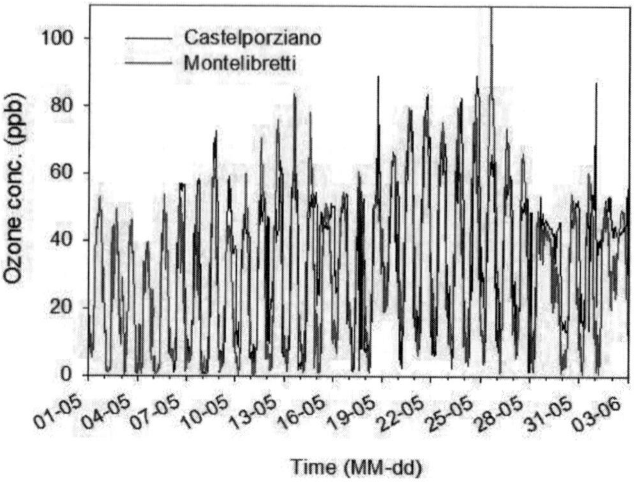

Figure 6. Daily profiles of ozone levels in Castelporziano (ppbv), cen-tral Italy and at the EMEP station in Montelibretti, located NE of Castelporziano, 20 km downwind Rome, inside the Tiber valley. 30 min means are reported for the period from 1 May to 4 June 2007.

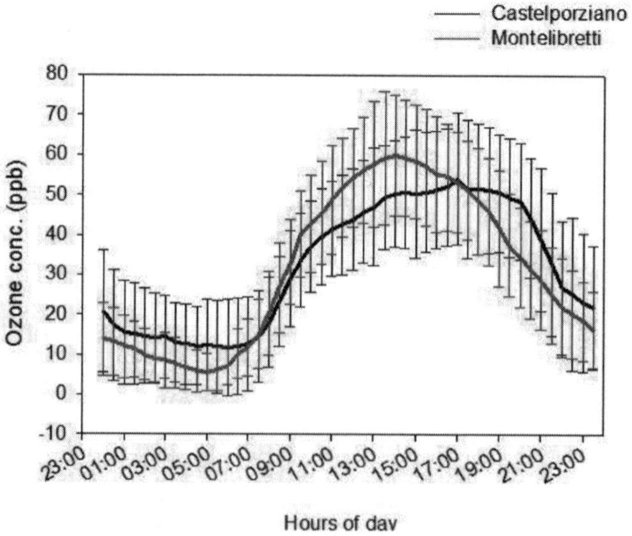

Figure 7. Half-hour averaged concentrations of ozone in Castelporziano (black line) and in Montelibretti (red line) for the period 1 May–4 June 2007. Mean ±SD are reported (n=42 days).

Ozone data collected in Castelporziano and Montelibretti during the campaign (Fig. 6) show that differences in ozone concentrations between the two sites occurred only when high-pressure conditions were generated over the area. This conclusion is supported by the polar plots reported in Fig. 5b. During low pressure episodes, the ozone levels in the two sites were almost the same. This suggests that, during the campaign, stability conditions that characterized hot and sunny days did not last long enough to generate accumulation of large amounts ozone over the sea. Consequently, levels measured in Castelporziano never reached those measured in summer, when values up to 90–95 ppbv have been recorded (Kalabokas et al., 1997). The observation that the mean daily ozone profile recorded in Castelporziano does not show the same bell-shaped profile of temperature and solar radiation which characterizes the suburban site of Montelibretti (Fig. 7), indicates that transport from aged air masses accumulated over the sea was the main source for this pollutant over the coast. Data in Fig. 7 also show that night time levels of ozone were often higher in Castelporziano than in Montelibretti. Under stable atmospheric conditions such difference may be explained by the higher NO emission from traffic over the Tiber valley than near the coast, as O3 is more efficiently removed from the lower layer by reaction with NO. As indicated by Fig. 5b, ozone advection caused by the land breeze could have also played a role in determing the ozone levels after 03:00 a.m.

Vegetation Characteristics

The dune vegetation of the experimental site is composed of patches of Mediterranean "maquis" and "garigue." Following the nomenclature suggested by Pignatti (Pignatti et al., 2001), we named "garigues" the Erico-Rosmarinetum formation, characterized by the abundant presence of low-shrubs species as Rosmarinus officinalis, Erica multiflora, Arbutus unedo, Phillyrea latifolia and Cistus incanus albidus. Many of these species are typically located over Regosols soils rich of $CaCO_3$ from deposited bioclasts. The woody species of this formation are rarely taller than two meters (Table 1). The Mediterranean "maquis" is a variation of the Quercetumilicis (Pignatti et al., 2001), which is dominated by Quercus ilex from the early developing stages. The early stages of "maquis" are formed by a mixture of Quercus ilex and other co-dominant species: Erica arborea, Phillyrea latifolia, Juniperus spp. and Arbutus unedo, with canopies rarely taller than five meters. The increase of below-ground marine water level could lead to a progressive replacement of high "maquis" with plant communities more typical of a degraded "garigue" (e.g. the Erico-Rosmarinetum), characterized by low shrubs adapted to very alkaline pH soils. Pignatti et al. (2001) hypothesized that vegetation patches of the two formations (garigue and maquis) observed in the experimental area correspond to an uneven distribution of sea shells which originated the bioclasts. Future studies should test this hypothesis.

As shown in Table 1, Quercus ilex was the tallest plant at the site, reaching the maximum of 3 m height. Arbutus unedo has the largest cover (21.9%), followed by Rosmarinus officinalis (17.2%), Quercus ilex (14.6%), Phillyrea latifolia (14%) and Erica multiflora (12.3%). The summed percentage coverage shows that the "garigue" (48.2% of the site) was slightly more widespread than the "maquis" (42.1% of the site). In terms of LAI, species as Erica arborea (4.5 m2 m−2), Arbutus unedo (2.9 m2 m−2), Quercus ilex (2.9 m2 m−2), Rosmarinus officinalis (2.5 m2 m−2) mostly contributed to the average LAI of the site, which was 2.3 m2 m−2. Only 5% of the experimental site was constituted by bare soil, and another 5% was covered with dead plant materials, i.e. fallen branches and leaves of annual herbs.

Table 2. Measurements of isoprenoids emitted by the different plant species of the dune vegetation of the experimental area of Castelporziano, Rome, central Italy. Photosynthesis and stomatal conductance are also reported. Missing values are undetected compounds. All values indicates means (n=4), standard deviations are reported in parenthesis for isoprene, total of monoterpenes, photosynthesis and stomatal conductance.

Isoprenoids (ug(C) g^{-1} DWh^{-1})	*Quercus ilex (mature leaves)*	*Quercus ilex (young leaves)*	*Cistus spp. (control)*	*Cistus spp. (irrigated)*	*Arbutus unedo*	*Phyllirea latifolia*	*Rosmarinus officinalis*	*Erica multiflora*	*Juniperus oxycedrus*
Isoprene	0.33(0.54)	0.31 (0.4)	0.05(0.04)	0.38(0.91)	0.02 (0.02)	0.05(0.04)	0.02(0.01)	1.48(0.4)	0.02(0.01)
Tricyclene	0.01	0.02	–	–	–	–	0.12	–	0.02
alpha-Thujene	0.05	0.07	...	...	...	...	0.12	..	0.01
alpha-Pinene	2.39	3.58	0.18	0.31	0.04	0.23	2.44	0.29	5.54
Camphene	0.09	0.14	0.01	0.02	...	0.01	2.81	0.14	0.08
6-metil-5 metil Eptane	0.39	0.36	0.36	0.31	0.14	0.32	0.56	0.54	0.37
Sabinene	1.66	2.29	0.14	0.12	0.02	0.09	0.06	0.01	0.10
beta-Pinene	1.65	2.43	0.12	0.17	0.03	0.11	4.09	0.26	0.19
Myrcene	0.39	0.65	0.03	0.06	0.01	0.06	0.54	0.05	0.58
alpha-Phellandrene	0.01	0.01	...	...	...	...	0.02	..	0.38
D-3-Carene	–	–	–	–	–	–	–	–	–
alpha-Terpinolene	0.01	0.02	..	..	...	...	0.04	...	0.01
para-Cymene	0.02	0.02	–	0.01	–	0.01	0.07	0.01	0.04
1,8-Cineol	0.10	0.34	0.01	0.02	–	0.02	4.94	0.58	0.15
beta-Phellandrene	0.09	0.08	0.01	0.04	0.01	0.01	0.03	–	1.50
Limonene	0.22	0.26	0.09	0.28	0.05	0.09	1.41	0.18	0.19
cis-b-Ocimene	1.77	0.03	0.04	0.23	0.00	0.03	0.25	0.03	0.02
trans-b-Ocimene	0.52	0.02	0.02	0.07	0.01	0.02	0.06	0.16	0.09
gamma-Terpinolene	0.02	0.04	..	..	...	...	0.10	0.01	0.01
Terpinolene	0.01	0.02	–	0.01	–	–	0.04	0.01	0.11
Linalool	–	0.01	–	–	–	–	0.07	0.01	0.00
4-Terpineol	...	...	–	–	–	–	0.18	0.03	0.00
a+g-Terpineol	–	0.02	–	–	–	–	0.47	0.09	0.13
trans-b-Caryophyllene	...	..	...	...	...	...	0.46	..	...
Total of Monoterpenes	**9.39** (4.5)	**10.4** (1.91)	**1.01** (0.7)	**1.66** (1.00)	**0.29** (0.20)	**0.98** (0.30)	**18.89** (5.00)	**2.40** (1.00)	**9.57** (2.50)
Photosynthesis (μmol m^{-2} s^{-1})	**5.28** (0.75)	**3.2** (0.60)	**10.41** (1.09)	**14.35** (1.25)	**9.22** (0.97)	**5.35** (2.61)	**8.58** (1.00)	**0.48** (0.10)	**6.73** (1.20)
Stomatal conductance (mol m^{-2} s^{-1})	**0.03** (0.001)	**0.038** (0.01)	**0.1245** (0.02)	**0.204** (0.04)	**0.0994** (0.02)	**0.073** (0.05)	**0.111** (0.06)	**0.012** (0.008)	**0.056** (0.02)

Changes in quality and depth of the water table of the sandy dune ecosystem can strongly affect biosphere-atmosphere exchanges. The waterpools present in the experimental site have an important ecological role, since they preserve a high floral and faunal biodiversity. However, a dramatic decrease of the water table level was observed in the last years (Bucci, 2006). This process is very common in the Mediterranean coastal areas, being caused mostly by increasing water demand by agricultural activities and coastal inurbation. As a consequence, marine water infiltrates on the freshwater table, reaching more superficial levels and salinizing the root environment of dune and retrodune plants. If this occurs, shrubs species belonging to the Erico-Rosmarinetum with a more superficial root system, could be favoured with respect to woody species. Models predict a complete salinization of the soil by 2050 (Bucci, 2006) for the coastal area of Castelporziano. We measured the depth of the water table two times, in the middle of the test site. The water table depth was found to be 210 and 290 cm, respectively, when measured on 27 May and 4 August. It is therefore hypothesized that the sinking depth of the water table could be the main cause of abiotic stress during the dry periods. A detailed study on water use under drought stress for the most representative species of the dune vegetation is proposed by Mereu et al., 2009. A strong physiological dependence of schlerophyllous plants on the freshwater table level was also observed, in our same experimental site, with a study of isotopic fractionation in xylematic water (Alessio et al., 2004). In particular, the shrubs examined in our study, and particularly A. unedo, P. latifolia and, Q. ilex, showed negative values of 18O2 values of xylematic water that are associated with high discrimination (Δ) values, even under very dry conditions. This indicates low long-term water use efficiency, especially when compared to the opposite behaviour of some other dune species (e.g. Smilax aspera). Interestingly, no species of the coastal dune system used marine water or mixtures of marine and freshwater (Alessio et al., 2004), indicating that in this site marine intrusion was absent.

Physiology

Plants showed an excellent physiological status during the experimental period. Rates of photosynthesis and stomatal conductance measured in mid May (Table 2) were comparable to data reported in the literature for unstressed leaves and higher than rates measured at the same experimental site during previous campaigns held in full summer (Manes et al., 1997b).

Since the measured ozone levels often exceeded 40 ppb during the day-time (Figs. 5, 6, 7), and this is considered a threshold after which plants can be damaged by the pollutant (UNECE 2004), we also checked for possible ozone injuries to vegetation. No reduction of chlorophyll content, photosynthesis, chlorophyll fluorescence, nor visible injuries that could be directly attributed to ozone damage were found (data not shown). Most of the schlerophyllous species growing in the experimental site (Quercus ilex, Arbutus unedo, Pistacia lentiscus) are known to be resistant to chronic or prolonged ozone exposures (Vitale et al., 2008; Nali et al., 2004). This can be attributed to: i) low stomatal opening which reduces the amount of ozone taken up by leaves (Loreto and Fares, 2007), especially when stomata are further shut down because of concurrent drought stress; ii) thick cuticular waxes and mesophylls that reduce non-stomatal ozone entry, and increase the chance of ozone reaction inside mesophyll before reaching target sensible organelles (Loreto and Velikova, 2001); iii) detoxification of ozone by reaction of the pollutant with BVOC (especially isoprenoids) emitted by plants. Loreto and Fares (2007) demonstrated that ozone damage is considerably reduced in high isoprenoid emitters, although ozone uptake is higher in these plants, and products of putative reactions of ozone with isoprenoids were not described in planta. Since many Mediterranean species do emit high levels of isoprenoids (see below and also Kesselmeier and Staudt (1999) for a comprehensive review), isoprenoids and other reactive molecules may be key players in driving ozone uptake and efficient detoxification thus limiting dangerous effects of pollution in peri-urban areas.

Isoprenoid Emission

As also shown by photosynthesis measurements, plants were actively growing during the experimental campaign. Growth occurs only when environmental constraints are low, and, in dry ecosystems of the Mediterranean area, it takes place mostly in two flushes during spring and fall (Reichstein et al., 2002). High photosynthesis is expected to drive high emissions of BVOC during spring, as some of these compounds, such as the main volatile isoprenoids, are formed by carbon directly shunted from photosynthetic carbon metabolism (Sharkey and Yeh, 2001). However, a strong seasonality has been observed for many BVOC, constraining biosynthesis and emissions of these compounds during spring. Isoprene emission is under developmental control, being very low in expanding leaves, and uncoupled from photosynthesis development (Sharkey and Loreto, 1993). A seasonal pattern of monoterpene emissions was also observed to depend diurnally and seasonally from light and temperature (Sabillon and Cremades, 2001) and phenology (Staudt et al., 1997, 2000; Ciccioli et al., 2003). This is attributed to slow development of the capacity to synthesize isoprenoid synthases, the enzymes catalyzing the formation of isoprene (Wiberley et al., 2005) or monoterpenes (Fischbach et al., 2002). Thus, a low emission of isoprene and monoterpenes may be expected early in the season. However, Wiberley et al. (2005) pointed out that the development of the trait (i.e. the transcriptional and translational limitation to isoprene synthase biosynthesis) is in turn controlled by growth temperature. Therefore, in the Mediterranean area, whose climate is characterized by already rather high spring temperatures, a sustained emission of isoprenoids could be seen already in spring.

Table 2 shows that emission of isoprene was low in the Mediterranean dunal ecosystem, confirming data previously obtained by Owen et al. (1997) at the same site, and by the whole previous campaign in the Mediterranean area (BEMA, 1997). However, monoterpene emission by some dune plants was relevant, despite the early season of measurement. Quercus ilex, in particular, was confirmed to be a high monoterpene emitter, with α and β-pinene being the most emitted compounds. The total emission of monoterpenes by Q. ilex leaves was around 10 μg(C) g–1DW h–1, a rate comparable to what was found also in previous field measurements (Loreto et al., 2001a). Monoterpene emissions were about

30% lower than those reported by Bertin and Staudt (1996) and Kesselmeier et al. (1997) in Castelporziano, but these other measurements were run in full summer (i.e. with higher temperatures and more complete leaf development) and on a different site, characterized by more structured soil and higher water availability.

In our measurements, both mature (second year) leaves, and young, still expanding leaves, emitted similar rates of total monoterpenes (Table 2), which suggests that the rate of biosynthesis and emission is not under developmental control, and that monoterpene emission is also rapidly induced in young leaves that were grown at rather elevated spring temperatures, confirming the indications of Wiberley et al. (2005). Interestingly, however, the composition of the emitted monoterpene blend was different in mature and developing Q. ilex leaves, with trans and cis-β-ocimene only emitted by mature leaves. We surmise that ocimene biosynthesis is under developmental control. Trans-β-ocimene seasonal emission was observed in pines (Loreto et al., 2000; Staudt et al., 2000). Emission of trans-β-ocimene and other acyclic monoterpenes may be also induced by herbivore feeding (Heil and Silva Bueno, 2007), and may therefore reveal induction caused by past stress occurrence in mature leaves. However, trans and cis-β-ocimene emissions were also measured in Q. ilex leaves that were grown in absence of abiotic and biotic stress (Loreto et al., 1996).

Rosmarinus officinalis and Juniperus oxycedrus also emitted monoterpenes at high rates, but with different emission patterns. β-Pinene and 1,8-cineol were the main compounds emitted by R. officinalis, whereas the emission of J. oxycedrus was mainly characterized by α-pinene. The total monoterpenes emission by rosemary leaves was 18.89 μg(C) g^{-1}DW h^{-1}, by far larger than reported by Hansen et al. (1997). Contrary to Q. ilex, R. officinalis and J. oxycedrus store monoterpenes in large pools in specialized glandular organs (Ormeno et al., 2007; Salido et al., 2002). The emission by monoterpene-storing species is generally not light-dependent, and a long time is needed to extinguish the pool once biosynthesis has been turned off (Guenther et al., 1993). Llusi´a and Penuelas (1998, 2000) observed that monoterpene emission by monoterpene storing species is more dependent on temperature than in non-storing species, which may drive relevant summer emissions of monoterpenes, even when photosynthesis is environmentally constrained. These authors also found an accumulation of monoterpenes in storage organs of Mediterranean species at increasing drought conditions, with maximal levels in the autumn. These findings explain why in our measuring period we did not observe the maximal monoterpene emission, and indicate that higher emissions can be detected during warmer periods, when the pools of monoterpenes are completely filled. However, the temperature dependence of monoterpenes stored in pools is not always that high. For instance in Pinus pinea emission of stored monoterpenes is less dependent on temperature than emission of non-stored monoterpenes (Staudt et al., 1997). In addition, it should be mentioned that rough handling can drive large emissions from storage pools (Loreto et al., 2000). We paid special attention to avoid breakage of storage pools in our measurements, but cannot exclude altogether that some of the abundant emission observed were also contributed by unwanted mechanical stress (see also Davison et al., 2009).

Owen et al. (1997), reported Cistus spp. to be weak isoprene and monoterpenes emitters in the Castelporziano ecosystem. However, we found values of monoterpene emission (1.35 µg(C) g−1DW h−1) much larger than previously reported. Llusi´a and Penuelas (2000) detected in Cataluna high emissions from Cistus incanus, more in line with our indications. Monoterpene emission by Cistus incanus is probably under seasonal control. The emission rates detected in our campaign from Cistus incanus leaves were more similar to rates measured in October than to those measured in May in Cistus incanus plants growing in the Pianosa island (Baraldi et al., 2001), but Llusi´a and Penuelas (2000) also measured maximal emissions in May. The Pianosa experiment indicated a strong dependency of Cistus incanus emission on water availability, as October sampling was done after a period of rainfalls (Baraldi et al., 2001). Indeed, we observed a further increase of the emission of both isoprene (>80%) and monoterpenes (>40% with respect to controls) in Cistus incanus plants that were artificially irrigated during our experiment. Since stomata opening should not regulate isoprenoid emissions, with few exceptions (Niinemets et al., 2004), our finding indicates a strong control of water availability over the synthesis of isoprenoids in this species. Recent research outlined that isoprenoid biosynthesis is resistant to drought stress (Pegoraro et al., 2004; Brilli et al., 2007), but after a threshold level of stress the isoprenoid formation is also inhibited (Brilli et al., 2007). This could be the case of Cistus spp. which are water-spending plants characterized by low stomatal control on water content, and very large changes of leaf water status. The "water-spending" strategy of this species seem to be successful in an environment subjected to fast changes of soil water content, as in sandy soils. However, the impact of these metabolic fluctuations on secondary metabolism leading to isoprenoid formation remains unknown.

A detectable emission of isoprene and monoterpenes was also observed from Erica multiflora leaves (1.48 and 2.4 µg(C) g−1DW h−1, respectively). Again, this is a higher emission than previously measured by Owen et al. (1997) at the same site. However, strong isoprene-emitters are characterized by rates of emissions ten times higher than in monoterpene emitters. Thus, on the basis of our spring measurements and of the summer measurements of Owen et al. (1997), E. multiflora can only be categorized as a low isoprene emitter.

The other two screened species emitted low amount of monoterpenes. The weak emission of monoterpenes by Arbutus unedo (0.29 µg(C) g−1DW h−1), was also observed by Owen et al. (1997), and Pio et al. (1993). Alessio et al. (2004) also reported inconspicuous emissions of isoprenoids by Arbutus unedo and Phillyrea latifolia. It is therefore confirmed that these plants do not have the capacity to produce and emit relevant amounts of isoprenoids.

After measuring the basal emission at leaf level of each representative species of the stand (Table 2), an upscaling procedure was developed to estimate isoprenoid emission at ecosystem level. The species representativeness, and the species-specific leaf biomass over the experimental site (the parameters shown in column 4 and 5 of Table 1) were considered. The basal emission of isoprene for the whole stand (normalized at 30°C and a light intensity of 1000 μmol photons m–2s–1, according to Guenther's algorithm, 1995) was 0.31 μg(C) g–1DW h–1. This emission is lower than the emission suggested for Mediterranean ecosystems by Guenther et al. (1995) (16 μg(C) g–1DW h–1) and by Owen et al. (1997) (14.88 μg(C) g–1DW h–1). It should be mentioned that these two studies concentrated on Mediterranean ecosystems more abundant of isoprene emitting species (Myrtus communis and Cistus spp.) than the dune ecosystem investigated here.

The same averaging procedure was adopted for monoterpenes, and yielded a basal emission rate of 4.7 μg(C) g–1DW h–1. For monoterpenes, Guenther et al. (1995) suggested a basal emission value of 1.2 μg(C) g–1 DWh–1, and a similar indication was given by Owen et al. (1997) (2.2 μg(C) g–1 DWh–1). Thus it is concluded that the Mediterranean dune vegetation is an important source of monoterpene emission, and a less important source of isoprene emission than other Mediterranean ecosystems. Emissions are also likely different in the macchia ecosystems growing on a less sandy soil, under different water limitations, especially during summer. Our estimate about stand emission rate may be lower than actual if integrated over the whole vegetative period. It has been shown that isoprene emission is controlled by seasonal temperatures (Sharkey et al., 1999) and a similar control may also occur in monoterpene-emitters (Ciccioli et al., 2003). Unless emissions are drastically restrained by stress episodes (Baraldi et al., 2001; Loreto et al., 2001b) emissions of dune vegetation might therefore increase when the stand is exposed to high summer temperatures for long periods.

Conclusions

The experimental area of Castelporziano is an ideal site to test the interactions between biosphere and atmosphere in an environment that is made fragile by heavy anthropogenic pressure and by climate change drivers. This situation is unfortunately typical of the entire Mediterranean coastal area, where plant ecosystems are particularly perturbed by a combination of anthropic impacts and natural climate changes (van Der Meulen and Salaman, 1997; IPCC, 2007).

The interdisciplinary effort of the ACCENT-VOCBAS campaign was concentrated on studying the source strength of BVOC emitted by Mediterranean dune vegetation, and the impact of BVOC on the physical and chemical properties of the atmosphere. We characterized the weather, climate and vegetation properties at the site, providing indispensable information for the correct interpretation of the results obtained by teams participating to the campaign. We showed that the campaign was run under generally good weather conditions which made easier measurements of chemical species at ecosystem level, and inspection of their sources. Our measurements of ozone concentrations also showed a regular trend that could be interpreted on the basis of the available climatic information. The sea-land breeze circulation system was often activated, transporting precursors and products of photochemical pollution along the Thyrrenian coast of Latium. The levels of ozone advected from the sea were lower than those usually observed in the middle of the summer season when photochemical smog episodes often occur. The ozone profiles were characterized by a top-flat peak reaching the maximum values late in the afternoon.

Plants were in good physiological conditions during the campaign, and showed BVOC emission factors higher than previously reported and not constrained by stress effects. The dune vegetation was dominated by monoterpene emitting species, while isoprene emitters were scarcely represented. Monoterpene emission was already quite high during spring, when the biosynthesis of volatile isoprenoids is believed to be limited by developmental causes both in species that do not store or do store these compounds. It is suggested that during summer, the emission may further increase as temperature rises, except than in water-spending plants such as Cistus incanus.

These measurements might be of interest to construct inventories and models of isoprenoid emissions at the ecosystem level, since the Mediterranean ecosystem may escape the generalizations that volatile isoprenoids are dominated by isoprene emissions and are under strong developmental control. Finally, we set the background to upscale measurements from leaf to whole ecosystem level, and to interpret concurrent measurements of BVOC fluxes (Davison et al., 2009), ozone fluxes (Gerosa et al., 2009), particle formation.

Acknowledgements.

The field campaign was supported by the VOCBAS and ACCENT/BIAFLUX programmes. We would also like to express our gratitude to the Scientific Committee of the Presidential Estate of Castelporziano and to its staff. We are particularly grateful to GianTommaso Scarascia Mugnozza and to Aleandro Tinelli. Dr Daniele Cecca assisted during the campaign, and to Roberto Moretti, provided environmental data. Ermenegildo Magnani, Giuseppe Santarelli and Leandro Brunacci assisted with the preparation of the campign and helped with field measurements. Riccardo Valentini and Luca Belelli of the University of Tuscia are kindly acknowledged for providing environmental data from the Carboeurope-IP station. Cinzia Perrino of CNR is also acknowledged for providing data on ozone concentration at the EMEP station of Montelibretti.

Edited by: A. Arneth

References

1. Alessio, G. A., De Lillis, M., Brugnoli, E., and Lauteri, M.: Water sources and water-use efficiency in Mediterranean Coastal Dune Vegetation, Plant Biol., 6, 350–357, 2004.
2. Atkinson, R. and Arey, J..: Gas-phase tropospheric chemistry of biogenic volatile organic compounds: a review, Atmos. Environ., 37(2), 197–219, 2003.
3. Bagnouls, F. and Gaussen, H.: Les climats biologiques et leur classification, Ann. Geogr., 66, 193–220, 1957.
4. Baldi, M., Colacino, M., and Dalu, G. A.: Isola di Calore a Brezza di Mare in area urbana: il caso dell'Area Romana, 1o Italian Symposium on the Strategies and Techniques for the Monitoring of the Atmosphere, P. Ciccioli Editor, Societ`a Chimica Italiana, Rome, Italy, 268–278, 1993.
5. Baraldi, R., Rapparini, F., Loreto, F., Pietrini, F., and Di Marco G.: Emissione di composti organici volatili dalla vegetazione della macchia mediterranea, Il progetto PianosaLab, Forum, Ed. Uni. Udinese, Udine, Italy, 51, 99, 2001.
6. BEMA: An European Commission project on biogenic emissions in the Mediterranean area, edited by: Seufert, G., Atmos. Env., 31, 1–256, 1997.
7. Bernetti, G.: La vegetazione forestale del bacino del Mediterraneo e le altre vegetazioni di tipo Mediterraneo, Italia Forestale e Montana LII, 6 469–471, 1997.
8. Bertin, N. and Staudt, M.: Effect of water stress on monoterpene emissions from young potted holm oak (Quercus ilex L.) trees, Oecologia, 107, 456–462, 1996.
9. Blasi, C.: Carta del fitoclima del Lazio (scala 1:250000) – Regione Lazio, Ass. agricoltura e foreste, caccia e pesca, usi civici, Universit`a di Roma "La Sapienza," Dipartimento di Biologia Vegetale, Roma, Italia, 1993.
10. Brilli, F., Barta, C., Fortunati, A., Lerdau, M., Loreto, F., and Centritto, M.: Response of isoprene emission and carbon metabolism to drought in white poplar saplings, New Phytol., 175, 244–254, 2007.

11. Bucci, M.: Stato delle risorse idriche, in: Il sistema ambientale della tenuta presidenziale di Castelporziano: Ricerche sulla complessit`a di un ecosistema forestale costiero mediterraneo, edited by: Accademia Nazionale delle Scienze detta dei Quaranta, 2006.
12. Cantuti, V., Ciccioli, P., Cecinato, A., Brancaleoni, E., Brachetti, A., Frattoni, M., and Di Palo V.: PAN nella valle del Tevere, 1o Italian Symposium on the Strategies and Techniques for the Monitoring of the Atmosphere, P. Ciccioli Editor, Societ`a Chimica Italiana, Rome, Italy, 137–145, 1993.
13. Chameides, W. L., Lindsay, R. W., Richardson, J., and Kiang C. S.: The role of biogenic hydrocarbons in urban photochemical smog: Atlanta as a case study, Science, 241, 1473–1475, 1988.
14. Ciccioli, P., Brancaleoni, E., Frattoni, M., Cucinato, A., and Brachetti A.: Ubiquitous occurrence of semi-volatile carbonyl compounds in tropospheric samples and their possible sources, Atmos. Environ., 27(12), 1891–1901, 1993.
15. Ciccioli, P., Brancaleoni, E., and Frattoni, M.: Reactive Hydrocarbons in the atmosphere at urban and regional scale, in Reactive hydrocarbons in the atmosphere, edited by: Hewitt, N. C., Academic Press, 159–207, 1999.
16. Ciccioli, P., Brancaleoni, E. and Frattoni, M.: Sampling of atmospheric volatile organic compounds (VOCs) with sorbent tubes and their analysis by GC-MS, In Environmental Monitoring Handbook, edited by: Burden, F. R., Mc Kelvie, I., Forstner, U., and Guenther, A., Mc Graw-Hill, New York, USA, 21—85, 2002.
17. Ciccioli, P., Brancaleoni, E., Frattoni, M., Marta, S., Brachetti, A., Vitullo, M., Tirone, G., and Valentini, R.: Relaxed eddy accumulation, a new technique for measuring emission and deposition fluxes of volatile organic compounds by capillary gas chromatography and mass spectrometry, J. Chromatog., A, 985, 283–296, 2003.
18. Ciccioli, P., and Mannozzi, M.: High molecular weight carbonyls in Volatile Organic Compounds in the Atmosphere, R. Koppmann Editor, Blackwell Publishing Ltd, Oxford, 292-334, 2007.
19. Claeys, M., Wang, W., Ion, A. C., Kourtchev, I., Gelencs´erb, A., and Maenhaut, W.: Formation of secondary organic aerosols from isoprene and its gas-phase oxidation products through reaction with hydrogen peroxide, Atmos. Environ., 38, 4093–4098, 2004.
20. Davison, B., Taipale, R., Langford, B., Misztal, P., Fares, S., Matteucci, G., Loreto, F., Cape, J. N., Rinne, J., and Hewitt, C. N. : Concentrations and fluxes of biogenic volatile organic compounds above a Mediterranean macchia ecosystem in Western Italy, Biogeosciences Discuss., 6, 2183–2216, 2009.

21. Finlayson-Pitts, B. J. and Pitts Jr., J. N.: Chemistry of the lower and upper atmosphere, Theory experiments and applications, Academic Press, San Diego, USA, 1999.

22. Fischbach, J., Staudt, M., Zimmer, I., Rambal, S., and Schnitzler, J. P.: Seasonal pattern of monoterpene synthase activities in leaves of the evergreen tree Quercus ilex, Physiol. Plant., 114, 3, 354– 360, 2002.

23. Francaviglia, R., Gataleta, L., Marchionni, M., Trinchera, A., Aromolo, R., Benedetti, A., Nisini, L., Morselli, L., Brusori, B., and Olivieri, P.: Soil quality and vulnerability in a Mediterranean natural ecosystem of Central Italy, in: Il sistema ambientale della tenuta presidenziale di Castelporziano: Ricerche sulla complessit`a di un ecosistema forestale costiero mediterraneo, edted by: Accademia Nazionale delle Scienze dctta dci Quaranta, 2006.

24. Fuentes, J. D., Wang, D., and Gu, L.: Seasonal variations of isoprene emissions from a boreal forest, J. Appl. Meteorol., 38, 855–869, 1999.

25. Gariazzo, C., Silibello, C., Finardi, S., Radice, P., Piersanti, A.,Calori, G., Cucinato, A., Perrino, C., Nussio, F., Cagnoli, M., Pelliccioni, A., Gobbi, G. P., and Di Filippo, P.: A gas/aerosol air pollutants study over the urban area of Rome using a comprehensive chemical transport model, Atmos. Environ., 41, 7286–7303, 2007.

26. Georgiadis, T., Giovanelli, G., and Fortezza, F.: Vertical layering of photochemical ozone during land-sea breeze transport, Il Nuovo Cimento, 17, 371–375, 1994.

27. Gerosa, G., Finco, A., Mereu, S., Marzuoli, R., and Ballarin-Denti, A.: Interactions among vegetation and ozone, water and nitrogen fluxes in a coastal Mediterranean maquis ecosystem. Biogeosciences Discuss., 6, 1453–1495, 2009.

28. Guenther, A. B., Zimmerman, P. R., Harley, P. C., Monson, R. K., and Fall, R.: Isoprene and monoterpene emission rate variability: model evaluations and sensitivity analyses, J. Geophys. Res., 98D, 12609–12617, 1993.

29. Guenther, A., Hewitt, C. N., Erickson, D., Fall, R., Geron, C., Graedel, T., Harley, P., Klinger, L., Lerdau, M., McKay, W. A., Pierce, T., Scholes, B., Steinbrecher, R., Tallamraju, R., Taylor, J., and Zimmerman, P.: A global model of natural volatile organic compounds emissions, J. Geophys. Res., 100, 8873–8892, 1995.

30. Guenther, A., Karl, T., Harley, P., Wiedinmyer, C., Palmer, P. I., and Geron, C.: Estimates of global terrestrial isoprene emissions using MEGAN (Model of Emissions of Gases and Aerosols from Nature), Atmos. Chem. Phys., 6, 3181–3210, 2006, http://www.atmos-chem-phys.net/6/3181/2006/.

31. Hansen, U., Van Eijk, J., Bertin, N., Staudt, M., Kotzias, D., Seufert, G., Fugit, J. L., Torres, L., Cecinato, A., Brancaleoni, E., Ciccioli, P., and Bomboi, T.: Biogenic emissions and CO2 gas exchange investigated on four Mediterranean shrubs, Atmos. Environ., 31, 157–167, 1997.

32. Heil, M. and Silva Bueno, J. C.: Within-plant signaling by volatiles leads to induction and priming of an indirect plant defense in nature, PNAS, 104(13), 5467–5472, 2007.

33. Kalabokas, P., Bartzis, J.C., Bomboi, T., Ciccioli, P., Cieslik, S., Dlugi, R., Foster, P., Kotzias, D., and Steinbrecher, R.: Ambient atmospheric trace gas concentrations and meteorological parameters during the first BEMA measuring campaign on May 1994 at Castelporziano, Italy, Atmos. Environ., 31, 67–77, 1997.

34. Karl, T.G., Spirig, C., Rinne, J., Stroud, C., Prevost, P., Greenberg, J., Fall, R., and Guenther, A.: Virtual disjunct eddy covariance measurements of organic compound fluxes from a subalpine forest using proton transfer reaction mass spectrometry, Atmos. Chem. Phys., 2, 279–291, 2002, http://www.atmos-chem-phys.net/2/279/2002/.

35. Kesselmeier, J. and Staudt, M.: Biogenic Volatile Organic Compound (VOC): An overview on emission, physiology and ecology, J. Atm. Chem., 33, 23–88, 1999.

36. Iinuma, Y., M¨uller, C., Berndt, T., B¨oge, O., Claeys, M., and Herrmann, H.: Evidence for the existence of organosulfates from beta-pinene ozonolysis in ambient secondary organic aerosol, Environ. Sci. Technol., 19, 6678–6683, 2007.

37. INFC: Guida alla classificazione della vegetazione forestale. Inventario Nazionale delle Foreste e dei Serbatoi Forestali di Carbonio. MiPAF – Direzione Generale Risorse Forestali Montane Idriche Corpo Forestale dello Stato, CRA-ISAFA, Trento, 2003.

38. IPCC: Climate change 2007: contribution of the three Working Groups to the fourth assessment report of the Intergovernmental Panel on Climate change, Cambridge University Press, 2007.

39. Lammel, G. and Cape, G. N.: Nitrous acid and nitrates in the atmosphere, Chem. Soc. Rev., 25(5), 361–369, 1996.

40. Lindinger, W., Hansel, A., and Jordan, A.: On-line monitoring of volatile organic compounds at pptv levels by means of Proton-Transfer-Reaction Mass Spectrometry (PTR-MS), Medical applications, food control and environmental research, Int. J. Mass Spectrom. Ion Proc., 173, 191–241, 1998.

41. Loreto, F., Ciccioli, P., Brancaleoni, E., Cecinato, A., Frattoni, M., and Sharkey, T.: Different sources of reduced carbon contribute to form three classes of terpenoid emitted by Quercus ilex L. Leaves, Proc. Natl. Acad. Sci., USA, 93, 9966–9969, 1996.

42. Loreto, F. and Delfine, S.: Emission of isoprene from salt-stressed Eucalyptus globulus leaves, Plant Physiol., 123, 1605–1610, 2000.

43. Loreto, F., Nascetti, P., Graverini, A., and Mannozzi, M.: Emission and content of monoterpenes in intact and wounded needles of the Mediterranean pine Pinus pinea, Funct. Ecol., 14, 589–595, 2000.

44. Loreto, F., Ferranti, F., Mannozzi, M., Maris, C., Nascetti, P., and Pasqualini, S.: Ozone quenching properties of isoprene and its antioxidant role in plants, Plant Physiol., 126, 993–1000, 2001a.

45. Loreto, F. and Velikova, V.: Isoprene produced by leaves protects the photosynthetic apparatus against ozone damage, quenches ozone products, and reduces lipid peroxidation of cellular membranes, Plant Physiol., 127(4), 1781–1787, 2001b.

46. Loreto, F., Fischbach, R. J., Schnitzler, J. P., Ciccioli, P., Brancaleoni, E., Calfapietra, C., and Seufert, G.: Monoterpene emission and monoterpene synthase activities in the Mediterranean evergreen oak Quercus ilex L. grown at elevated CO2 concentrations, Global Change Biol., 7, 709–717, 2001b.

47. Loreto, F. and Fares, S.: Is ozone flux inside leaves only a damage indicator? Clues from volatile isoprenoid studies, Plant Physiol.., 143, 1096–1100, 2007.

48. Loreto, F., Barta, C., Brilli, F., and Nogu`es, I.: On the induction of volatile organic compound emissions by plants as consequence of wounding or fluctuations of light and temperature, Plant, Cell Environ., 29, 1820–1828, 2006.

49. Llusi`a, J. and Pe˜nuelas, J.: Changes in terpene content and emission in potted Mediterranean woody plants under severe drought, Can. J. Bot., 76, 8, 1366–1373, 1998.

50. Llusi`a, J. and Pe˜nuelas, J.: Seasonal patterns of terpene content and emission from seven Mediterranean woody species in field conditions, Am. J. Bot., 87, 133–140, 2000.

51. Manes, F., Grignetti, A., Tinelli, A., Lenz, R., and Ciccioli, P.: General features of the Castelporziano test site, Atmos. Environ., 31, 19–25, 1997a.

52. Manes, F., Seufert, G., and Vitale, M.: Ecophysiological studies of Mediterranean plant species at the Castelporziano estate, Atmos. Environ., 31, 51–60, 1997b.

53. Mastrantonio, G., Viola, A.P., Argentini, S., Hocco, C., Giannini, L., Rossini, L., Abbate, G., Ocone, B., and Casonato, M.: Observation of sea breeze effects in Rome, Bound.-Lay. Meteorol., 71, 67–80, 1994.

54. Mereu, S., Salvatori, E., Fusaro, L., Gerosa, G., Muys, B., and Manes, F.: A whole plant approach to evaluate the water use of mediterranean maquis species in a coastal dune ecosystem, Biogeosciences Discuss., 6, 1713–1746, 2009.

55. Mooney, H. A. and Dunn, E. L.: Photosynthetic Systems of Mediterranean-Climate Shrubs and Trees of California and Chile, The American Naturalist, 104, 447–453, 1970.

56. Maun, M.A.: Adaptations enhancing survival and establishment of seedlings on coastal sand dunes, Plant ecology, 111, Vol. I, 15735052, 1994.

57. Millan-Millan, M., Salvador, R., Mantella, E., and Artinano, A.: Meteorology of photochemical air pollution in Southern Europe: experimental results from EC research projects, Atmos. Environ., 30, 2583–2593, 1998.

58. Mooney, H. A. and Dunn, E. L.: Convergent evolution of Mediterranean-climate evregreen sclerophyllous shrubs, Evolution, 24, 292–303, 1970.

59. Niinemets, U., Loreto, F., and Reichstein, M.: Physiological and physico-chemical controls on foliar volatile organic compound emissions, Trends Plant Sci., 9, 180–186, 2004.

60. Mitrakos, K.: A theory for Mediterranean plant life, Acta Oecol. Plant., 1, 245–252, 1980.

61. Ormeno, E., Fernandez, C., and Mevy, J. P.: Plant coexistence alters terpene emission and content of Mediterranean species, Phytochemistry, 68, 6, 840–852, 2007.

62. Owen, S., Boissard, S., Street, R. A., Duckham, S. C., Csiky, O., and Hewitt, N.: Screening of 18 mediterranean plant species for volatile organic compound emission, Atmos. Environ., 31, 101– 117, 1997.

63. Nali, C., Paoletti, E., Marabottini, R., Della Rocca, G., Lorenzini, G., Paolacci, A., Ciaffi, M., and Badiani, M.: Ecophysiological and biochemical strategies of response to ozone in Mediterranean evergreen broadleaf species, Atmos. Environ., 38, 15, 2247– 2257, 2004.

64. Pegoraro, E., Rey, A., Bobich, E. G., Barron-Gafford, G., Grieve, K. A., Malhi, Y., and Murthy, R.: Effect of elevated CO2 concentration and vapour pressure deficit on isoprene emission from leaves of Populus deltoides during drought, Funct. Plant Biol., 31, 1137–1147, 2004.

65. Pignatti, S., Bianco, P. M., Tescarollo, P., and Scarascia Mugnozza, G. T.: La vegetazione della tenuta di Castelporziano, In: Il sistema ambientale della tenuta di

presidenziale di Castelporziano, Accademia Nazionale dei Quaranta, "scritti e documenti" XXVI, Rome, Vol. II, 441–709, 2001.

66. Pinzari, F., Trinchera, A., Benedetti, A., and Sequi, P.: Use of biochemical indexes in the Mediterranean environment: comparison among soils under different forest vegetation, J. Microbiol. Meth., 36, 21–28, 1999.
67. Pio, C. A., Nunes, T. V., and Brito, S.: Volatile hydrocarbon emissions from common and native species of vegetation in Portugal. Air Pollution Research Report 47, Joint Workshop CEC/BIATEX of EUROTRAC, General assessment of biogenic emissions and deposition of nitrogen compounds, sulphur compounds and oxidants in Europe, 291–298, 1993.
68. Reichstein, M., Tenhunen, J. D., Roupsard, O., Orcival, J. M., Rambal, S., Miglietta, F., Peressotti, A., Pecchiari, M., Tirone, G., and Valentini, R.: Severe drought effects on Ecosystem CO2 and H2O fluxes at three Mediterranean evergreen sites: revision of current hypothesis?, Global Change Biol., 8, 999–1017, 2002.
69. Rinne, J., Guenther, A., Warneke, C., de Gouw, J. A., and Luxembourg, S. L.: Disjunct eddy covariance technique for trace gas flux measurements, Geophys. Res. Lett., 28, 3139–3142, 2001.
70. Rosenstiel, T. N., Potosnak, M. J., Griffin, K. L., Fall, R., and Monson, R. K.: Increased CO2 uncouples growth from isoprene emission in an agriforest ecosystem, Nature, 421, 256–259, 2003.
71. Sabillon, D. and Cremades, L. V.: Diurnal and seasonal variation of monoterpene emission rates for two typical Mediterranean species (Pinus pinea and Quercus ilex) from field measurements—relationship with temperature and PAR, Atmos. Environ., 35, 26, 4419–4431, 2001.
72. Salido, S., Altarejos, J., Nogueras, M., S′anchez, A., Pannecouque, C., Witvrouw, P., and De Clercq, E.: Chemical studies of essential oils of Juniperus oxycedrus ssp. Badia, J. Ethnopharmacol., 81, 129–134, 2002.
73. Sharkey, T. D., and Yeh, S.: Isoprene emission from plants, Annual Rev. Plant Phys. Plant Mol. Biol., 52, 407–436, 2001.
74. Sharkey, T. D. and Loreto, F.: Water stress, temperature, and light effects on the capacity of isoprene emission and photosynthesis of kudzu laves, Oecologia, 95, 328–333, 1993.
75. Sharkey, T. D., Singsaas, E. L., Lerdau, M. T., and Geron, C. D.: Weather effect on isoprene emission capacity and applications in emission algorithms, Ecol. Appl., 9, 1132–1137, 1999.

76. Staudt, M., Bertin, N., Hansen, U., Seufert, G., Ciccioli, P., Foster, P., Frenzel, B., and Fugit, J. L.: Seasonal and diurnal patterns of monoterpene emissions from Pinus Pinea L., Atmos. Environ., 32, 145–156, 1997.

77. Staudt, M., Bertin, N., Frenzel, B., and Seufert, G.: Seasonal variation in amount and composition of monoterpenes emitted by young Pinus pinea trees – Implications for emission modelling, J. Atmos. Chem., 35, 77–99, 2000.

78. Thompson, J. D.: Plant Evolution in the Mediterranean, Oxford University Press, ISBN 0198515332, 2005.

79. Trinchera, A., Pinzari, F., Fiorelli, F., Marchionni, M., and Benedetti, A. Duna antica e duna recente: due ecosistemi a confronto, Bollettino della Societ`a italiana di Scienze del Suolo, 48, 399–416, 1998.

80. UNECE: Revised manual on methodologies and criteria for mapping critical levels/loads and geographical areas where they are exceeded, www.icpmapping.org, 2004.

81. van der Meulen, F. and Salman, A. H. P. M.: Management of Mediterranean coastal dunes, Ocean Coast. Manage., 30, 177– 195, 1996.

82. Verheggen, B., Weingartner, E., Baltensperger, U., Metzger, A., Duplissy, J., Dommen, J., and Pr′ot, A. S. H.: Aerosol Formation ev^from Isoprene: Determination of Particle Nucleation and Growth Rates., Nucl. Atmos. Aerosol., 989–993, doi:10.1007/978-14020-6475-3, 2007.

83. Vitale, M., Salvatori, E., Loreto, F., Fares, S., and Manes, F.: Physiological responses of Quercus ilex leaves to water stress and acute ozone exposure under controlled conditions, Water Air Soil Poll., 189(1–4), 113–125, 2008.

84. Vitale, M., Matteucci, G., Fares, S., and Davison, B.: A process-based model to estimate gas exchange and monoterpene emission rates in the mediterranean maquis – comparisons between modelled and measured fluxes at different scales, Biogeosciences Discuss., 6, 1747-1776, 2009.

85. Vickers, C. E., Possel, M., Cojocariu, C. I., , Velokova, V. B., , Wornkitkul, J. L., Ryan, A., Mullineaux, P. M, and Hewitt, N.: Isoprene synthesis protects transgenic tobacco plants from oxidative stress, Plant Cell Environ., 32, 520–531, 2009.

86. Wiberley, A. E., Linskey, A. R., Falbel, T. G., and Sharkey, T. D.: Development of the capacity for isoprene emission in kudzu, Plant Cell Environ., 28, 898–905, 2005.

Molecular Adaptation During Adaptive Radiation in the Hawaiian Endemic Genus Schiedea

Maxim V. Kapralov and Dmitry A. Filatov

ABSTRACT

Background

"Explosive" adaptive radiations on islands remain one of the most puzzling evolutionary phenomena. The rate of phenotypic and ecological adaptations is extremely fast during such events, suggesting that many genes may be under fairly strong selection. However, no evidence for adaptation at the level of protein coding genes was found, so it has been suggested that selection may work mainly on regulatory elements. Here we report the first evidence that positive selection does operate at the level of protein coding genes during rapid adaptive radiations. We studied molecular adaptation in Hawaiian

endemic plant genus Schiedea (Caryophyllaceae), which includes closely related species with a striking range of morphological and ecological forms, varying from rainforest vines to woody shrubs growing in desert-like conditions on cliffs. Given the remarkable difference in photosynthetic performance between Schiedea species from different habitats, we focused on the "photosynthetic" Rubisco enzyme, the efficiency of which is known to be a limiting step in plant photosynthesis.

Results

We demonstrate that the chloroplast rbcL gene, encoding the large subunit of Rubisco enzyme, evolved under strong positive selection in Schiedea. Adaptive amino acid changes occurred in functionally important regions of Rubisco that interact with Rubisco activase, a chaperone which promotes and maintains the catalytic activity of Rubisco. Interestingly, positive selection acting on the rbcL might have caused favorable cytotypes to spread across several Schiedea species.

Significance

We report the first evidence for adaptive changes at the DNA and protein sequence level that may have been associated with the evolution of photosynthetic performance and colonization of new habitats during a recent adaptive radiation in an island plant genus. This illustrates how small changes at the molecular level may change ecological species performance and helps us to understand the molecular bases of extremely fast rate of adaptation during island adaptive radiations.

Introduction

The most dramatic "bursts" of adaptive radiation often occur within confined geographical regions (e.g. oceanic islands or inland freshwater lakes; e.g. [1]). Although island adaptive radiations may be viewed as extreme examples of evolutionary diversification, it is thought that major adaptive radiations in the history of our planet have been following the same evolutionary processes as island endemic radiations. Thus, islands may be viewed as evolutionary laboratories one can use to understand general processes of adaptation and speciation [1], [2].

"Explosive" island adaptive radiations are accompanied by tremendous phenotypical and ecological changes that suggest many genes might be under fairly strong positive selection. However, we do not know how natural selection works at the molecular level during adaptive radiation events. Island habitats are always limited in area, so island populations are limited in size. In addition, many island species are thought to evolve via colonization of new islands or habitats ("island hopping speciation," [3]). Such colonization events should lead to a drastic reduction in population size. As the efficacy of selection is proportional to the product of the selective coefficient and the effective population size [4], the relatively small effective population size of island species should result in a reduced efficacy of natural selection. In small populations (e.g. in endemic island species), the dynamics of non-synonymous mutations is dominated by drift and the fixation probabilities of deleterious and advantageous mutations are expected to be approximately equal [5]. Few studies have investigated the action of selection at the molecular level during island adaptive radiations [6], [7]. These studies indicated some increase in non-synonymous (dN) to synonymous (dS) substitution rates in the Hawaiian silversword alliance, which may reflect slight relaxation of purifying selection in small island populations, but no convincing evidence of positive selection has been reported.

The small size of island populations may also limit the genetic variability required for natural selection to work, so the fast rate of phenotypic diversification on islands is quite surprising. Interspecific hybridization may be a possible source of additional genetic variation within species [2], [8], [9]. Closely related species are often cross-compatible and there are numerous examples of interspecific hybridization in plants and animals [10]. Even with low rates of introgression positively selected alleles can spread across several species [11]. Occasional events of interspecific hybridization allow adaptive radiations to be considered as metapopulations, where adaptive mutations may spread across several species significantly accelerating the adaptation process [12]. However, it is not known how common such sharing of adaptive mutations by several species might be.

In this paper we report the analysis of positive selection at the molecular level and the spread of adaptive alleles across several species in Hawaiian endemic plant genus Schiedea (Caryophyllaceae), which represents one of the largest plant adaptive radiations on Hawaii. Schiedea comprises 32 living (and at least two extinct) species adapted to a wide range of habitats from wet rainforest to dry desert-like conditions of coastal cliffs [13]. Among Schiedea's most prominent evolutionary transitions have been remarkable changes in its growth habit, ranging from rainforest vines and perennial herbs through mesic and dry forest subshrubs and shrubs to cliff-dwelling shrubs [13]. The latter are particularly notable for a lineage within the Caryophyllaceae family which contains mainly herbaceous annuals and perennials.

Schiedea species from contrasting environments (e.g. rainforest vs. coastal cliffs) are dramatically different from each other in many physiological traits [13]. In particular, there are substantial differences in photosynthetic performance in Schiedea, suggesting that some of the protein coding genes involved in photosynthesis could be under positive selection. This motivated us to choose two chloroplast "photosynthetic" genes, psbA and rbcL, for phylogeny-based analysis of positive selection. The first gene, psbA, encodes photosystem II reaction center protein D1. Photosystem II is the first link in the chain of photosynthesis, it captures photons and uses the energy to extract electrons from water molecules [14]. The second gene, rbcL, encodes large subunits of Ribulose-1, 5-bisphospate (RuBP) carboxylase/oxigenase (Rubisco; EC 4.1.1.39) which catalyzes the first step in net photosynthetic CO2 assimilation and photorespiratory carbon oxidation [15]. "The most abundant protein in the world," Rubisco comprises about 40–50% of all soluble proteins in green plant tissues and is responsible for almost all carbon fixation on Earth. Despite its critical importance for life on our planet, this protein is notoriously inefficient in its function, creating a bottleneck, which limits plant growth [15]. This makes rbcL a likely target of positive selection, as any improvements in its function may drastically change plant growth rate. For comparison we also studied a non-photosyntetic chloroplast gene, matK, that encodes a protein of unknown function which is hypothesized to be involved in splicing in the chloroplast genome [16], [17].

Below we demonstrate that one of the studied photosynthetic genes, rbcL, evolved under positive selection during adaptive radiation in Schiedea. The differences in amino acid sequence among different Schiedea species could possibly account for the observed differences in photosynthetic performance and may have helped the genus to colonise a new habitat–dry sunny slopes and cliffs. Interestingly, the positive selection at Schiedea rbcL may have caused adaptive chloroplast haplotypes to spread across several Schiedea species, which are known to occasionally form hybrids in the wild [13]. This supports the view that sharing of adaptive mutations by several species may play a significant role in plant adaptive evolution [2], [9], [12].

Results

Positive Selection in Schiedea rbcL

Phylogenetic maximum likelihood analysis of selection at the molecular level assumes that the phylogeny of an analyzed gene is correct [18]. The published phylogeny of the genus Schiedea is based on morphology and ITS and ETS sequences and is relatively well established [13]. However, individual genes may have gene trees that differ from the species tree due to horizontal gene transfer and lineage sorting [2], [19], [20]. Thus, in order to conduct phylogenetic maximum likelihood analysis of selection in chloroplast genes it was essential to construct a robust gene tree of the Schiedea chloroplast DNA. For this purpose we sequenced fragments of psbA, rbcL and matK protein coding genes, as well as noncoding trnL intron, psbA-trnK and trnL-trnF intergenic spacers and trnS-trnG region (in total, 5.3 kb per individual, Table 1) from all 27 Schiedea species used in this study (Fig. 1). As expected for a non-recombining chloroplast genome, the phylogenies based on individual chloroplast regions were consistent with each other, allowing us to concatenate the datasets, which resulted in a fairly well resolved phylogeny with high bootstrap support (25 haplotypes for concatenated dataset, Fig. 1). Interestingly, the Schiedea chloroplast gene tree (Fig. 1) substantially differed from the accepted ITS+ETS+morphology based phylogeny of the genus [13]. The plausible explanation for the observed discordance of the phylogenies as well as for rather "shallow" clades III and IV in the cpDNA tree (Fig. 1) could be the transition of cytotypes via interspecific hybridization and further fixation of favorable haplotypes within multiple species. Indeed, all the Schiedea species from cpDNA clade III live on the same island, Kaua'i, while the species from cpDNA clade IV inhabit several younger islands (O'ahu, Maui, Moloka'i, Lana'i, Hawai'i) that were connected to each other at various points in history of the archipelago. Thus, it seems likely that the cpDNA clades III and IV represent chloroplast hapoltypes that have spread across several species in Kaua'i, or younger islands. Strong positive selection in the Schiedea photosynthetic gene rbcL described below may have caused the cytotypes to spread across several species.

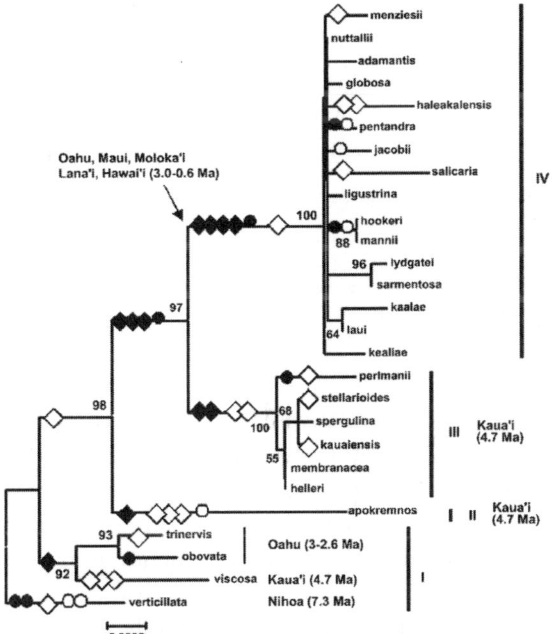

Figure 1. Neighbor joining tree of 27 Schiedea species constructed using three coding and four non-coding cpDNA regions. Numbers above branches are bootstrap support values (%). Non-synonymous (diamonds) and synonymous (circles) substitutions are shown for rbcL (black filled symbols) and matK (white symbols). The four clades are marked by Roman numbers.

Table 1. Investigated Chloroplast DNA Regions of 27 Schiedea Species

Region	Length (indel length), bp	Haplotypes	Mutations (singletons)
matK	1077 (9)	17	25 (19)
psbA	723 (0)	8	8 (5)
rbcL	1362 (0)	9	18 (5)
psbA-trnK intergenic spacer	545 (52)	14	16 (13)
trnL gene, intron	611 (24)	9	8 (4)
trnL-trnF intergenic spacer	404 (0)	9	15 (10)
trnS-trnG intergenic spacer+*trnG* gene	638 (0)	16	28 (22)
Concatenated dataset	5360 (85)	25	118 (78)

One out of the three investigated protein coding genes–psbA–appeared to be under strong purifying selection in Schiedea, showing no non-synonymous and only eight synonymous substitutions (Table 2). Moreover, when we compared Schiedea's psbA with homologs from Silene latifolia (Caryophyllaceae, Caryophyllales; GenBank AB189069) and Chenopodium rubrum (Chenopodiaceae, Caryophyllales; GenBank Y14732) all 47 observed mutations again appeared to be synonymous.

Table 2. Non-synonymous and Synonymous Substitutions Rates in Three Investigated Chloroplast Genes of 27 Schiedea Species

Gene	Codons	cDNA % [a]	*N* [b]	*S* [c]	d*N* [d]	d*S* [e]	d*N*/d*S* [f]
matK	356	70.8	19 (14)	6 (5)	0.02317	0.03234	0.72
psbA	241	68.3	0 (0)	8 (5)	0.00000	0.05295	0.00
rbcL	454	95.4	10 (0)	8 (5)	0.01386	0.02789	0.50

[a]Proportion of the whole gene length based on coding sequences annotated in GenBank.
[b]The number of non-synonymous mutations (singletons).
[c]The number of synonymous mutations (singletons).
[d]Non-synonymous divergence estimated by PAML.
[e]Synonymous divergence estimated by PAML.
[f]The ratio of non-synonymous to synonymous substitutions rates.

Both matK and rbcL showed relatively high dN/dS averaged across the whole Schiedea phylogeny–0.72 and 0.50 respectively (Table 2). However, the distribution of non-synonymous substitutions on the phylogenetic tree for rbcL and all other regions was remarkably different (Table 2). While all ten non-synonymous substitutions in rbcL occurred in the internal branches, non-synonymous substitutions in matK as well as synonymous mutations in rbcL, matK and psbA and mutations in all non-coding regions appeared mainly in the terminal branches (Table 2; Fig 1). 2×2 contingency tests demonstrate that an unusually large number of amino acid replacements in the Schiedea rbcL occurred on the internal branches of the tree in comparison to synonymous mutations in rbcL as well as to non-synonymous and synonymous mutations in matK and psbA and mutations in non-coding regions (Table 3).

Table 3. Comparisons of Mutation Numbers on Internal and External Branches of the Schiedea Chloroplast Gene Tree

Gene				*RbcL*		*matK*		*psbA*	*psbA-trnK*	*trnL*	*trnL-trnF*	*trnS-trnG*
	N/S [a]			*N*	*S*	*N*	*S*	*S*	*S*	*S*	*S*	*S*
		IntM [b]		**10**	**3**	**5**	**1**	**3**	**3**	**4**	**5**	**6**
			ExtM [b]	**0**	**5**	**14**	**5**	**5**	**13**	**4**	**10**	**22**
rbcL	*N*	**10**	**0**	-	8.65**	14.25***	12.12***	8.65**	16.25***	6.43*	11.11***	18.66***
rbcL	*S*	**3**	**5**		-	0.34	0.73	0.00	0.08	0.25	0.04	0.86
matK	*N*	**5**	**14**			-	0.23	0.34	0.28	1.42	0.20	0.15
matK	*S*	**1**	**5**				-	0.73	0.01	1.66	0.58	0.07
psbA	*S*	**3**	**5**					-	0.08	0.25	0.04	0.86
psbA-trnK	*S*	**3**	**13**						-	2.52	0.86	0.05
trnL	*S*	**4**	**4**							-	0.61	0.05
trnL-trnF	*S*	**5**	**10**								-	0.73
trnS-trnG	*S*	**6**	**22**									-

Mutation numbers for internal and external branches are given in bold; χ^2 values for 2×2 contingency tests of the number of mutations on the internal vers. external branches are given at the intersection of the column and row for the respective gene and mutation type.

[a] *S*–synonymous or non-coding mutations; *N*–non-synonymous mutations.

[b] IntM–mutations number on internal branches; ExtM–mutations number on external branches.

P values: * $P<0.02$; ** $P<0.01$; *** $P<0.001$.

To test for the presence of codons under positive selection in matK and rbcL we used likelihood ratio tests (LRTs) to compare the nested models allowing for variation in dN/dS ratio across codons [21]. In this analysis we compared the following pairs of models implemented in codeml program [18]: M1a/M2a [22], M7/M8 [21] and M8a/M8 [23]. Model M1a allows two classes of sites: one class with dN/dS varying freely between 0 and 1, and another one with dN/dS = 1 [22]. Model 2a has an additional class of sites, which can accommodate codons with dN/dS>1 [21]. The model M2a fits rbcL data significantly better than model M1a ($\chi 2$ = 12.88, P = 0.0016, df = 2), while there is no significant difference in fit of the two models to matK data ($\chi 2$ = 1.82, P = 0.4025, df = 2). In another nested pair of models M7 assumes that all codons have dN/dS distributed according to discrete beta distribution between 0 and 1, while model M8 allows for an additional class of codons with dN/dS>1 [21]. The comparison of these two models in a LRT is a test for the presence of a class of codons with dN/dS>1 [21]. The model M8 fits rbcL data significantly better than model M7 ($\chi 2$ = 13.25, P = 0.0013, df = 2), while there is no significant difference in fit of the two models for matK data ($\chi 2$ = 1.87, P = 0.3926, df = 2). Under the model M8 about 4% of codons in rbcL fall into the positively selected class, which had dN/dS = 13.92. A more stringent test for positive selection compares models M8 and M8a, which is the same as model M8, but the class of codons with dN/dS>1 in M8 is forced to have dN/dS = 1 in M8a. This LRT specifically tests whether the dN/dS for codons falling into this class is significantly larger than unity [23]. While in M8-M8a comparison there was no significant difference in fit for matK data ($\chi 2$ = 1.82,

P = 0.1773, df = 1), application of this test to Schiedea rbcL demonstrated that this gene does have codons with dN/dS significantly larger than unity ($\chi2$ = 12.87, P = 0.0003, df = 1), providing strong evidence for positive selection.

Amino Acid Substitutions in Schiedea RbcL

The summary of the amino acid substitutions in Schiedea rbcL and their possible effects is presented in Table 4. Throughout the text amino acid positions are numbered according to spinach rbcL for which protein cristal structure is available [24]. Nine out of ten amino acid substitutions occurred on the branches of the phylogeny predating the split of the clades III and IV or leading to the clades III and IV (Fig. 1).

Table 4. Physical Properties of Amino Acid Substitutions in Schiedea's RbcL

Codon No[a]	Amino acid changes	Type of changes[b]	ΔH[c]	ΔP[d]	ΔV[e]	IS[f]	ΔG[g], kcal/mol	SA[h], %	Location of residue[j]
Branch within the clade I									
145	Ile⇒Val	HN⇒HN	−0.3	0.7	−19	D	−0.51	0.7	α-helix D
Branches leading to the clades II & III (two independent mutations)									
86	His⇒Tyr	UB⇒UR	1.9	−4.2	23	D	−1.62	47.3	β-strand C
Branch leading to the clades III & IV									
230	Ala⇒Thr	HN⇒UP	−2.5	0.5	26	S	0.70	58.1	α-helix 2
326	Ile⇒Val	HN⇒HN	−0.3	0.7	−19	S	0.30	0.0	β-strand 6
449	Cys⇒Ser	HP⇒UP	−3.3	3.7	−13	n.a.	n.a.	20.9	α-helix G
Branch leading to the clade III									
23	Thr⇒Asn	UP⇒UP	−2.8	3.0	3	D	−0.74	28.1	N-terminal-domain above β-strand A
Branch leading to the clade IV									
354	Thr⇒Ile	UP⇒HN	5.2	−3.4	31	D	−1.52	3.6	β-strand G
363	Tyr⇒Phe	UR⇒HR	4.1	−1.0	−6	D	−2.92	8.0	below β-strand H
367	Ser⇒Pro	UP⇒HN	−0.8	−1.2	17	S	3.68	32.4	β-strand H
470	Gln⇒Glu	UP⇒UA	0.0	1.8	−5	n.a.	n.a.	81.0	the carboxyl terminus

[a]Codon numbering is based on the sequence of the spinach [24].
[b]Side chain type changes. Types abbreviations: A–acidic (negatively charged); B–basic (positively charged); H–hydrophobic; N–nonpolar aliphatic; P–polar uncharged; R–aromatic; U–hydrophilic [42].
[c]Hydropathicity difference [43].
[d]Polarity difference [44].
[e]Van der Waals volume difference [42].
[f]IS–impact on overall stability: D–destabilising; S–stabilising [39].
[g]Predicted free energy changes [39].
[h]Solvent accessibility [39].
[j][15,26].

For four out of ten amino acid substitutions in Schiedea rbcL (residues 86, 230, 326 and 449) a Bayesian posterior probability of positive selection larger than 0.99 was shown by the Bayes Empirical Bayes analysis implemented in the PAML package [25]. Three of these residues (positions 230, 326 and 86) reside in regions that play key role in the functioning of Rubisco enzyme.

Replacement Ala230⇒Thr230 occurred on the branch leading to the clades III and IV of the Schiedea chloroplast gene tree (Fig. 1). Residue 230 interacts with the βA-βB loop of small subunit [26]. This residue 230 is highly solvent accessible (about 60% of total surface area; Table 4) and has a hydrogen bond with residue 10 of the small subunit of Rubisco. Replacement Ala230⇒Thr230 significantly decreases hydrophobicity of the residue that has a stabilizing effect in this position (Table 4).

Replacement Ile326⇒Val326 happened on the branch leading to the clades III and IV of the gene tree (Fig. 1). Residue 326 has six internal contacts and located inside of the protein molecule in the β-strand flanking loop 6, a flexible element that folds over substrate during catalysis and plays a key role in discriminating between CO2 and O2 in competing RuBP carboxylation and oxygenation reactions of Rubisco [15], [26]. Although Ile and Val have similar properties, Val is smaller and such replacement should increase the overall molecule stability (Table 4).

Interestingly, replacement His86⇒Tyr86 happened twice independently in Schiedea phylogeny–on the branches leading to the clade II and to the clade III (Fig. 1). Residue 86 is highly solvent accessible (about 50% of total surface area; Table 4) and may be a part of the Rubisco activase recognition region located in the N-terminal domain [15], [27]. The activase recognition region provides a physical contact between Rubisco and Rubisco activase, an ATP-dependent enzyme that releases tight-binding sugar phosphates from the Rubisco active site and facilitates conversion of Rubisco from the closed to the open conformation. Rubisco activase plays a vital role in the response of photosynthesis to temperature [27]. Properties of His and Tyr are very different: after His86⇒Tyr86 replacement hydrophobicy and volume of residue increased and polarity decreased dramatically making the molecule more tight while less soluble (Table 4). Although this is predicted to decrease stability of the molecule (Table 4), this analysis was done without taking interaction with Rubisco activase into account (for which no protein structure is available). Decreased polarity and increased hydrophobicy of the residue interacting with Rubisco activase may result in tighter binding. Thus, His86⇒Tyr86 replacement is likely to affect physical interaction of Rubisco with Rubisco activase. Although residue 86 is one of the most variable positions in the large subunit (up to 11 different amino acids across the 499 plant species; [26]), His86⇒Tyr86 replacement is very rare, considering that about 76% from 491 flowering species have His86, but only two species (<1%) have Tyr86 [15].

Furthermore, residue 86 is not the only one of Schiedea rbcL replacements that may be involved into Rubisco-Rubisco activase interactions. The critical residues for these interactions identified so far are immediately adjacent to the active site (Fig. 2, [27]), as well as residues in strand G (particularly strand 6)–strand H region and carboxyl terminus [15]. Based on published data [15], [27] and structural modelling we found that apart from residue 86 six other residues out of ten replacements in Schiedea rbcL could be involved in Rubisco-Rubisco activase interactions (residues 23 and 326 are close to the active site; residues 354, 363,

367 belong to strand G–strand H region and residue 470 is close to the carboxyl terminus). Six out of ten detected amino acid mutations while residing far from each other in the amino acid sequence appeared relatively close in the tertiary structure and could potentially influence each other: in the tertiary structure the average distance between the residues 86, 145, 326, 354, 363 and 367 is 15.1Å; between the residues 145, 326, 354 and 363 is 12.2Å; and between the residues 145, 354 and 363 is 9.2Å). The proximity of these replacements in the tertiary protein structure suggests that several mutations may have a cumulative effect that affects overall properties of Rubisco activase interaction region in Schiedea's Rubisco (Fig. 2).

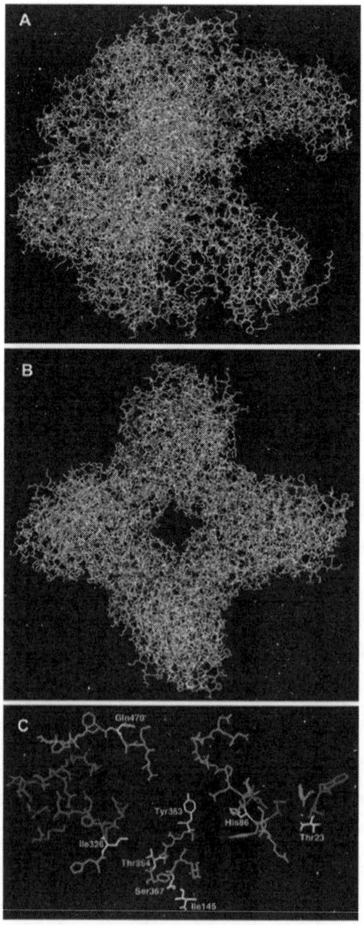

Figure 2. The structure of Rubisco enzyme in two projections (A, B) and (C) the residues that belong to active site (coloured red) and that are involved in interactions with Rubisco activase (coloured green; after [15], [24], [26], [27]). Eight residues replaced in Schiedea (positions 23, 86, 145, 326, 354, 363, 367, 470) are coloured yellow.

Discussion

We demonstrated that the rbcL gene, encoding the large subunit of Rubisco enzyme, might have been under strong positive selection during recent adaptive radiation in Hawaiian Schiedea. Rubisco catalyzes the first step in net photosynthetic CO_2 assimilation and photorespiratory carbon oxidation. The enzyme is subject to competitive inhibition by O_2, inactivation by loss of carbamylation, and dead-end inhibition by RuBP, that makes Rubisco inefficient as a catalyst for the carboxylation of RuBP and limiting for photosynthesis and plant growth [15]. Thus, even small improvements in efficiency of this enzyme may provide significant physiological advantage.

In land plants Rubisco is composed of eight large subunits (LSUs) encoded by the chloroplast rbcL gene and eight small subunits (SSUs) encoded by a family of rbcS nuclear genes [28], [29]. By directed mutagenesis in Rhodospirillum rubrum, Synechococcus, Chlamydomonas, and tobacco it has been shown that even single mutations can positively or negatively change stability or substrate specificity of Rubisco [15], [30]. The most dramatic changes in Rubisco performance are inducted by replacements in the active site and in the regions providing interactions between LSUs and SSUs, and between Rubisco and Rubisco activase [15], [30].

Most amino acid replacements in Schiedea rbcL (residues 23, 86, 326, 354, 363, 367 and 470) reside in regions influencing interactions with Rubisco activase, a chaperone which promotes and maintains the catalytic activity of Rubisco [15], [27]. Rubisco activase plays a vital role in the response of photosynthesis to temperature [27], thus molecular adaptation of Rubisco-Rubisco activase interactions may have played an important role in adaptation of Schiedea species to dry sunny conditions. Furthermore, five of the replaced residues (86, 326, 354, 363 and 367) are close to each other (distances<20Å) in the Rubisco tertiary structure, suggesting possible cumulative effect. Sequencing and investigation of Schiedea's Rubisco activase might be of considerable interest for future studies of possible coevolution of Rubisco and Rubisco activase in Schiedea.

The distribution of rbcL amino acid replacements in the Schiedea cpDNA phylogeny corroborates their possible functional importance. Non-synonymous mutations favored by positive selection are expected to be more common at the internal branches relative to terminal branches [31]. Indeed all amino acid replacements in Schiedea appeared in the internal branches (Fig. 1), a pattern significally different from ones of rbcL synonymous substitutions as well as from non-synonymous and synonymous mutations in other investigated cpDNA regions (Table 3).

The possible changes of Rubisco properties in Schiedea predicted from structural modeling match well with the observed difference in rates of photosynthesis [13] between "basal" and "advanced" species (roughly corresponding to clades I+II and III+IV, respectively) as well as with wide distribution of "advanced" rbcL haplotypes within Schiedea. The "basal" Schiedea species inhabit mesic or wet shady forests, while most species in the "advanced" clades (section Schiedea) colonised dry sunny habitats, such as coastal cliffs. Given the importance of Rubisco enzyme performance for plant growth and the significant effect of mutations affecting the contacts with Rubisco activase [15], [30], the His86⇒Tyr86 and other replacements in Schiedea rbcL may have played an important role in colonisation of dry habitats during recent adaptive radiation in Schiedea. Molecular adaptation in photosynthetic Rubisco enzyme represents the first known case of adaptation at the protein level during a recent adaptive radiation and reveals molecular bases of physiological and ecological evolution during rapid radiations in island endemics.

Positive selection on rbcL may possibly be the cause of fixation of two chloroplast haplotypes in virtually all species of Schiedea sensu stricto (clades III and IV on Fig. 1) and hence the main reason for cytonuclear discordance. This hypothesis is corroborated by the geographical pattern of Schiedea cpDNA haplotype distribution, where clades represent islands, rather than recognized Schiedea sections (Fig. 1).

Despite remarkable morphological and ecological divergence, natural interspecific hybrids have been found for many Schiedea species and the ability to cross-hybridize with each other has been shown for virtually all Schiedea species in green-house experiments [13]. However, strong geographical isolation between and within the islands makes interspecies contacts quite rare. Indeed, a previous DNA diversity study has demonstrated that isolation between the populations of S. globosa from different islands is much stronger than one between the populations from the same island [32]. Thus, it is quite likely that only genes under strong positive selection can spread across Schiedea species.

Positive selection on non-recombining chloroplast DNA is expected to lead to a spread of the selected chloroplast haplotype across several species, causing phylogenetic cytonuclear discordance. Cytonuclear discordance promoted by interspecific hybridization has been found in many adaptive radiations including Darwin's finches, African cichlids, Lake Baikal sculpins and Hawaiian silversword alliance (reviewed in [2]). Complete plastom and mitochondrion replacements via interspecific introgression have been documented for various plant and animal groups (reviewed in [33]). Most authors typically attribute the occurrence of introgression to demographic events and chance fixation, whereas relatively few suggest positive selection as a possible cause [34], [35]. The adaptive amino acid replacements in Schiedea rbcL occurred on the branches leading to the clades III and IV of the chloroplast gene tree. The spread of advantageous rbcL alleles across many Schiedea species inhabiting the same island (or a group of previously connected islands) argues in favor of positive selection as a main cause of cytonuclear discordance and suggests that sharing of adaptive mutations by several closely related species may be an important factor in adaptive evolution in small populations within confined geographical regions, such as oceanic islands or big lakes.

Materials and Methods

Isolation and Sequencing of Schiedea Genes

Genomic DNA was isolated from fresh leaf material using magnetic beads-based Plant DNA Charge Switch Kit (Invitrogen) in accordance with manufacturer protocol. For PCR amplification of all regions except trnS-trnG we used BioMix Red (Bioline) with the following PCR conditions: one cycle of 95°C, 2 min, 55°C, 30 s, 72°C, 4 min followed by 36 cycles of 93°C, 30 s, 53°C, 30 s, 72°C, 3.5 min. For PCR amplification of trnS-trnG region we used Protocol 1 from [36]. The PCR products were extracted from the agarose gels using the Qiagen gel extraction kit. Sequencing was performed using ABI BigDye v3.1 system on an ABI3700 automated sequencing machine. Sequence chromatograms were checked and corrected, and the contigs were assembled and aligned using ProSeq3 software [37]. All polymorphic sites were checked against original sequence chromatograms and doubtful regions were resequenced; obtained sequences were compared with homologues from GenBank and ORFs integrity was confirmed for protein coding sequences; all indels were removed before further analyses. Novel sequences have been submitted to GenBank under accession numbers DQ907721-DQ907909.

Statistical Tests for Positive Selection

The neighbor-joining trees for every investigated chloroplast region as well as for three concatenated datasets (noncoding regions, coding regions, all regions) were created using MEGA v3.1 [38]. The topologies of all obtained trees were similar and for further phylogenetic analyses of positive selection in Schiedea's chloroplast protein coding genes we used the unrooted tree based on concatenated dataset of all regions.

We used the codeml program in the PAML v.3.14 [18] package to estimate the non-synonymous divergence (dN), synonymous divergence (dS), and their ratio (dN/dS) in model 0, that allows for a single dN/dS value throughout the whole phylogenetic tree. Further, codeml was used to perform likelihood ratio tests (LRTs) for rate heterogeneity and positive selection among amino acid sites. We applied models of codon evolution which allow for variation in dN/dS among codons but assume the same distribution in all branches of the phylogeny. We performed three LRTs for positive selection: M1a–M2a LRT, M7–M8 LRT and M8a–M8 LRT [21]–[23]. For all LRTs, the first model is a simplified version of the second, with fewer parameters, and is thus expected to provide a poorer fit to the data (lower maximum likelihood). The M1, M7 and M8a models are the null models without positive selection (no codons with dN/dS>1) and the M2 and M8 models are the alternative models with positive selection. The significance of the LRTs was calculated assuming that twice the difference in the log of maximum likelihood between the two models is distributed as a chi-square distribution with the degrees of freedom (df) given by the difference in the numbers of parameters in the two nested models. For both M1a–M2a and M7–M8 comparisons we used df = 2 [21], [22]. It was argued that for M8a–M8 comparisons the appropriate test would use a 50:50 mixture of df = 0 and df = 1 [23], however we assumed df = 1 for this test, which is conservative [22].

To identify amino acid sites potentially under positive selection, the parameter estimates from M8 model were used to calculate the posterior probabilities that an amino acid belongs to a class with dN/dS>1 using the Bayes Empirical Bayes approaches implemented in PAML [25].

Structural Analysis of Rubisco

We used spinach Rubisco protein structure [24] to infer the possible effect(s) of mutations at the residues identified as being under positive selection in Schiedea. The divergence between rbcLs of spinach and Schiedea at the amino acid level is between 3.2% and 4.4%, depending on the Schiedea species. Furthermore, the ancestral states of eight out of ten replacements found in Schiedea rbcL are identical to corresponding residues in spinach, making it appropriate to use protein structure obtained for spinach. Rubisco structural data for spinach (1RBO) were obtained from the RCB Protein Data Bank (http://www.rcsb.org/pdb). The solvent accessible surface areas for individual amino acids in the structure and the impact of single replacements on overall structural stability were analyzed using CUPSAT software [39; http://cupsat.uni-koeln.de]. The structural contacts for individual amino acids in the structure were analyzed using DeepView–Swiss-PdbViewer v. 3.7 [40; http://www.expasy.org/spdbv/] and STING Report [41; http://trantor.bioc.columbia.edu/SMS/].

Acknowledgements

We are grateful to Dr S. Weller for providing us Schiedea leaf material, to Dr Z. Yang and Dr M. Anisimova for advice on PAML, to Dr K. Futterrer for help and advice with protein structure analysis and to Dr G. Muir for suggestions and correction of the manuscript.

Author Contributions

Conceived and designed the experiments: DF. Performed the experiments: MK. Analyzed the data: MK. Wrote the paper: DF MK

References

1. Schluter D (2000) The ecology of adaptive radiation. Oxford: Oxford University Press.
2. Seehausen O (2004) Hybridization and adaptive radiation. Trends Ecol Evol 19: 198–207.
3. Hollocher H (1996) Island hopping in Drosophila: patterns and processes. Philos Trans R Soc Lond B Biol Sci 351: 735–743.
4. Kimura M (1983) The neutral theory of molecular evolution. Cambridge: Cambridge University Press.
5. Ohta T (1992) The nearly neutral theory of molecular evolution. Annu Rev Ecol Syst 23: 263–286.
6. Barrier M, Robichaux RH, Purugganan MD (2001) Accelerated regulatory gene evolution in an adaptive radiation. Proc Natl Acad Sci USA 98: 10208–10213.
7. Remington DL, Purugganan MD (2002) GAI Homologues in the Hawaiian Silversword Alliance (Asteraceae-Madiinae): molecular evolution of growth regulators in a rapidly diversifying plant lineage. Mol Biol Evol 19: 1563–1574.
8. Stebbins GL Jr (1959) The role of hybridisation in evolution. Proc Am Philos Soc 103: 231–251.
9. Morjan CL, Riesberg LH (2004) How species evolve collectively: implications of gene flow and selection for the spread of advantageous alleles. Mol Ecol 13: 1341–1356.
10. Arnold ML (1997) Natural hybridisation and evolution. New York: Oxford University Press.

11. Slatkin M (1976) The rate of spread of advantageous alleles in a subdivided population. In: Karlin S, Nevo E, editors. Population Genetics and Ecology. New York: Academic Press. pp. 767–780.
12. Lewontin RC, Birch LC (1966) Hybridization as a source of variation for adaptation to new environments. Evolution 20: 315–336.
13. Wagner WL, Weller SG, Sakai A (2005) Monograph of Schiedea (Caryophyllaceae subfam. Alsinoideae). Syst Bot Monographs 72: 1–169.
14. Ferreira KN, Iverson TM, Maghlaoui K, Barber J, Iwata S (2004) Architecture of the photosynthetic oxygen-evolving center. Science 303: 1831–1838.
15. Spreitzer RJ, Salvucci ME (2002) RUBISCO: structure, regulatory interactions, and possibilities for a better enzyme. Annu Rev Plant Biol 53: 449–475.
16. Neuhaus H, Link G (1987) The chloroplast tRNALys (UUU) gene from mustard (Sinapsis alba) contains a class II intron potentially coding for a maturase related polypeptide. Current Genetics 11: 251–257.
17. Vogel J, Hübschmann T, Börner T, Hess WR (1997) Splicing and intron-internal RNA editing of trnK-matK transcripts in barley plastids: support for matK as an essential splice factor. J Mol Biol 270: 179–187.
18. Yang Z (1997) PAML: a program package for phylogenetic analysis by maximum likelihood. Comput Appl Biosci 13: 555–556.
19. Maddison WP (1997) Gene trees in species trees. Syst Biol 46: 523–536.
20. Machado CA, Hey J (2003) The causes of phylogenetic conflict in a classic Drosophila species group. Proc Biol Sci 270: 1193–1202.
21. Yang Z, Nielsen R, Goldman N, Pedersen AM (2000) Codon-substitution models for heterogeneous selection pressure at amino acid sites. Genetics 155: 431–449.
22. Wong WSW, Yang Z, Goldman N, Nielsen R (2004) Accuracy and power of statistical methods for detecting adaptive evolution in protein coding sequences and for identifying positively selected sites. Genetics 168: 1041–1051.
23. Swanson WJ, Nielsen R, Yang Q (2003) Pervasive adaptive evolution in mammalian fertilization proteins. Mol Biol Evol 20: 18–20.
24. Knight S, Andersson I, Branden CI (1990) Crystallographic analysis of ribulose 1,5-bisphosphate carboxylase from spinach at 2.4Å resolution. J Mol Biol 215: 113–160.
25. Yang Z, Wong WS, Nielsen R (2005) Bayes empirical Bayes inference of amino acid sites under positive selection. Mol Biol Evol 22: 1107–1118.

26. Kellogg EA, Juliano ND (1997) The structure and function of RuBisCO and their implications for systematic studies. Am J Bot 84: 413–428.

27. Portis AR Jr (2003) Rubisco activase–Rubisco's catalytic chaperone. Photosynthesis Res 75: 11–27.

28. Dean C, Pichersky E, Dunsmuir P (1989) Structure, evolution, and regulation of RbcS genes in higher plants. Annu Rev Plant Physiol Plant Mol Biol 40: 415–439.

29. Roy H, Andrews TJ (2000) Rubisco: assembly and mechanism. In: Leegood RC, Sharkey TD, von Caemmerer S, editors. Photosynthesis: Physiology and Metabolism. Dordrecht: Kluwer. pp. 53–83.

30. Satagopan S, Spreitzer RJ (2004) Substitutions at the Asp-473 latch residue of Chlamydomonas ribulosebisphosphate carboxylase/oxygenase cause decreases in carboxylation efficiency and CO2/O2 specificity. J Biol Chem 279: 14240–14244.

31. Ruiz-Pesini E, Mishmar D, Brandon M, Procaccio V, Wallace DC (2004) Effects of purifying and adaptive selection on regional variation in human mtDNA. Science 303: 223–226.

32. Filatov DA, Burke S (2004) DNA diversity in Hawaiian endemic plant Schiedea globosa. Heredity 92: 452–458.

33. Avise JC (2004) Molecular markers, natural history, and evolution. Sunderland: Sinauer Ass. 684 p.

34. Ballard JWO, Kreitman M (1995) Is mitochondrial DNA a strictly neutral marker? Trends Ecol Evol 10: 485–488.

35. Doiron S, Bernatchez L, Blier PU (2002) A comparative mitogenomic analysis of the potential adaptive value of arctic charr mtDNA introgression in Brook charr populations (Salvelinus fontinalis Mitchill). Mol Biol Evol 19: 1902–1909.

36. Shaw J, Lickey EB, Beck JT, Farmer SB, Liu W, et al. (2005) The tortoise and the hare II: relative utility of 21 noncoding chloroplast sequences for phylogenetic analysis. Amer J Bot 92: 142–166.

37. Filatov DA (2002) PROSEQ: A software for preparation and evolutionary analysis of DNA sequence data sets. Mol Ecol Notes 2: 621–624.

38. Kumar S, Tamura K, Nei M (2004) MEGA3: Integrated software for molecular evolutionary genetics analysis and sequence alignment. Briefings Bioinformatics 5: 150–163.

39. Parthiban V, Gromiha MM, Schomburg D (2006) CUPSAT: prediction of protein stability upon point mutations. Nucleic Acids Res 34: W239–242.

40. Guex N, Peitsch MC (1997) SWISS-MODEL and the Swiss-PdbViewer: An environment for comparative protein modeling. Electrophoresis 18: 2714–2723.

41. Neshich G, Mancini AL, Yamagishi ME, Kuser PR, Fileto R, et al. (2005) STING Report: convenient web-based application for graphic and tabular presentations of protein sequence, structure and function descriptors from the STING database. Nucleic Acids Res 33: 269–274.

42. Nelson DL, Cox MM (2005) Lehninger principles of biochemistry,. New York: WH Freeman and Company.

43. Kyte J, Doolittle RF (1982) A simple method for displaying the hydropathic character of a protein. J Mol Biol 157: 105–132.

44. Grantham R (1974) Amino acid difference formula to help explain protein evolution. Science 185: 862–864.

45. Cuénoud P, Savolainen V, Chatrou LW, Powell M, Grayer RJ, et al. (2002) Molecular phylogenetics of Caryophyllales based on nuclear 18S rDNA and plastid rbcL, atpB, and matK DNA sequences. Amer J Bot 89: 132–144.

46. Kadereit G, Borsch T, Weising K, Freitag H (2003) Phylogeny of Amaranthaceae and Chenopodiaceae and the evolution of C4 photosynthesis. Int J Plant Sci 164: 959–986.

47. Taberlet P, Gielly L, Pautou G, Bouvet J (1991) Universal primers for amplification of three non-coding regions of chloroplast DNA. Plant Mol Biol 17: 1105–1109.

Analysis of the Chloroplast Protein Kinase Stt7 during State Transitions

Sylvain Lemeille, Adrian Willig, Nathalie Depège-Fargeix, Christian Delessert, Roberto Bassi and Jean-David Rochaix

ABSTRACT

State transitions allow for the balancing of the light excitation energy between photosystem I and photosystem II and for optimal photosynthetic activity when photosynthetic organisms are subjected to changing light conditions. This process is regulated by the redox state of the plastoquinone pool through the Stt7/STN7 protein kinase required for phosphorylation of the light-harvesting complex LHCII and for the reversible displacement of the mobile LHCII between the photosystems. We show that Stt7 is associated with photosynthetic complexes including LHCII, photosystem I, and the cytochrome b_6f complex. Our data reveal that Stt7 acts in catalytic amounts. We also provide evidence that Stt7 contains a transmembrane region that separates its catalytic kinase domain on the stromal side from its N-terminal end in the thylakoid *lumen with two conserved Cys that are critical for its activity and state transitions. On the basis of these data, we propose that the activity of Stt7 is regulated through its transmembrane domain and that a disulfide bond between the two lumen Cys is essential for its activity. The high-light–induced reduction of this bond may occur through a transthylakoid thiol–reducing pathway driven by the ferredoxin-thioredoxin system which is also required for cytochrome b_6f assembly and heme biogenesis.*

Author Summary

To grow optimally, photosynthetic organisms need to constantly adjust to changing light conditions. One of these adjustments, called state transitions, allows light energy to be redistributed between the two photosynthetic reaction center complexes in a cell's chloroplasts. These complexes act in concert with other components of the photosynthetic machinery to turn light energy into cellular energy. A key component in the regulation of state transitions is the chloroplast protein Stt7 (also known as STN7), which can modify other proteins by adding a phosphate group. When light levels change, the oxidation level of a pool of another chloroplast component, plastoquinone, changes, which in turn activates Stt7, inducing it to phosphorylate specific proteins of the light-harvesting complex of one reaction center. As a result, a portion of this light-harvesting complex is transferred from one photosynthetic reaction center to the other, thereby optimizing photosynthetic efficiency. Here, we have addressed the configuration of Stt7 within the thylakoid membrane of the chloroplast and the molecular mechanisms underlying its activation. Our data reveal that the level of Stt7 protein changes drastically under specific environmental conditions, that the protein does not need to be present in a one-to-one ratio with its targets for activity, and that it associates directly with a number of components of the photosynthetic machinery. The protein-modifying domain of Stt7 is exposed to the outer side of the thylakoid membrane, whereas the domain critical for regulation of its activity lies on the inner side of the thylakoid membrane. These results shed light on the molecular mechanisms that allow photosynthetic organisms to adjust to fluctuations in light levels.

Introduction

Photosynthetic organisms are constantly subjected to changes in light conditions. These organisms have developed different mechanisms to rapidly acclimate to this changing environment. At one extreme, when the absorbed light excitation energy vastly exceeds the assimilation capacity of the photosynthetic apparatus, these organisms need to protect themselves. Excess light energy is dissipated as heat through nonphotochemical quenching, which involves conformational changes in the light-harvesting system of photosystem II [1]. In contrast, under low light, photosynthetic organisms optimize the absorption capacity of their antenna systems. This is especially true when changes in light quality occur that lead to the preferential stimulation of either photosystem II (PSII) or photosystem I (PSI), which are linked through the photosynthetic electron transport chain. Under these conditions, balancing of the light excitation energy between the antenna systems of PSII and PSI occurs through a process called state transitions [2–4]. Upon preferential excitation of PSII, the plastoquinone pool is reduced, a process that favors binding of plastoquinol to the Qo site of the cytochrome b6f complex and leads to the activation of a thylakoid protein kinase required for the phosphorylation of the light-harvesting system of PSII (LHCII) [5,6]. In the green alga Chlamydomonas reinhardtii, the LHCII protein set consists of Type I (Lhcbm3, Lhcbm4, Lhcbm6, Lhcbm8, and Lhcbm9), Type II (Lhcbm5), Type III (Lhcbm2 and Lhcbm7), and Type IV (Lhcbm1) proteins, and of CP26 and CP29 [7]. Because of their nearly identical sequences and size, several of these Lhcbm proteins cannot be distinguished by SDS-PAGE. Most of them fractionate into two major bands called P11/P13 (Type I) and P17 (Type III). CP29, Lhcbm5, P11, P13, and P17 are phosphorylated during a state 1 to state 2 transition [7–9]. Although CP29 and Lhcbm5 are mobile during state transitions, it is not yet clear which among the other LHCII proteins of C. reinhardtii are mobile [10,11]. The phosphorylation of LHCII is followed by a displacement of LHCII from PSII to PSI, thus increasing the size of the PSI antenna at the expense of the PSII antenna and rebalancing the excitation energy between both photosystems. Binding of the mobile LHCII to PSI requires the PsaH subunit [12]. This state corresponds to state 2. The process is reversible as preferential excitation of PSI leads to the dephosphorylation of LHCII by unknown phosphatases and its return to PSII (state 1).

State transitions can be induced, not only by changes in light conditions, but also through changes in cellular metabolism. Thus, in C. reinhardtii, the process can be triggered when the level of ATP is low or when the cells are grown in the dark under anaerobiosis. These conditions lead to the influx of reducing equivalents into the plastoquinone pool and to the activation of the LHCII kinase [13]. Moreover, transition from state 1 to state 2 in C. reinhardtii is associated with a switch from linear to cyclic electron transfer [14,15]. In this alga, the major role of state transitions appears to be ATP homeostasis. In land plants, the LHCII kinase can also be activated in the dark by the addition of sugar compounds [16]. Recent studies further confirm that state transitions are not limited to the balancing of excitation energy between the photosystems but that they also play a major role in the adjustment of the light reactions with carbon metabolism [17].

In C. reinhardtii, transition from state 1 to state 2 causes the displacement of 80% of LHCII from PSII to PSI, as deduced from the measurements of the quantum yield of PSI and PSII charge separation [18]. In contrast in land plants, only 15% of LHCII is mobile during state transitions, although this process is associated with considerable structural rearrangements of the thylakoid membranes [19]. The large size of the mobile LHCII antenna leads to significant changes in fluorescence yield during state transitions in C. reinhardtii, a feature that has been exploited for the screening of mutants deficient in this process [9,20]. Such a screen has revealed the existence of the thylakoid Ser-Thr protein kinase Stt7 (AA063768) [21]. Mutants deficient in this kinase are deficient in LHCII phosphorylation and fail to undergo a transition from state 1 to state 2. The Stt7 protein kinase is associated with the thylakoid membrane and contains a potential transmembrane domain upstream of the catalytic kinase domain. In Arabidopsis thaliana, the ortholog STN7 (NP_564946) is also specifically involved in LHCII phosphorylation and state transitions [21,22]. At this time, it is not yet clear whether Stt7 and STN7 act directly on LHCII or whether they act further upstream as part of a kinase cascade. The cytochrome b6f complex plays a key role in the activation of the kinase [4]. It is thus very likely that Stt7/STN7 interacts directly with this complex. If LHCII is the substrate of Stt7/STN7, an interaction between the two is expected. Here, we have used coimmunoprecipitations and pull-down experiments to show that interactions of this type do indeed occur. Our data reveal that Stt7 acts in catalytic amounts. We also show that Stt7 contains a transmembrane region with the catalytic domain on the stromal side of the thylakoid membrane and the N-terminal region in the lumen. This domain appears to play a key role in the regulation of the kinase activity.

Results

Stt7 is Associated with a Large Molecular Weight Complex

To understand how the Stt7 kinase functions, we first estimated its abundance, in particular its molar ratio compared with the cytochrome b_6f complex under state 2 conditions. An antibody was raised against Stt7 and the amount of Stt7 was estimated using recombinant Stt7 protein for calibration. A similar calibration was performed with the cytochrome b_6f complex. This analysis revealed that the molar ratio between Stt7 and the cytochrome b_6f complex is 1:20, clearly indicating that Stt7 is present at substoichiometric levels compared to the photosynthetic complexes.

The Stt7/STN7 protein kinase has been shown to be associated with the thylakoid membrane [21]. However, it is not known whether Stt7/STN7 acts singly or whether it is associated with other proteins in a larger complex whose composition might change during state transitions. To test this possibility, thylakoid membranes from the stt7 mutant complemented with Stt7 containing a haemagglutinin (HA)-tag at its C-terminal end (Stt7-HA) were isolated. Membranes were prepared from cells in state 1 obtained by illumination in low light (6 μmol m−2 s−1) in the presence of DCMU (3-(3,4-dichlorophenyl)-1,1-dimethylurea) for 30 min and in state 2 by incubating the cells under anaerobic conditions for 30 min in the dark. The occurrence of state transitions was verified by measuring the change in maximum fluorescence (Fmax). The thylakoid membranes were solubilized with n-dodecyl-β-maltoside and fractionated by sucrose density gradient centrifugation. Individual fractions of the two gradients from Stt7-HA thylakoid membranes were separated by PAGE and then tested by immunoblot analysis using antibodies directed against HA, Cytf, PsaA, D1, CP26, CP29, and Lhcbm5 (Figure 1A). Under both state 1 and state 2 conditions, the Stt7 protein kinase was associated with a large complex that partly overlaps with the high molecular weight fractions of PSI and the cytochrome b6f complex but not with PSII (Figure 1A). No major changes in the distribution of the thylakoid complexes were observed under the state 1 and state 2 conditions used. A similar distribution of the complexes was found in the stt7 mutant, confirming that the high molecular weight LHC complexes are not only formed under state 2, but also under state 1 conditions (Figure 1B).

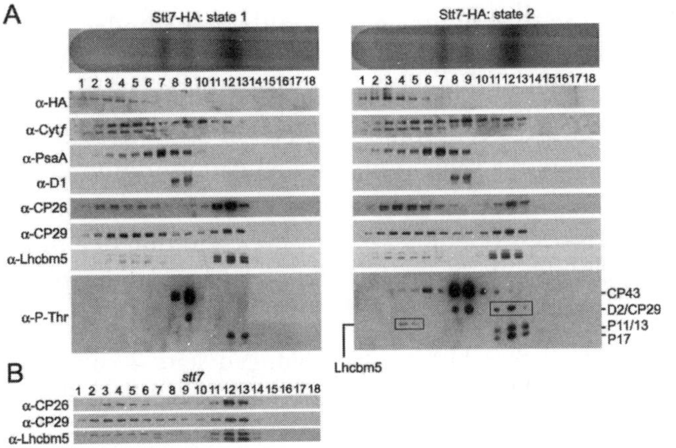

Figure 1. Fractionation of Stt7 and Chlorophyll-Protein Complexes by Sucrose Density Gradient Centrifugation

(A) Thylakoid membranes from Stt7-HA cells in state 1 and state 2 were solubilized with n-dodecyl-β-maltoside, and the chlorophyll-protein complexes were separated by centrifugation on a sucrose density gradient. Proteins from the fractions of the gradients corresponding to Stt7-HA were separated by SDS-PAGE. Immunoblotting was performed with the indicated antibodies; α-P-Thr, anti–phospho-Thr. The identity of the phosphorylated proteins in the lower panel was determined by immunoblotting with antisera from the indicated proteins. In the immunoblot with a-P-Thr, signals corresponding to CP29 and Lhcbm5 are framed.

(B) Immunoblots prepared as in (A) except that the thylakoid membranes from stt7 were used. P11/13 and P17 are the major LHCII proteins of C. reinhardtii and correspond to Type I and Type III LHCII proteins, respectively.

Although the level of Stt7-HA was between 25%–50% compared to Stt7 in wild-type cells, state transitions proceeded to the same extent as in the wild type (unpublished data). Moreover, the amount of photosynthetic complexes was the same in Stt7-HA, stt7, and wild-type cells. As expected, immunoblots with an anti–P-Thr antiserum revealed increased phosphorylation of several proteins in state 2, notably the major LHCII proteins P11, P13, and P17 in the wild-type strain (Figure 1A). Moreover, a weak phosphorylation signal corresponding to Lhcbm5 was detected under state 2 conditions in the same high molecular weight fractions containing PSI. Although an increase of phosphorylation was also observed for the PSII core proteins CP43 and D2 under state 2 conditions, in other experiments, no significant increase of phosphorylation of these proteins was detected between state 1 and state 2.

The Level of Stt7 Decreases under Prolonged State 1 Conditions

The immunoblots in Figure 1A indicate that the levels of Stt7 are significantly higher in state 2 than in state 1. To examine this further, cells containing Stt7-HA were grown for 2 h under state 2 conditions, and growth was continued for 4 h either under state 2 or state 1 conditions (Figure 2). At different time points, aliquots of cells were processed for immunoblot analysis with HA and Cytf antibodies. Whereas the level of Stt7 remained the same under continuous state 2 conditions, its level decreased gradually 1 h after the shift to state 1 conditions. It decreased 4-fold after 2 h and more than 20-fold after 4 h (Figure 2C). After shifting the cells to state 2 conditions, the level of Stt7 increased 2-fold after 2 h but did not reach its initial value (unpublished data). Addition of cycloheximide did not affect the decline of Stt7, indicating that no newly synthesized protease is involved in this process (Figure 2D). However, addition of a protease inhibitor mixture to the cells abolished the degradation of Stt7 under state 1 conditions (Figure 2E).

The degradation of Stt7-HA under prolonged state 2 conditions monitored by immunoblotting with the HA antiserum could be due to the removal of the HA tag from Stt7. To test this possibility, we repeated this experiment with the Stt7-HA strain by using both Stt7 and HA antiserum. In both cases, a decline of the Stt7 kinase was confirmed under state 2 conditions. We checked that this decrease also occurs with untagged Stt7. To identify the type of proteases involved, different protease inhibitors were tested under the same conditions as above: ACA (ε-aminocaproic acid) (Sigma) (50 mM), AEBSF (4-(2-aminoethyl)-benzenesulfonyl fluoride hydrochloride) (Roche) (5 mM), NEM (N-ethylmaleimide) (Sigma) (10 mM), phenylmethanesulfonyl fluoride (Sigma) (5mM), and EDTA (50 mM). Leupeptin and NEM were also used at different concentrations. Samples were taken and analyzed at different time points. Whereas NEM completely prevented the breakdown of Stt7, the serine protease inhibitors ACA and AEBSF had no effect, and EDTA enhanced this process. Other inhibitors of cysteine proteases besides NEM, such as E64 and leupeptin, prevented Stt7 degradation under prolonged state 1 conditions. Although no convincing proof of chloroplast Cys proteases has been reported, their existence cannot be excluded. A bioinformatic search for Cys proteases in plastids of A. thaliana did not reveal any convincing candidate.

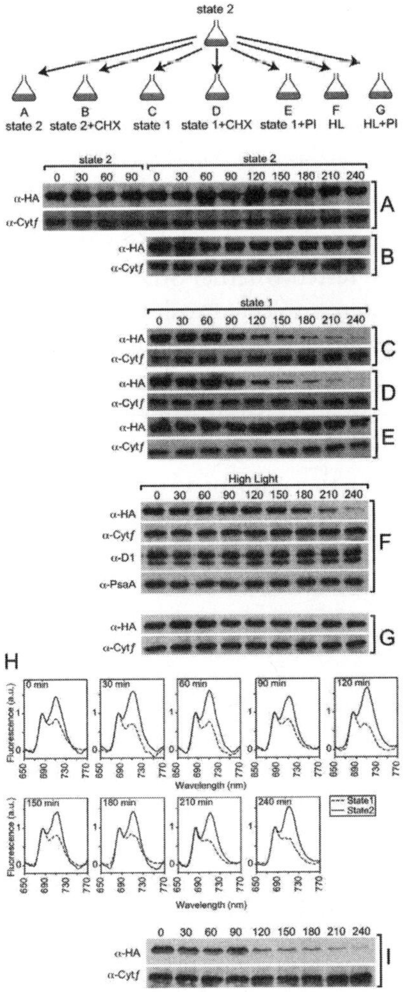

Figure 2. Levels of Stt7 Decrease under Prolonged State 1 Conditions and High Light Cells were grown for 120 min under state 2 conditions and subsequently maintained in state 2 (dark + anaerobiosis) or subjected to

state 1 conditions (light + strong aeration) or to high light (900 μmol photons m^{-2} s^{-1}) for 240 min. At various time intervals, total proteins were extracted and the amount of Stt7-HA was measured by immunoblotting with HA antiserum. Cytf was used as a loading control. Top: outline of the experimental conditions (for details, see Materials and Methods).

(A) Cells maintained under continuous state 2 conditions.

(B) Same as (A), but with addition of cycloheximide after the initial culture was maintained in state 2 for 120 min.

(C) Cells in state 2 were subjected to state 1 conditions after 120 min.

(D) Same as (C), but with addition of cycloheximide after 120 min.

(E) Same as (C) but with addition of protease inhibitors after 120 min.

(F) Cells in state 2 were subjected to high light. The levels of PSII and PSI were assessed by immunoblotting with antibodies against D1 and PsaA, respectively.

(G) Same as (F), but with addition of leupeptin. Total proteins were examined by immunoblotting with HA and Cytf antibodies.

(H) Cells shifted to state 1 conditions as shown in (C) were tested for transition from state 1 to state 2 at different times. In this case, transition to state 2 was induced by adding 5 μM FCCP and assayed after 15 min by measuring low-temperature fluorescence emission spectra for each time point.

(I) Stt7 levels of the cells used in (H) were determined by immunoblotting with HA antiserum.

As high-light treatment is known to lead to the inactivation of the LHCII protein kinase [23,24], we tested whether Stt7 was stable under these conditions. Cells adapted to state 2 were subjected to high-light treatment (900 μmol m−2 s−1), and the Stt7 levels were determined by immunoblotting at different times. Under these conditions, a steady decrease of Stt7 was observed (Figure 2F), which could be fully prevented by addition of leupeptin (Figure 2G). Under the same light regime, the level of PSII and of PSI were nearly unaffected (Figure 2F). However, measurements of the ratio of variable (Fv) over maximal florescence (Fv/Fmax) revealed that this ratio decreased from 0.7 to 0.2, indicating photo-damage to PSII without apparent decrease of D1 protein level.

Stt7 Acts in Catalytic Amounts

We took advantage of the decrease in Stt7 under prolonged state 1 conditions to test whether the cells are still able to switch from state 1 to state 2 under these conditions. Cells were first grown under state 1 conditions as shown in Figure 2. After different time periods, cells were collected and assayed for state transitions using the uncoupler FCCP (carbonyl cyanide p-fluoromethoxyphenylhydrazone), a known inducer of transition to state 2 [13]. The fluorescence emission spectra at low temperature and the level of Stt7 were measured under state 1 and state 2 conditions. Figure 2H shows that transition to state 2 occurred readily in these cells collected after 4 h under state 1 conditions, although the amount of Stt7 protein kinase was decreased 20-fold (Figure 2I). This indicates that the Stt7 kinase acts in catalytic amounts.

Stt7 Interacts with the Cytochrome b_6f Complex, LHCII, and PSI

Activation of the LHCII kinase depends critically on the cytochrome b_6f complex [4]. Moreover, this kinase is likely to interact with its putative substrate LHCII. To test whether the Stt7 kinase interacts with the cytochrome b_6f complex and/or LHCII, thylakoid membranes from the Stt7-HA strain in state 1 or state 2 were solubilized with dodecyl maltoside and immunoprecipitated with HA antiserum. The immunoprecipitates were fractionated by SDS-PAGE and immunoblotted with antibodies against several known thylakoid proteins. Figure 3A shows that the LHCII proteins P13, P11, P17, CP29, CP26, and Lhcbm5 were coimmunoprecipitated with the Stt7 kinase. The signal obtained with CP29 was only detectable under state 2 conditions. Signals were also observed with Cytf and Rieske protein as well as with the PSI subunit PsaA. In contrast, no interaction was observed between Stt7 and D1 from PSII and with subunit α of ATP synthase (Figure 3A), indicating that the interactions detected between Stt7-HA and the other photosynthetic proteins are specific. Furthermore, no signal was observed with the untagged wild-type strain (Figure 3A). Reciprocal immunoprecipitations with PsaA, Cytf, and P11 antibodies confirmed the interaction of these proteins with Stt7 (Figure 3B).

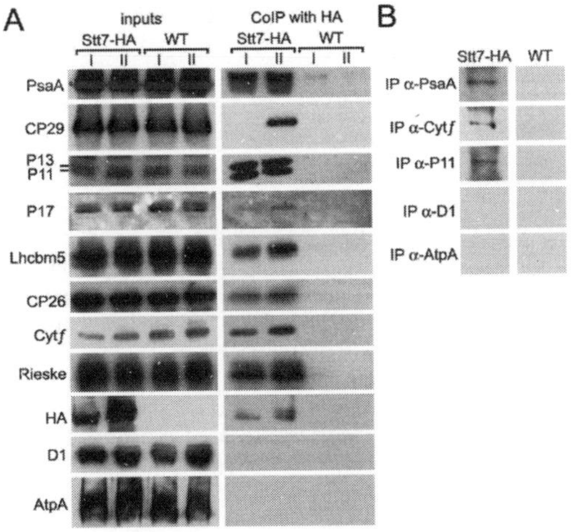

Figure 3. Coimmunoprecipitations of Stt7 with LHCII, the Cytochrome b6f Complex, and PSI

(A) Right panel: thylakoid membranes from the stt7 mutant strain complemented with Stt7-HA and wild-type (WT) cells in state 1 and state 2 were solubilized with n-dodecyl-β-maltoside, and proteins were coimmunoprecipitated (CoIP) with HA antibodies. The immunoprecipitates were separated by SDS-PAGE and immunoblotted with the antibodies indicated. Left panel: the inputs are shown and represent 10% of the samples used for the coimmunoprecipitations.

(B) Thylakoid membranes from the Stt7-HA or wild-type (WT) strain were processed as above and coimmunoprecipitated with PsaA, Cytf, D1, and AtpA antibodies. The immunoprecipitates were separated by SDS-PAGE and immunoblotted with HA antibodies.

To further test the interaction of Stt7 with the cytochrome b6f complex, the Stt7-HA strain was transformed with petA containing a His tag at its 3′-end [25]. After testing the homoplasmic state of the transformed strain for petA-His, thylakoid membranes were isolated, solubilized, and the cytochrome b6f complex was purified on a Ni-NTA column. After washing the column, immunoblotting of the eluted fraction revealed that a small portion of Stt7 kinase was associated with the tagged cytochrome b6f complex, but not with the untagged strain (Figure 4A). To determine which of the subunits of the cytochrome b6f complex interacts with Stt7, a pull-down assay with GST-Stt7 and solubilized purified cytochrome b6f complex was performed. The results (Figure 4B) show that the Rieske protein, but not Cytf, could be eluted from the GST-Stt7 column, indicating that Stt7 interacts with the Rieske protein.

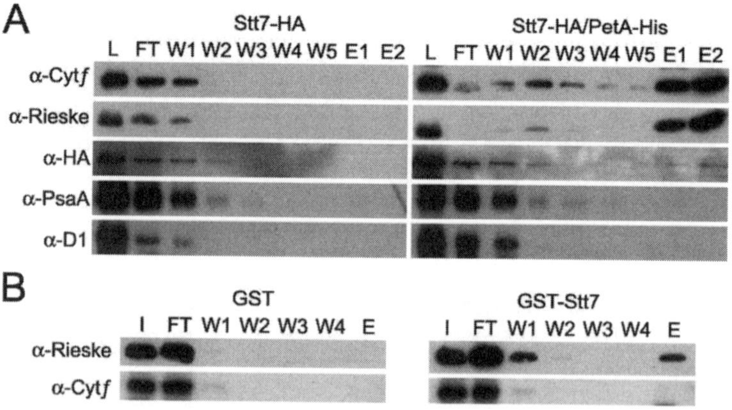

Figure 4. Stt7 Cofractionates with the Cytochrome b6f Complex

(A) Thylakoid membranes from strains containing Stt7-HA and Stt7-HA with His-tagged Cytf (PetA-His) were solubilized with n-dodecyl-β-maltoside and fractionated by Ni-NTA chromatography. E1, E2 eluates (300 mM imidazole) were analyzed by PAGE and immunoblotting with the indicated antibodies. FT, flow-through; L, loading (10 mM imidazole); W1–W5, washes (20 mM imidazole).

(B) Stt7 interacts with the Rieske protein. GST and GST-Stt7 fusion protein were incubated separately with purified solubilized cytochrome b6f complex from wild-type cells. The eluates (E) (100 mM DTT, 2% SDS) were analyzed by PAGE and immunoblotting with Rieske and Cytf antibodies. FT, flow-through; I, input; W1–W4, washes (PBS).

Orientation of the Stt7 Kinase in the Thylakoid Membrane

Although a putative transmembrane domain within Stt7 is predicted by several algorithms [21], the presence of four Pro residues within this domain raises some questions, and two models need to be considered. In the first, the N-terminal end of Stt7 would be separated from the large catalytic domain of Stt7 by a transmembrane domain. In the second model, Stt7 could be localized entirely on one side of the thylakoid membrane. To distinguish between these possibilities, the Stt7 protein was tagged with FLAG and HA at its N-and C-terminal ends, respectively. It should be noted that FLAG-Stt7-HA is not functional but that it stably accumulates in the thylakoid membranes (see below). Because the presumed substrates of the Stt7 kinase are localized on the stromal side of the thylakoid membrane, it is expected that the kinase domain of Stt7 is also located on the stromal side. This was tested by isolating intact thylakoid membranes from the strains containing Stt7-HA or FLAG-Stt7-HA and by subjecting the thylakoid membranes to mild digestion with protease V8. The resulting protein extracts were then examined by PAGE and immunoblotting using antibodies directed against HA, FLAG, PsaD, and OEE2. PsaD is known to be partially exposed to the stromal side, whereas OEE2 is entirely located on the lumenal side of the thylakoid membrane. OEE2 and the main body of PsaD are thus expected to be protected from any external protease. Under conditions (75 μg/ml V8) in which proteolysis mildly affected the OEE2 protein and led to partial digestion of PsaD, the level of Stt7 was significantly decreased as measured with HA antibodies, confirming that the kinase domain is located on the stromal side of the membrane (Figure 5A). Similar results were obtained with Stt7-HA (unpublished data). In contrast when antibodies against FLAG were used, products of smaller size were detected, indicating that the N-terminal end of Stt7 is localized on the lumenal side of the thylakoid membrane (Figure 5A). Sonication of the thylakoid membranes followed by V8 protease treatment revealed that the levels of protected fragments detected with the FLAG antibodies were significantly reduced as also observed with the lumenal protein OEE2 (Figure 5A). Sonication also led to enhanced degradation of the HA-tag of Stt7, presumably because the domain of the kinase exposed to the stroma was more accessible to the protease under these conditions and/or the thylakoid membrane was damaged. In contrast, no difference was observed for PsaD with and without sonication.

The orientation of Stt7 was further tested with the yeast split-ubiquitin system [26]. The C-terminal fragment of ubiquitin (Cub) was fused to either the N- or C-terminal end of Stt7 and expressed in yeast together with the N-terminal end of ubiquitin (Nub) fused to the C-terminal end of the endoplasmic reticulum (ER) protein Alg5, which places Nub on the cytoplasmic side of the membrane. Ubiquitin was reconstituted only with the construct in which Cub was fused to the C-terminal end of Stt7 but not when it was fused to the N-terminal end (Figure 5B). Because membrane proteins usually insert with their lumen domain in the periplasmic space of yeast, the two-hybrid results are fully compatible with the topology of Stt7 derived from the protease protection studies.

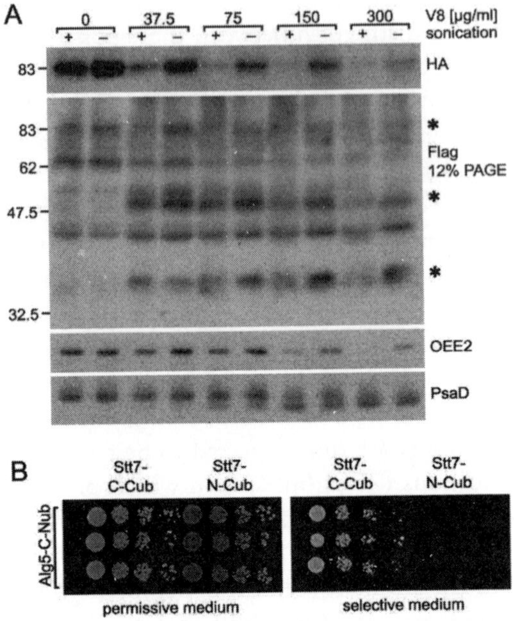

Figure 5. Stt7 Contains a Transmembrane Region with Its N-Terminal End in the Thylakoid Lumen

(A) Thylakoid membranes from a strain containing FLAG-Stt7-HA were subjected to increasing concentrations of V8 protease at room temperature, subjected to PAGE, and immunoblotted with HA, FLAG, PsaD, and OEE2 antibodies. An asterisk (*) indicates degradation products of Stt7 revealed with the FLAG antiserum.

(B) Split-ubiquitin assays were performed with fusions of the C-terminal half (Cub) of ubiquitin with either the N- or C-terminal end of Stt7 and the ER protein Alg5 fused at its C-terminal end with the N-terminal half of ubiquitin (Nub). Yeast colonies were plated on permissive (-L-W) and selective (-Ade-H-L-W) media with 10-fold and two 5-fold serial dilutions, respectively.

Mutations in the N-Terminal Region of Stt7 Abolish Kinase Activity

The transmembrane domain of Stt7 near its N-terminal end is preceded by a region that contains two conserved Cys separated by four residues that are also conserved in the orthologous STN7 kinase of Arabidopsis [21]. In land plants, it has been shown that under high light, the LHCII kinase is inactivated through the ferredoxin-thioredoxin system [23,24]. One possibility is that these conserved Cys are targets of this redox system. To test the role of these residues, the two Cys were changed individually to Ser or Ala by transforming the stt7 mutant with the Stt7-HA-C68S/A and Stt7-HA-C73S/A constructs. In all four cases, the mutant kinase accumulated as in Stt7-HA cells (Figure 6A). However, the low-temperature fluorescence emission spectra measured under conditions inducing state 1 or state 2 were nearly identical in these mutants, indicating that they are deficient in state transitions (Figure 6A). As expected, an increase of PSI fluorescence at

715 nm, which is characteristic for state 2, was detected in the rescued Stt7-HA strain. Moreover, the Cys mutants failed to phosphorylate LHCII under state 2 conditions (Figure 6B). Thus, the Cys residues are critical for kinase activity, although they are separated from the catalytic kinase domain by the transmembrane region.

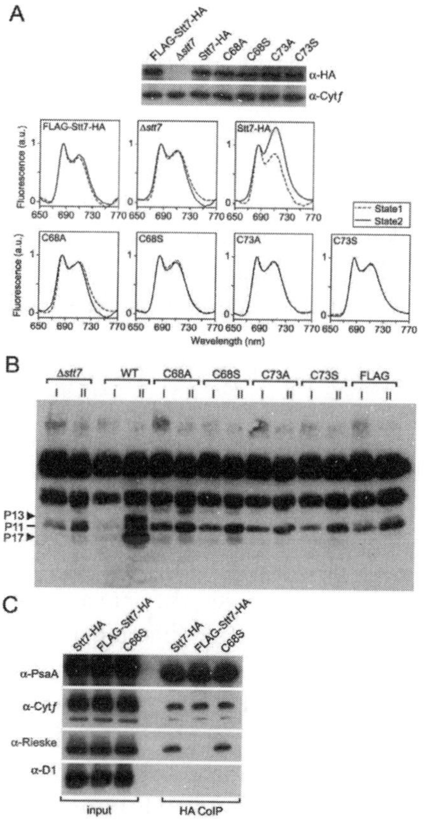

Figure 6. Analysis of Mutants of Stt7 Affected in the N-Terminal Region

Transformants of stt7 containing mutant forms of Stt7-HA with changes of Cys68 and Cys73 to Ala/Ser or with a FLAG-tag (FLAG-Stt7-HA) at the N-terminal end were used.

(A) Cys68 and Cys73 of Stt7 are essential for state transitions and LHCII phosphorylation. Fluorescence emission spectra at 70 K of wild type, stt7, and the different transformants under state 1 and state 2 conditions. Right: immunoblots of total extracts from these strains and Stt7-HA with HA antibodies.

(B) Cys68 and Cys73 are required for LHCII phosphorylation under state 2 conditions. Membrane proteins from the wild type and the indicated transformants under state 1 and state 2 conditions were fractionated by SDS PAGE and immunoblotted with an anti–P-Thr antiserum. The bands missing in the transformants are indicated by two arrows.

(C) Insertion of a FLAG tag at the N-terminal end of Stt7 specifically prevents its interaction with the Rieske protein. Thylakoids from Stt7-HA, FLAG-Stt7-HA, and Stt7-C68S were solubilized with n-dodecyl-β-maltoside and immunoprecipitated with HA antiserum. The immunoprecipitate was fractionated by PAGE and immunoblotted with the indicated antibodies.

The insertion of a FLAG-tag near the N-terminal end of mature Stt7 abolished state transitions and kinase activity but did not affect the stable accumulation of the protein (Figure 6A and 6B). Moreover, the presence of the FLAG tag specifically prevented the coimmunoprecipitation of Stt7 with the Rieske protein, but not with Cytf (Figure 6C). At first view, this appears to contradict the results of the pull-down experiment, which indicate that the Rieske protein, but not Cytf, interacts with Stt7. In the case of the pull-down experiment, recombinant Stt7 protein was used in which the transmembrane may not be correctly folded. This domain could be responsible for the binding to Cytf. In the case of the immunoprecipitation, solubilized thylakoid membranes were used in which the interaction between Stt7 and Cytf is preserved. The addition of the FLAG epitope appears to prevent proper interaction of Stt7 with the Rieske protein. In contrast, coimmunoprecipitations of Stt7-C68S occurred both with Cytf and the Rieske protein (Figure 6C). Taken together, these results indicate that the N-terminal end of Stt7 plays a crucial role in the activation of its kinase activity and that this domain may be involved in the interaction with the Rieske protein.

Discussion

The Stt7 Protein Kinase is Associated with a Large Molecular Weight Complex

State transitions lead to a considerable reorganization of the antenna systems of PSII and PSI in C. reinhardtii. Moreover, they are accompanied by large changes in the PSI-LHCI supercomplex [11,27]. Fractionation of solubilized thylakoid membranes by sucrose density gradient centrifugation revealed no major changes in the distribution of Stt7 and the LHCII proteins between PSII and PSI under state 1 and state 2 conditions. Interestingly, Stt7 is associated with a large molecular weight complex which cofractionates with the high molecular weight fractions of the cytochrome b_6f complex and PSI, but clearly not with PSII (Figure 1). These partial cofractionations of Stt7 with the cytochrome b_6f complex and PSI are compatible with the coimmunoprecipitation experiments (Figure 3). The exact composition of these complexes remains to be determined.

Our results on the fractionation of the complexes differ slightly from those of Takahashi et al. [11]. Whereas these authors found a clear shift of CP26, CP29, and Lhcbm5 towards the PSI region upon a transition from state 1 to state 2, which they attribute to the formation of a large PSI-LHCII supercomplex, we only detected a small increase in the ratio between high and low molecular weight fractions of the sucrose gradient in state 2 for CP26 and Lhcbm5 (Figure 1A). In the case of CP29, there was no significant difference in its distribution in state 1 and state 2. These differences may be due to the fact that state 2 and state 1 were induced through different means in our study and that of Takahashi et al. They used FCCP plus NaF and DCMU plus staurosporine for inducing state 2 and state 1, respectively, whereas we used anaerobiosis and vigorous aeration in the presence of DCMU. We further confirmed the presence of the LHCII proteins detected in the high molecular weight fractions under state 1 conditions in the stt7 mutant.

Although we could not detect major changes in the distribution of the LHC complexes in the sucrose density gradient under state 1 and state 2 conditions, there were changes in the phosphorylation patterns. A phosphorylated form of Lhcbm5 was detectable in the high molecular weight fractions under state 2, but not state 1 conditions (Figure 1A). Takahashi et al. [11] observed phosphorylated forms of CP29 and Lhcbm5 in this fraction. A phosphorylated form of CP29 in the high molecular weight fraction was also reported by Kargul et al. [10]. The differences between these studies might be partly accounted for by the different specificities of the anti–P-Thr antibodies used.

Stt7 is Less Abundant under State 1 Conditions

A striking feature is that the level of Stt7 is significantly lower in state 1 than in state 2. This was further investigated by a state 2–state 1 time course experiment (Figure 2). Within 2 h after shifting from state 2 to state 1 conditions, the level of Stt7 decreased to one fourth of state 2 levels. The level of Stt7 decreased further to 2%–5% of state 2 levels after 4 h in state 1, indicating that Stt7 is unstable under prolonged state 1 conditions. Thus, Stt7 is more abundant and stable under conditions where it is active. The decrease of Stt7 under state 1 conditions could be prevented by addition of inhibitors of Cys proteases to intact cells. It is therefore possible that under state 2 conditions, Stt7 is more protected from proteases either because of a posttranslational modification, e.g., phosphorylation, or of its association with other proteins. Although experimental evidence for the presence of cysteine proteases in chloroplasts is weak, their existence and role in Stt7 turnover cannot be excluded. It is possible that other proteases are involved in this process, such as the FtsH and Deg proteases, which are known to degrade thylakoid membrane proteins [28]. Some Deg proteases are indeed sensitive to NEM [29].

We note that the decrease in Stt7 occurs only after a prolonged period in state 1, whereas state transition is a short-term acclimation response that occurs within minutes. It is therefore possible that the control of the Stt7 level is part of a long-term acclimation response. There is indeed evidence for a role of the ortholog STN7 of A. thaliana in such a process [30].

The Level of Stt7 is Decreased under High Light

The level of Stt7 decreases under high light. Under the same conditions, PSII levels remained constant. In land plants, the LHCII kinase is known to be inactivated through the ferredoxin-thioredoxin system under high light [31]. It is thus conceivable that the kinase is less stable when it is maintained for a prolonged period in its inactive state, induced either by state 1 conditions or high light. Although the redox state of the plastoquinone pool is critical for activation of the kinase, it does not solely determine the level of Stt7. Under state 1 conditions, the plastoquinone pool is oxidized, whereas it is expected to be reduced under high light.

Stt7 Interacts with the Cytochrome b_6f Complex, LHCII, and PSI

A close interaction between Stt7 and the cytochrome b_6f complex is apparent from the coimmunoprecipitation results. This complex is known to play a critical role for the activation of the kinase during a state 1 to state 2 transition [32]. Mutants deficient in cytochrome b_6f complex fail to phosphorylate LHCII and are blocked in state 1 [32].

The original state transition model postulates that upon activation of the Stt7 kinase through the cytochrome b6f complex, the kinase is released from the complex to phosphorylate LHCII. However, we find that the association of the Stt7 kinase with the cytochrome b6f complex does not markedly change between state 1 and state 2. One possibility is that another downstream kinase is phosphorylated by Stt7, which in turn phosphorylates LHCII. Alternatively, the Stt7 kinase may be part of a large supercomplex that includes the cytochrome b6f complex and the PSII-LHCII complex. We have not been able to detect complexes of this kind, although it is possible that they are only formed transiently. The PetO subunit of the C. reinhardtii b6f complex is known to be phosphorylated during state transitions [33]. Despite several attempts, we were unable to detect any interaction between Stt7 and PetO based on coimmunoprecipitations or yeast two-hybrid screens. It remains to be seen whether the phosphorylation of PetO depends on Stt7.

To identify which subunit of the cytochrome b6f complex interacts with Stt7, pull-down experiments were performed. The Rieske protein was identified as an interactant (Figure 4B). Based on structural studies of the mitochondrial bc1 and of the chloroplast b6f complexes, electron transfer between plastoquinol at the Qo site and cytochrome f (Cytf) is mediated by the Rieske protein which moves from a proximal position when the Qo site is occupied by plastoquinol to a distal position when the Qo site is unoccupied [25,34,35]. It has been suggested that this dynamic behavior of the Rieske protein could be coupled to the activation of the Stt7 kinase [36–38]. Such a dynamic model is compatible with the low abundance of the Stt7 kinase with a molar ratio of 1:20 relative to the cytochrome b6f complex found in this study. Assuming one cytochrome b6f complex for one PSII core complex and an average of ten LHCII proteins per PSII reaction center [39], the molar ratio of Stt7 kinase to LHCII protein can be estimated at 1:200. At first sight, the kinase would have to phosphorylate several LHCII substrates and may undergo multiple rounds of activation. It is possible that the Stt7 kinase acts first on the LHCII located in the edges of the grana and that this phosphorylation induces extensive remodeling of the thylakoid membrane as reported recently during state transitions in A. thaliana [19]. These changes may facilitate the access of Stt7 to LHCII in the grana core. Alternatively, the observed strong LHCII phosphorylation under state 2 conditions could be due to very flexible movements of PSII-LHCII supercomplexes in the grana core and grana margins.

An intriguing component of the cytochrome b6f complex is its single chlorophyll a molecule whose chlorine ring lies between helices F and G of the PetD subunit, whereas the phytyl chain protrudes near the Qo site [25]. Interestingly, mutants affected in the binding site of chlorophyll a, besides having reduced cytochrome b6f turnover, also display a decreased rate of transition from state 1 to state 2 [36]. This region may thus either be an interaction site for the N-terminal region of Stt7 or it could act as a sensor for the presence of plastoquinol at the Qo site and initiate a signaling pathway through the chlorophyll a molecule towards the catalytic domain of Stt7 on the stromal side of the thylakoid membrane.

The coimmunoprecipitation experiments indicate that Stt7 is associated with LHCII in most cases under both state 1 and state 2 conditions. LHCII could be the direct substrate of Stt7 or, alternatively, Stt7 could be part of a multikinase complex, which ultimately phosphorylates LHCII. The only marked difference in coimmunoprecipitation between state 1 and state 2 was observed for CP29 (Figure 3). CP29 is particularly interesting. First, this monomeric LHCII together with CP26 and Lhcbm5 has been proposed to act as linker between the LHCII trimers and the dimeric PSII reaction center for the transfer of excitation energy [11,39–41]. Second, during transition from state 1 to state 2, CP29 undergoes hyperphosphorylation: in addition to Thr6 and Thr32, which are phosphorylated in state 1, Thr16 and Ser102 are phosphorylated in state 2 [10,42]. Third, electron microscopy (EM) analysis of PSII-LHCII complexes lacking CP29 could not distinguish between C2S2 and PSII monomeric complexes [39]. Fourth, in maize bundle sheath cells, which carry out mostly cyclic electron flow, the few remaining PSII complexes are monomeric [43]. Taken together, these results raise the possibility that the phosphorylation of CP29 in state 2 may act as a switch for cyclic electron flow with the detachment of CP29 from PSII and its monomerization.

The coimmunoprecipitation experiments also reveal an interaction of Stt7 with the PSI complex. This finding is surprising, as one would expect that the kinase acts on LHCII bound to PSII and that it would not interact with PSI. However, movement of PSI-LHCI complexes from the stromal lamellae to the grana margins following LHCII phosphorylation occurs in land plants, and it was proposed that the PSI absorption cross section is increased in this region through interaction of PSI-LHCI with the phosphorylated LHCII originating from the grana [44]. It is possible that the grana margins constitute a platform where the observed interactions of Stt7 with PSI could occur. It is not known whether the kinase is active or inactive when it is bound to PSI, and the role of this association remains to be determined.

The N-Terminal Part of Stt7 is Localized in the Thylakoid Lumen and is Essential for Kinase Activity

Analysis of the Stt7 amino acid sequence by bioinformatic means predicts the presence of a single transmembrane domain. However, this putative transmembrane domain contains four Pro residues that may prevent the formation of an α helix. It was therefore important to test the topology of Stt7 by experimental means using Stt7 tagged at its N-terminal end with FLAG and at its C-terminal end with HA. Using intact thylakoid membranes, we showed that whereas the C-terminal end of Stt7 is susceptible to protease digestion, the N-terminal end is protected, indicating that Stt7 indeed contains a transmembrane domain with the kinase domain on the stromal side and the N-terminal end in the lumen. The N-terminal region contains Cys68 and Cys73, which are conserved in the ortholog of Stt7 in land plants [21]. Among the seven Cys residues in Stt7, these are the only conserved Cys residues between Stt7 and STN7. In land plants, the LHCII kinase is inactivated by high-light treatment through the ferredoxin-thioredoxin system [31]. The two conserved Cys could therefore be the targets of this redox system and/or play a major role in the activation of the kinase. By changing either of the two Cys to Ala or Ser, the kinase was inactivated. There was no phosphorylation of LHCII under state 2 conditions, and the mutants were blocked in state 1. Possibly a disulfide bridge between these two Cys is required for kinase activity, or redox changes of this disulfide bridge are critical for its activity (Figure 7). The question arises how the redox state of these two Cys in the lumen is regulated through the redox state of the stromal compartment. Recently, at least two components of a transthylakoid thiol–reducing pathway have been identified in chloroplasts. The CcdA thiol disulfide transporter is a polytopic thylakoid protein with two highly conserved Cys in membrane domains, which is able to convey reducing power from the stroma to the lumen [45]. The second component is the Hcf164 thioredoxin-like protein, which acts as a thiol disulfide oxidoreductase [46,47]. This system, which is also conserved in bacteria, is thought to be involved in cytochrome c and cytochrome b_6f assembly but could also have additional roles. In this respect, given the tight physical association of Stt7 with the cytochrome b_6f complex and the requirement of this active complex for the activation of the kinase, it is tempting to propose that the redox state of the Cys68 and Cys73 couple is controlled through the same thioreduction system that operates in cytochrome b_6f assembly. These changes in redox state of Stt7 could in turn induce conformational changes of the kinase and affect both its activity and stability.

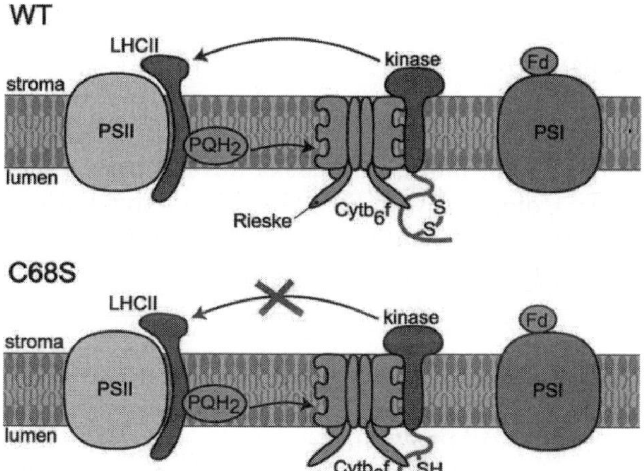

Figure 7. Model of Stt7 Kinase within the Thylakoid Membrane in Wild Type, Stt7-C68S, and FLAG-Stt7-HA. The N-terminal part of Stt7 interacts with the lumenal domain of the Rieske protein. A disulfide bridge between Cys68 and Cys73 may be required for the activity of Stt7 that is abolished in Stt7-C68S. Interaction of Stt7 with Rieske protein is prevented by the insertion of a FLAG-tag at the N-terminal end of Stt7. The catalytic domain of Stt7 is on the stromal side of the thylakoid membrane where it could phosphorylate LHCII. The HA-tag of the Stt7-HA is at is C-terminal end on the stromal side of the thylakoid membrane. Fd, ferredoxin; PQH_2, plastoquinol.

Materials and Methods

Strains and Media

Chlamydomonas reinhardtii wild-type and mutant cells were grown as described [48]. The stt7 mutant and stt7 complemented with Stt7-HA were used [21]. The HA-tag consists of six copies of the HA peptide YPYDVPDYA inserted at the C-terminal end of Stt7. In some experiments, the double-tagged FLAG-Stt7-HA was used with FLAG inserted after the Stt7 transit peptide and the HA-tag at the C-terminal end of Stt7. Strains were maintained on Tris-acetate-phosphate (TAP) medium at 25 °C in dim light (10 µmol m^{-2} s^{-1}). The stt7 mutant strain complemented with Stt7-HA was also transformed with the ph6FA1 plasmid [25] in order to obtain a strain expressing Stt7-HA and cytochrome f tagged with a His-tag. Homoplasmicity of this strain was checked by PCR.

Tagging of Stt7

Stt7-HA consists of six copies of the HA epitope YPYDVPDYA inserted at the C-terminal end of Stt7. The molecular mass of this HA tag is 8.4 kDa and that of Stt7-HA is 85 kDa. Compared with Stt7-HA, FLAG-Stt7-HA contains in addition the FLAG tag RDYKDHDGDYKDHDIDYKDDDDKS with a molecular mass of 3 kDa inserted after the 41 amino acid transit peptide of Stt7.

Analysis of Thylakoid Membranes in State 1 and State 2

Cultures were grown in TAP medium to a density of 2×10^6 cells/ml. Cells were subsequently concentrated 10-fold in HSM medium. State 1 was induced by incubating cells in 10^{-5} M DCMU (3-(3,4-dichlorophenyl)-1,1-dimethylurea) in dim light (10 µmol m^{-2} s^{-1}) under strong aeration, and state 2 was obtained by incubating cells under anaerobic conditions in the dark. Cells were harvested at 6,000g for 10 min and resuspended in buffer at a density of 108 cells/ml and broken in a French press at 1,200 psi. Thylakoid membranes (0.8 mg/ml) were prepared as described [11] and solubilized with 0.9% n-dodecyl-β-maltoside for 30 min in ice, then 0.5 ml were layered on a sucrose density gradient (0.1–1.3 M sucrose in 5 mM Tricine-NaOH [pH 8.0], 0.05% n-dodecyl-β-maltoside) and centrifuged at 280,000g for 16 h in a SW40 Beckman rotor. After centrifugation, the gradient was divided into 18 fractions that were analyzed by SDS/PAGE and immunoblotting. The antibodies used were against HA, FLAG, Cytf, PsaA, D1, CP26, CP29, Lhcbm5, and P-Thr.

Protein Stability Measurements During State Transitions

Cultures were grown in TAP medium to a density of 2 × 10^6 cells/ml. After a 10-fold concentration, cells were incubated in HSM medium under anaerobic conditions in the dark for 2 h. Cells were then maintained under state 2 conditions or cultured under dim light under strong aeration (state 1 conditions). In some cases, cycloheximide (10 μg/ml) or protease inhibitor cocktail (Roche) (2× concentration recommended by the manufacturer) were added prior to the onset of state 1 conditions.

Fluorescence Measurements

Chlorophyll fluorescence emission spectra were recorded with a Jasco FP-750 spectrofluorimeter using intact cells at a concentration of 10^6 cells/ml frozen in liquid nitrogen. The excitation light had a wavelength of 435 nm, and emission was detected from 650 to 800 nm. Fv and Fmax measurements at room temperature were performed with a Hansatech PAM fluorimeter.

Immunoblot Analysis of Proteins Phosphorylated During State Transitions

Proteins from thylakoid membranes isolated from cells in either state 1 or state 2 were separated on a 15% SDS-polyacrylamide 6 M urea gel and transferred to nitrocellulose membranes. The membranes were blocked with bovine serum albumin and incubated with rabbit anti-phosphothreonine antibody (Cell Signaling Technology).

Immunoprecipitations and Pull-Down Experiments

Thylakoid membranes (0.8 mg/ml) were solubilized with n-dodecyl-β-maltoside for 30 min in ice, and nonsolubilized material was removed by centrifugation at 12,000g for 10 min at 4 °C. Fifty microliters of anti–HA-affinity matrix was added to 0.3 mg of chlorophyll of solubilized membranes and incubated overnight at 4 °C. The beads were washed five times in TBS-BSA (100 mM Tris/HCl [pH 7.5], 150 mM NaCl, 0.05% BSA), and the bound proteins were eluted in 40 μl of 2× SDS loading buffer (100 mM Tris/HCl [pH 6.8], 4% SDS, 0.2% bromophenol blue, 20% glycerol) for 30 min at room temperature. The immunoprecipitated proteins were analyzed by immunoblotting with antibodies against subunits of the photosynthetic complexes.

The full-length Stt7 cDNA was cloned in the pGEX-4T-1 expression vector (Amersham Pharmacia Biotech). Proteins were expressed in Escherichia coli as GST fusion and purified with the GST fusion system kit (Amersham Pharmacia Biotech). Two micrograms of GST fusion protein were immobilized on glutathione Sepharose 4B beads (incubation for 2 h at 4 °C) and mixed with 50 µg of solubilized thylakoid extract in 0.5 ml of PBS. After overnight incubation at 4 °C on a rotary shaker, beads were washed four times with the same buffer, resuspended in 30 µl of SDS gel-loading buffer (100 mM DTT, 2% SDS), and the eluate was fractionated by SDS-PAGE.

Proteolysis of Thylakoid Membranes

Thylakoids were washed three time in 50 mM Hepes (pH 7.4), 0.3 M sucrose [49] without protease inhibitors and resuspended in NH_4HCO_3 25 mM. Samples were sonicated in a waterbath sonicator with 1-min sonication and 30-s cooling five times. Endoproteinase GLU-C (Sigma) was added at indicated concentrations to thylakoid membranes at a chlorophyll concentration of 0.3 mg/ml at room temperature for 15 min. The digestion was arrested by TCA precipitation and samples were resuspended in 1× loading buffer containing 8 M Urea. Further analyses of the samples were performed by SDS-PAGE and immunoblotting.

Isolation of Cytochrome b_6f Complex

All steps were performed at 4 °C with 1 mM AEBSF (Roche) as protease inhibitor in all buffers. After solubilization with 0.9% n-dodecyl-β-maltoside, 640 µg of thylakoid membranes (chlorophyll equivalent) were incubated 2 h with Ni-NTA matrix (Qiagen) in the presence of 200 mM NaCl and 10 mM imidazole. The matrix was washed five times in washing buffer (TBS, 200 mM NaCl, 20 mM imidazole), and the bound proteins were eluted in 2× 1 ml of elution buffer (TBS, 250 mM NaCl, 300 mM imidazole). The eluted proteins were analyzed by immunoblotting with antibodies against subunits of the photosynthetic complexes.

Split-Ubiquitin System

The split-ubiquitin experiments were performed using the DUALmembrane kit 3 from Dualsystems Biotech (http://www.dualsystems.com). Cub fused to the artificial transcription factor LexA-VP16 was fused to either the N-terminal end of Stt7 (minus the chloroplast signal peptide) or to its C-terminal end (including STE leader sequence) according to the manufacturer's protocol using SfiI restriction sites. These constructs were then tested for their ability to release LexA-VP16 when coexpressed with an integral membrane protein (Alg5) fused at its C-terminal to the N-terminal half of ubiquitin (Nub). If both Cub and Nub are located in the cytoplasm, spontaneous reassociation will occur, and ubiquitin-specific proteases will be recruited, releasing LexA-VP16, which will in turn activate the reporter genes (Ade and His).

Acknowledgements

We thank N. Roggli for artwork, Stefano Cazzaniga for technical assistance, Michel Schneider for the bioinformatic analysis, F.-A. Wollman for the cytochrome f, subunit V, and Rieske antisera, and M. Goldschmidt-Clermont for critical reading of the manuscript.

Author Contributions

SL and JDR conceived and designed the experiments. SL, AW, NDF, and CD performed the experiments. SL, AW, NDF, CD, RB, and JDR analyzed the data. SL, AW, NDF, RB, and JDR contributed reagents/materials/analysis tools. SL and JDR wrote the paper.

Funding. This work was supported by the National Centre of Competence in Research (NCCR) Plant Survival and by grants 3100-0667763.02 and 3100A0-117712 from the Swiss National Foundation. The funders had no role in study design, data collection and analysis, decision to publish, or preparation of the manuscript.

Competing interests. The authors have declared that no competing interests exist.

References

1. Holt NE, Fleming GR, Niyogi KK (2004) Toward an understanding of the mechanism of nonphotochemical quenching in green plants. Biochemistry 43: 8281–8289.
2. Allen JF (1992) Protein phosphorylation in regulation of photosynthesis. Biochim Biophys Acta 1098: 275–335.
3. Rochaix JD (2007) Role of thylakoid protein kinases in photosynthetic acclimation. FEBS Lett 581: 2768–2775.
4. Wollman FA (2001) State transitions reveal the dynamics and flexibility of the photosynthetic apparatus. EMBO J 20: 3623–3630.

5. Vener AV, van Kan PJ, Rich PR, Ohad II, Andersson B (1997) Plastoquinol at the quinol oxidation site of reduced cytochrome bf mediates signal transduction between light and protein phosphorylation: Thylakoid protein kinase deactivation by a single-turnover flash. Proc Natl Acad Sci USA 94: 1585–1590.
6. Zito F, Finazzi G, Delosme R, Nitschke W, Picot D, et al. (1999) The Qo site of cytochrome b6f complexes controls the activation of the LHCII kinase. EMBO J 18: 2961–2969.
7. Minagawa J, Takahashi Y (2004) Structure, function and assembly of Photosystem II and its light-harvesting proteins. Photosynth Res 82: 241–263.
8. De Vitry C, Wollman FA (1988) Changes in phosphorylation of thylakoid membrane proteins in light-harvesting complex mutants from Chlamydomonas reinhardtii. Biochim, Biophys Acta 933: 444–449.
9. Fleischmann MM, Ravanel S, Delosme R, Olive J, Zito F, et al. (1999) Isolation and characterization of photoautotrophic mutants of Chlamydomonas reinhardtii deficient in state transition. J Biol Chem 274: 30987–30994.
10. Kargul J, Turkina MV, Nield J, Benson S, Vener AV, et al. (2005) Light-harvesting complex II protein CP29 binds to photosystem I of Chlamydomonas reinhardtii under State 2 conditions. FEBS J 272: 4797–4806.
11. Takahashi H, Iwai M, Takahashi Y, Minagawa J (2006) Identification of the mobile light-harvesting complex II polypeptides for state transitions in Chlamydomonas reinhardtii. Proc Natl Acad Sci U S A 103: 477–482.
12. Lunde C, Jensen PE, Haldrup A, Knoetzel J, Scheller HV (2000) The PSI-H subunit of photosystem I is essential for state transitions in plant photosynthesis. Nature 408: 613–615.
13. Bulté L, Gans P, Rebeille F, Wollman FA (1990) ATP control on state transitions in Chlamydomonas. Biochim Biophys Acta 1020: 72–80.
14. Finazzi G, Barbagallo RP, Bergo E, Barbato R, Forti G (2001) Photoinhibition of Chlamydomonas reinhardtii in State 1 and State 2: damages to the photosynthetic apparatus under linear and cyclic electron flow. J Biol Chem 276: 22251–22257.
15. Finazzi G, Rappaport F, Furia A, Fleischmann M, Rochaix JD, et al. (2002) Involvement of state transitions in the switch between linear and cyclic electron flow in Chlamydomonas reinhardtii. EMBO Rep 3: 280–285.
16. Hou CX, Rintamaki E, Aro EM (2003) Ascorbate-mediated LHCII protein phosphorylation–LHCII kinase regulation in light and in darkness. Biochemistry 42: 5828–5836.

17. Tikkanen M, Piippo M, Suorsa M, Sirpio S, Mulo P, et al. (2006) State transitions revisited-a buffering system for dynamic low light acclimation of Arabidopsis. Plant Mol Biol 62: 779–793.

18. Delosme R, Olive J, Wollman FA (1996) Changes in light energy distribution upon state transitions: an in vivo photoacoustic study of the wild type and photosynthesis mutants from Chlamydomonas reinhardtii. Biochim Biophys Acta 1273: 150–158.

19. Chuartzman SG, Nevo R, Shimoni E, Charuvi D, Kiss V, et al. (2008) Thylakoid membrane remodeling during state transitions in Arabidopsis. Plant Cell 20: 1029–1039.

20. Kruse O, Nixon PJ, Schmid GH, Mullineaux CW (1999) Isolation of state transition mutants of Chlamydomonas reinhardtii by fluorescence video imaging. Photosynthesis Res 61: 43–51.

21. Depège N, Bellafiore S, Rochaix JD (2003) Role of chloroplast protein kinase Stt7 in LHCII phosphorylation and state transition in Chlamydomonas. Science 299: 1572–1575.

22. Bellafiore S, Barneche F, Peltier G, Rochaix JD (2005) State transitions and light adaptation require chloroplast thylakoid protein kinase STN7. Nature 433: 892–895.

23. Rintamaki E, Martinsuo P, Pursiheimo S, Aro EM (2000) Cooperative regulation of light-harvesting complex II phosphorylation via the plastoquinol and ferredoxin-thioredoxin system in chloroplasts. Proc Natl Acad Sci USA 97: 11644–11649.

24. Martinsuo P, Pursiheimo S, Aro EM, Rintamäki E (2003) Dithiol oxidant and disulfide reductant dynamically regulate the phosphorylation of light-harvesting complex II proteins in thylakoid membranes. Plant Physiol 133: 37–46.

25. Stroebel D, Choquet Y, Popot JL, Picot D (2003) An atypical haem in the cytochrome b(6)f complex. Nature 426: 413–418.

26. Johnsson N, Varshavsky A (1994) Split ubiquitin as a sensor of protein interactions in vivo. Proc Natl Acad Sci USA 91: 10340–10344.

27. Subramanyam R, Jolley C, Brune DC, Fromme P, Webber AN (2006) Characterization of a novel Photosystem I-LHCI supercomplex isolated from Chlamydomonas reinhardtii under anaerobic (State II) conditions. FEBS Lett 580: 233–238.

28. Adam Z, Clarke AK (2002) Cutting edge of chloroplast proteolysis. Trends Plant Sci 7: 451–456.

29. Helm M, Luck C, Prestele J, Hierl G, Huesgen PF, et al. (2007) Dual specificities of the glyoxysomal/peroxisomal processing protease Deg15 in higher plants. Proc Natl Acad Sci USA 104: 11501–11506.

30. Bonardi V, Pesaresi P, Becker T, Schleiff E, Wagner R, et al. (2005) Photosystem II core phosphorylation and photosynthetic acclimation require two different protein kinases. Nature 437: 1179–1182.

31. Rintamaki E, Salonen M, Suoranta UM, Carlberg I, Andersson B, et al. (1997) Phosphorylation of light-harvesting complex II and photosystem II core proteins shows different irradiance-dependent regulation in vivo. Application of phosphothreonine antibodies to analysis of thylakoid phosphoproteins. J Biol Chem 272: 30476–30482.

32. Wollman FA, Lemaire C (1988) Studies on kinase-controlled state transitions in photosystem II and b6f mutants from Chlamydomonas reinhardtii which lack quinone-binding proteins. Biochim Biophys Acta 933: 85–94.

33. Hamel P, Olive J, Pierre Y, Wollman FA, de Vitry C (2000) A new subunit of cytochrome b6f complex undergoes reversible phosphorylation upon state transition. J Biol Chem 275: 17072–17079.

34. Breyton C (2000) Conformational changes in the cytochrome b6f complex induced by inhibitor binding. J Biol Chem 275: 13195–13201.

35. Zhang Z, Huang L, Shulmeister VM, Chi YI, Kim KK, et al. (1998) Electron transfer by domain movement in cytochrome bc1. Nature 392: 677–684.

36. de Lacroix de Lavalette A, Finazzi G, Zito F (2008) b6f-Associated chlorophyll: structural and dynamic contribution to the different cytochrome functions. Biochemistry 47: 5259–5265.

37. Finazzi G, Zito F, Barbagallo RP, Wollman FA (2001) Contrasted effects of inhibitors of cytochrome b6f complex on state transitions in Chlamydomonas reinhardtii: the role of Qo site occupancy in LHCII kinase activation. J Biol Chem 276: 9770–9774.

38. Gal A, Zer H, Ohad I (1997) Redox-cntrolled thylakoid protein phosphorylation. News and views. Plant Physiol 100: 869–885.

39. Dekker JP, Boekema EJ (2005) Supramolecular organization of thylakoid membrane proteins in green plants. Biochim Biophys Acta 1706: 12–39.

40. Dainese P, Bassi R (1991) Subunit stoichiometry of the chloroplast photosystem II antenna system and aggregation state of the component chlorophyll a/b binding proteins. J Biol Chem 266: 36–42.

41. Harrer R, Bassi R, Testi MG, Schafer C (1998) Nearest-neighbor analysis of a photosystem II complex from Marchantia polymorpha L. (liverwort), which contains reaction center and antenna proteins. Eur J Biochem 255: 196–205.

42. Turkina MV, Kargul J, Blanco-Rivero A, Villarejo A, Barber J, et al. (2006) Environmentally modulated phosphoproteome of photosynthetic membranes in the green alga Chlamydomonas reinhardtii. Mol Cell Proteomics 5: 1412–1425.

43. Bassi R, Marquardt J, Lavergne J (1995) Biochemical and functional properties of photosystem II in agranal membranes from maize mesophyll and bundle sheath chloroplasts. Eur J Biochem 233: 709–719.

44. Tikkanen M, Nurmi M, Suorsa M, Danielsson R, Mamedov F, et al. (2008) Phosphorylation-dependent regulation of excitation energy distribution between the two photosystems in higher plants. Biochim Biophys Acta 1777: 425–432.

45. Page ML, Hamel PP, Gabilly ST, Zegzouti H, Perea JV, et al. (2004) A homolog of prokaryotic thiol disulfide transporter CcdA is required for the assembly of the cytochrome b6f complex in Arabidopsis chloroplasts. J Biol Chem 279: 32474–32482.

46. Lennartz K, Plucken H, Seidler A, Westhoff P, Bechtold N, et al. (2001) HCF164 encodes a thioredoxin-like protein involved in the biogenesis of the cytochrome b(6)f complex in Arabidopsis. Plant Cell 13: 2539–2551.

47. Motohashi K, Hisabori T (2006) HCF164 receives reducing equivalents from stromal thioredoxin across the thylakoid membrane and mediates reduction of target proteins in the thylakoid lumen. J Biol Chem 281: 35039–35047.

48. Harris EH (1989) The Chlamydomonas sourcebook : a comprehensive guide to biology and laboratory use. San Diego (California): Academic Press. 780 p.

49. Cline K, Werner-Washburne M, Andrews J, Keegstra K (1984) Thermolysin is a suitable protease for probing the surface of intact pea chloroplasts. Plant Physiol 75: 675–678.

A Rapid, Non-Invasive Procedure for Quantitative Assessment of Drought Survival Using Chlorophyll Fluorescence

Nick S. Woo, Murray R. Badger and Barry J. Pogson

ABSTRACT

Background

Analysis of survival is commonly used as a means of comparing the performance of plant lines under drought. However, the assessment of plant water status during such studies typically involves detachment to estimate water shock, imprecise methods of estimation or invasive measurements such as osmotic adjustment that influence or annul further evaluation of a specimen's response to drought.

Results

This article presents a procedure for rapid, inexpensive and non-invasive assessment of the survival of soil-grown plants during drought treatment. The changes in major photosynthetic parameters during increasing water deficit were monitored via chlorophyll fluorescence imaging and the selection of the maximum efficiency of photosystem II (F_v/F_m) parameter as the most straightforward and practical means of monitoring survival is described. The veracity of this technique is validated through application to a variety of Arabidopsis thaliana ecotypes and mutant lines with altered tolerance to drought or reduced photosynthetic efficiencies.

Conclusion

The method presented here allows the acquisition of quantitative numerical estimates of Arabidopsis drought survival times that are amenable to statistical analysis. Furthermore, the required measurements can be obtained quickly and non-invasively using inexpensive equipment and with minimal expertise in chlorophyll fluorometry. This technique enables the rapid assessment and comparison of the relative viability of germplasm during drought, and may complement detailed physiological and water relations studies.

Background

With the increasing demands of industrial, municipal and agricultural consumption on dwindling water supplies [1], the development of sustainable farming practices has taken higher priority. For this reason, advancement of the current understanding of plant responses to drought stress and the mechanisms involved has become a major target of research and investment, with the ultimate goal of developing crops with improved water use efficiencies and minimized drought-induced loss of yield [2,3]. On a multi-gene scale, analysis of quantitative trait loci allows identification of genetic regions responsible for control of complex responses such as the co-ordination of the whole-plant response to water deficit [4,5]. In parallel to this, as our comprehension of the molecular signaling events leading to drought responses has increased, genetic engineering techniques now also permit the manipulation of these response mechanisms through targeted overexpression or suppression of specific genes [3,6].

Irrespective of the method used to generate plants with altered drought responses, their performance under drought conditions must be evaluated in order to determine their effectiveness. This introduces a number of experimental decisions, not only with respect to the manner in which water deficit is applied, but also the means used to assess the drought stress response. In regards to the application of water deficit to small model plants such as Arabidopsis thaliana several alternative procedures are in common use, including the detachment of leaves or whole rosettes [7], air-drying of uprooted plants [8], or the transfer of specimens to solute-infused media [9]. Rosette detachment and uprooting are suitable for assessment of a plant's ability to resist rapid water loss using dehydration avoidance mechanisms, such as stomatal closure. In contrast, growth on solute-infused media allows exposure of specimens to a defined level of water deficit over a longer period of time, and thus is a valid means of evaluating adaptive responses [10]. Possibly the most straightforward and relevant application of drought stress is through experiments where water is withheld from soil-grown plants. Soil-drying techniques are generally regarded as the most practical means of approximating field drought conditions for laboratory-based research. However, their use introduces complicating factors such as variation in leaf or soil water loss rates due to differences in plant size and soil composition [10,11] and may necessitate the monitoring and adjustment or control of soil water content [12,13].

In order for soil-drying experiments to yield quantifiable comparisons between genotypes it is crucial that a suitable method of assessment be employed [11,14]. Measurements of stomatal conductance [15,16], leaf or soil water potential [12,17] or plant relative water content (RWC) [12] provide meaningful quantitative data and are necessary in a detailed physiological analysis of drought response characteristics. However, determination of leaf water potential or water content involves destructive analyses that may influence future measurements and may not accurately represent the plant as a whole. Physical disturbance to specimens is also typically unavoidable during analyses of transpiration and soil water content. The simplest assessment of viability in response to drought is the capacity of a plant to grow and remain alive under progressively increasing water deficit conditions, and thus it is common practice to utilize such survival assays to compare the drought performance of different plant lines. In such survival experiments, watering is resumed after the majority of specimens appear to have perished, and the percentage of surviving (viable) plants is presented as a measure of the drought tolerance of a line [7,18-20]. However, these survival studies rely on qualitative observation of physical symptoms of water deficit stress such as turgor loss, chlorosis, and other qualities that can vary greatly between specimens and are also sensitive to experimental conditions. Critically, the timing of rehydration presents a major problem; for instance, for plants that fail to recover upon rewatering, it is not be possible to determine retrospectively the time at which they perished. Thus, current laboratory-based techniques require either invasive or destructive measurements or are largely subjective and qualitative.

With respect to drought, the negative impact on photosynthesis is well-documented, with carbon assimilation declining progressively with increasing water deficit as a result of both stomatal and metabolic limitations [21-24]. Thus, non-invasive measurement of photosynthesis by chlorophyll a fluorometry [25,26] may potentially provide a means to determine plant viability and performance in response to drought. Measurement of chlorophyll fluorescence by probe-based systems has been utilized for non-invasive analyses of stress-induced perturbations to photosynthesis for several decades [27,28]. Indeed, dissection and analysis of the rapid polyphasic chlorophyll a fluorescence transient OJIP [29], a technique applied previously to measure tolerance to light [30] and chilling [31] stresses, was recently employed to assess the response of several barley cultivars to non-lethal drought stress [32]. The recent introduction of chlorophyll fluorescence imaging systems has allowed acquisition of fluorescence data from larger sample areas than probe-based systems [33,34], thereby enabling simultaneous measurement of several specimens and the identification of spatial heterogeneities in photosynthesis across whole leaves or rosettes. Such imaging techniques have also been successfully utilized to examine the impact of numerous environmental stresses [35], including cold [36,37], high light [38] and wounding [34].

In this article, we tested the response of major photosynthetic parameters to increasing water deficit in Arabidopsis with the objective of developing a rapid, reproducible, accurate and non-invasive method for monitoring plant viability in response to prolonged drought. We have developed a procedure that allows a quantitative and precise determination of viability in intact, drought-stressed Arabidopsis plants. The accuracy and general application of this technique has been demonstrated in different wild-type cultivars and in mutant lines that possess differences in drought performance or altered photosynthetic characteristics.

Results

Identification of Drought-Induced Changes in Photosynthetic Parameters in Arabidopsis Wild-Type Ecotypes

In order to identify a parameter suitable for monitoring survival in Arabidopsis in response to water deficit, an assessment of common photosynthetic parameters was performed spanning the duration of a prolonged, terminal drought treatment. To verify that any observed trends would be applicable across experiments involving Arabidopsis lines of different ecotypic backgrounds, three commonly-used species of Arabidopsis were examined: Columbia (Col), Landsberg erecta (Ler) and C24.

The maximum efficiency of photosystem II (Fv/Fm) and operating efficiency of photosystem II (ΦPSII) represent the capacity for photon energy absorbed by photosystem II (PSII) to be utilized in photochemistry under dark- and light-adapted conditions respectively [25,39]. As shown in Figures 1a and 1d, Fv/Fm did not vary from levels expected for plants under non-stressed conditions (~0.800) until late in the course of the treatment, when a slight decline (to 0.700–0.750) was observed. This was followed by a sudden and rapid decline to very low levels (0.100–0.250) over a 2–3-day period, after which very little change was noted. This decrease in Fv/Fm affected all rosette leaves and was readily discernible from false-colour images of Fv/Fm measurements (Figure 1d). For clarity, Figure 1a shows representative measurements from a single plant of each ecotype. ΦPSII levels under the growth illumination conditions were likewise stable until the latter stages of drought, at which time a rapid decline was observed (Figure 1b). This decline appeared to precede the decline in Fv/Fm by approximately one day; often ΦPSII fell to 50% or less of normal levels before an appreciable change in Fv/Fm was noted.

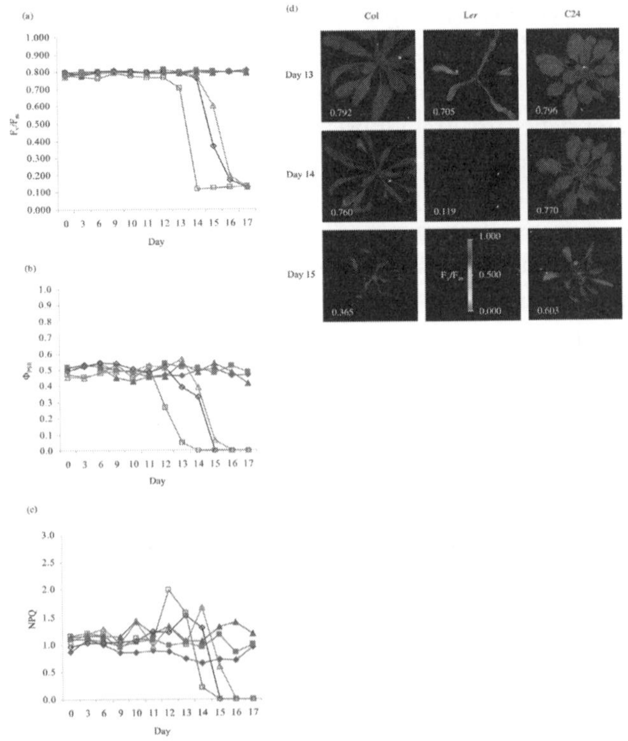

Figure 1. Measurements of (a) F_v/F_m, (b) Φ_{PSII} and (c) NPQ during progression of drought. Measurements are shown for Columbia (□), Landsberg (□) and C24 (Δ) plants; filled symbols represent controls, empty symbols represent drought-treated plants. For both control and drought-treated populations, n = 8 for each line; for clarity, only measurements from one control and one drought specimen of each line are displayed. (d) False-colour images of F_v/F_m measurements obtained from drought-affected specimens during late drought. The average F_v/F_m measurements of each plant are shown in the lower left corner of the respective images. Note that false-colour images were not generated at F_v/F_m values of < ~0.125; for details, refer to Experimental Procedures. The same individual specimens provided all the measurements presented in Figure 1a-d.

Under conditions where absorption of photons exceeds the capacity for their utilization in photochemical processes, excess excitation energy may be dissipated as thermal radiation via xanthophyll-mediated non-photochemical quenching (NPQ) [40]. NPQ did not show appreciable changes for most of the treatment, with values ranging from approximately 0.8–1.6 (Figure 1c). During late drought, NPQ levels tended towards the higher end of this range, around 1.6–1.8. This slight increase was followed by a more pronounced decrease to minimal levels, and eventually nil. A number of other photosynthetic parameters were also monitored, including the rate of photosynthetic electron transport (ETR) [39] and non-regulated energy dissipation (ΦNO) [41]. All parameters investigated underwent similar changes to those described above, remaining mostly constant before undergoing a sudden, catastrophic decline (or, in the case of ΦNO, a sudden increase) to critical levels. The rapid decline in photosynthetic parameters occurred concurrently with the appearance of physical symptoms of drought stress, including chlorosis of leaves and loss of turgor (Figure 1d). As Fv/Fm is the most readily measurable of these parameters, it was investigated further.

Correlation of the Decline in Fv/Fm with Decreased Plant Water Status and Viability

To determine if the rapid decline in F_v/F_m during late drought correlates with deterioration in plant water status, the RWC of drought-affected plants exhibiting signs of photosynthetic decline (F_v/F_m < 0.750) was determined (Figure 2). Well-watered plants had RWCs of 80–90% and F_v/F_m levels of ~0.800. Under drought conditions, for RWCs in the range of 20–80%, F_v/F_m varied between 0.700–0.750. Plants experiencing critical levels of water deficiency (RWC of 10–20%) displayed noticeably depressed F_v/F_m levels, in the range of 0.450–0.750. The close correlation between the sudden decline in F_v/F_m and critical levels of water deficit suggest that the rapid changes in F_v/F_m may be a useful indicator of terminal water loss, or loss of viability, at which point plants are unable to recover even if the soil is rehydrated. Association of this loss of viability with the decline of F_v/F_m beyond a 'threshold' value would provide a convenient, non-invasive means of identifying the time of death of plants subjected to drought.

To determine the threshold for viability, drought-treated Columbia, Landsberg and C24 plants exhibiting Fv/Fm measurements in the range 0.100–0.750 were rehydrated. None of the plants whose Fv/Fm measurements were less than the 33% of the mean Fv/Fm of watered control plants showed signs of recovery after 3 days, whereas the large majority (87%) of plants with Fv/Fm values above this threshold recovered following rehydration (Figure 3a, b). This visible recovery post-rehydration correlated with a gradual recovery in Fv/F= (Figure 3b).

For plants that showed no visible signs of recovery, Fv/Fm levels remained below 0.300. Thus, a threshold of 33% of the mean Fv/Fm of control plants provides a method to reliably identify non-viable specimens within a severely drought-affected population. The Fv/Fm threshold test provides a level of accuracy not possible through visual evaluation alone, as demonstrated in Figure 4. In this example, Fv/Fm measurements were performed on a subset of plants, all of which were classified visually as being dead (Figure 4a, b) despite the presence of viable specimens. Application of the threshold test correctly distinguished between the viable and non-viable plants, as confirmed through rehydration (Figure 4c).

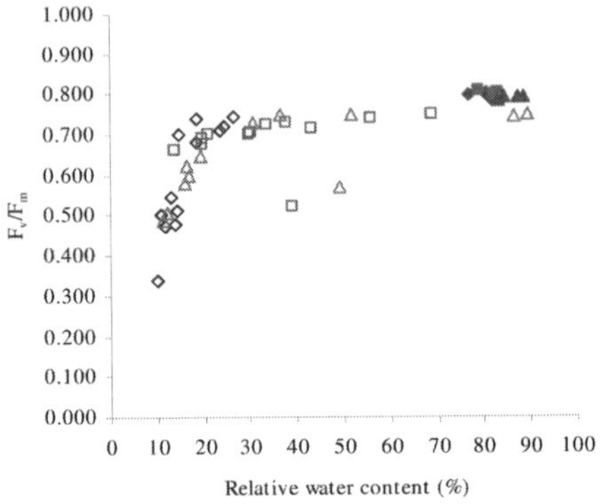

Figure 2. Relationship between F_v/F_m and plant relative water content. Measurements are shown for Columbia (□), Landsberg (□) and C24 (Δ) plants; filled symbols represent controls, empty symbols represent drought-treated plants. For control populations, n = 4 for each line; for drought-treated populations, n = 12 for each line. Data shown are representative of two separate experiments.

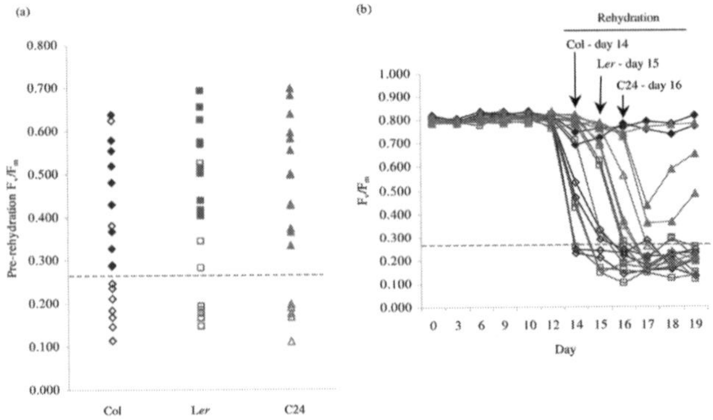

Figure 3. Validation of the F_v/F_m threshold test for viability. Drought-affected Columbia (□), Landsberg (□) and C24 (Δ) plants were rehydrated after their F_v/F_m levels were observed to fall below 0.750. Filled symbols represent plants that recovered within 3 days of rehydration, while empty symbols represent plants that failed to evidence signs of recovery following watering. The 33% threshold for a typical average control F_v/F_m of 0.800 is shown as a dotted line. (a) F_v/F_m measurements of individual specimens immediately prior to rehydration. For each line, n = 20. (b) Change in F_v/F_m of drought-treated plants following rehydration. Columbia, Landsberg and C24 plants were rewatered after 14, 15 and 16 days' drought respectively, as indicated by arrows. For each line, n = 6. The data presented in Figures 3a and 3b were obtained from separate experiments.

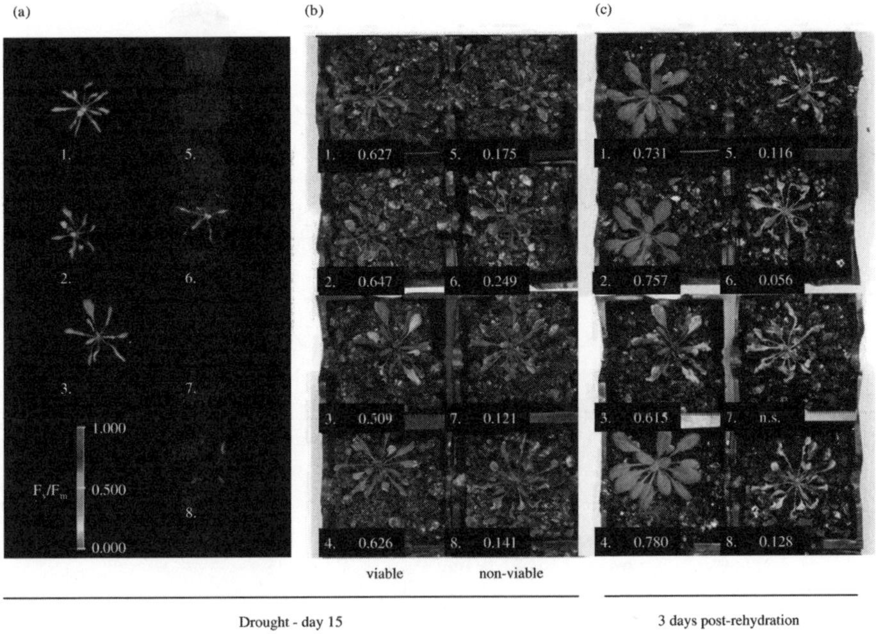

Figure 4. Visual estimation of drought survival. (a) False-colour representations of F_v/F_m measurements of Columbia plants following 15 days' drought treatment. The individual specimens were labeled 1 through 8, as indicated by the number below each plant. Note that false-colour images were not generated at F_v/F_m values of < ~0.125; for details, refer to Experimental Procedures. The image of plant #4 has been omitted for provision of the false-colour scale, however its F_v/F_m measurements were comparable to those of plant #1. (b) Photograph of the plants shown in (a). F_v/F_m measurements obtained from each plant are shown in the lower left corner of each punnet. The average F_v/F_m of control plants (not shown) was 0.800, providing a threshold F_v/F_m of 0.264. The 4 plants in the left column were classified as viable by application of the threshold test ($F_v/F_m > 0.264$), while the 4 plants in the right column were classified as non-viable ($F_v/F_m < 0.264$). (c) Photograph of the same 8 plants after watering was resumed for 3 days; n.s. = no signal detected.

Case Study: Measuring Drought Survival of Water Deficit-Tolerant Arabidopsis Mutants

To further appraise the precision of the threshold test for viability, it was utilized to perform an assessment of the survival during drought of an established water deficit-tolerant mutant, altered APX2 expression 8 (alx8; At5g63980) [42], and a drought-sensitive mutant, open stomata 1–2 (ost1-2; At4g33950) [43]. Monitoring of Fv/Fm levels and application of the threshold test (Figure 5a, b) permitted estimation of plant survival to a specific day (Figure 5c), with loss of viability confirmed via rehydration (data not shown). The experiment demonstrated that alx8 survived an average of 5.0 days longer than Columbia ($p < 0.0001$), while ost1-2 plants lost viability 1.4 days earlier than the Landsberg erecta wild-type parent ($p < 0.05$).

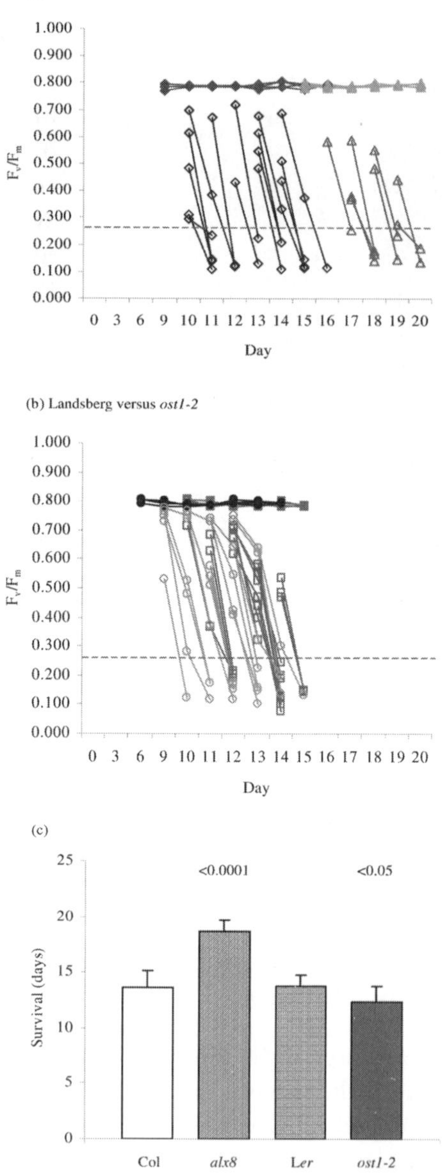

Figure 5. Drought survival analysis of alx8 and ost1-2 plants. (a, b) Application of the threshold test. The Fv/Fm measurements of individual (a) Columbia (□) and alx8 (Δ), and (b) Landsberg (□) and ost1-2 (O) specimens are shown. Filled symbols represent controls; empty symbols represent plants that failed to evidence signs of recovery within 3 days of rehydration. The 33% threshold for a typical average control F_v/F_m of 0.800 is shown as a dotted line. For control populations, n = 4 for each line; for drought-treated populations, n = 15 for Columbia, Landsberg and ost1-2, and n = 8 for alx8. (c) Comparison of drought survival times of alx8, ost1-2 and wild-type plants. Error bars indicate standard deviation. Pairwise t-tests were performed for the mutant lines against survival times of their corresponding wild-type (Columbia for alx8, Landsberg for ost1-2), yielding p-values as shown.

Case Study: Measuring Drought Survival of Photosynthetically-Impaired Arabidopsis Mutants

The use of the threshold test had now been validated on the common Columbia and Landsberg erecta ecotypes and on mutant plants with altered drought characteristics but comparable photosynthetic efficiencies. To determine whether the 33% F_v/F_m threshold test remained a valid predictor of viability when applied to Arabidopsis mutants with impaired photosynthetic activities, the drought survival of three variegated lines of Arabidopsis was evaluated. The yellow variegated 1, (var1-1; At5g42270) [44], yellow variegated 2 (var2-2; At2g30950) [45] and altered APX2 expression 13 (alx13) lines exhibit chlorotic sectoring and depressed photosynthetic efficiencies. Depending on the severity of chlorosis, the F_v/F_m values of control plants from the three mutant lines varied from 0.650–0.800, corresponding to threshold values in the range of 0.215–0.264. The threshold test was applied using the lower threshold values obtained from the mutant controls rather than the threshold of the non-chlorotic Columbia wild-type (Figure 6a–d). In this manner, survival times were estimated as shown in Figure 6e, with all plants failing to recover following rehydration.

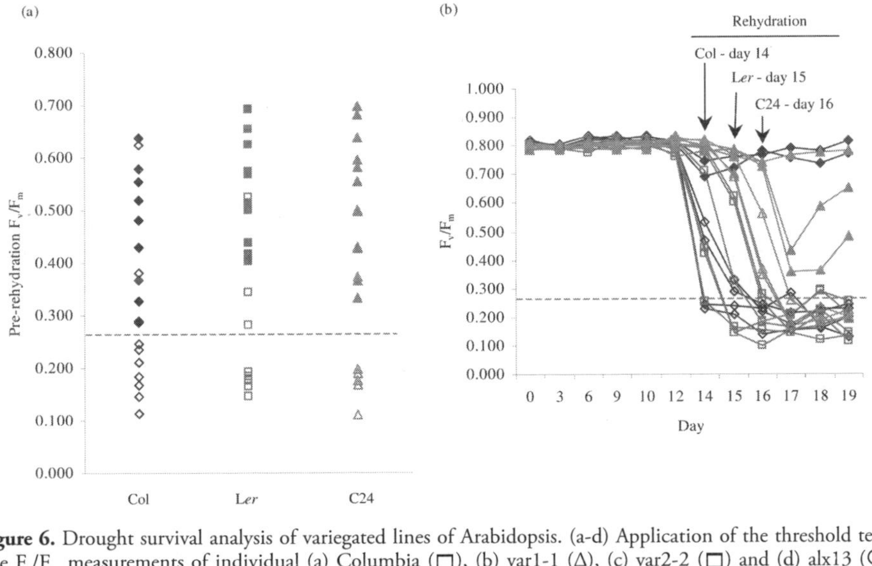

Figure 6. Drought survival analysis of variegated lines of Arabidopsis. (a-d) Application of the threshold test. The F_v/F_m measurements of individual (a) Columbia (□), (b) var1-1 (Δ), (c) var2-2 (□) and (d) alx13 (O) specimens immediately prior to rehydration are shown. Filled symbols represent controls; empty symbols represent plants that failed to evidence signs of recovery within 3 days of rehydration. The 33% threshold for each line is shown as a dotted line. For control populations, n = 7 for each line; for drought-treated populations, n = 16 for each line. For clarity, only measurements from 4 control plants are shown. (e) Comparison of drought survival times of variegated lines. Error bars indicate standard deviation. Pairwise t-tests were performed against survival times of wild-type Columbia plants, yielding p-values as shown; n.s. = not significant. Data shown are the combined results of two separate experiments.

Case Study: Comparison of a Traditional Rehydration Survival Test and the F_v/F_m Threshold Test

The threshold test was next applied to assess the drought survival of transgenic plants altered in the expression of an abiotic stress response transcription factor. The protein encoded by the HL-responsive gene zinc-finger of Arabidopsis 10 (ZAT10; At1g27730) has been shown to function as both a positive and negative regulator of a number of genes involved in the oxidative stress response and is implicated in the activation and suppression of several abiotic stress response pathways, including osmotic, heat and salinity stress [46]. However, overexpression of ZAT10 has been variously reported as either conferring a marked increase in drought resistance [47] or not affecting the drought response at all [46] when assessed using the traditional re-watering survival tests.

Two transgenic lines in which ZAT10 gene expression was suppressed via RNA interference (zat10(i)-1 and zat10(i)-3) and two lines in which ZAT10 was constitutively overexpressed under the direction of the cauliflower mosaic virus 35S promoter (35S:ZAT10-6 and 35S:ZAT10-14) were subjected to drought survival analysis via both traditional rehydration methods and our threshold test [48]. As shown in Table 1a, in a traditional rehydration test three zat10(i) plants were shown to survive 20 days' drought treatment whereas all Columbia wild-type and 35S:ZAT10 specimens had perished by this time. The inherent limitations of data obtained from this form of experiment make it difficult to draw substantive conclusions from these results as to whether this difference is significant and accurate. A threshold test survival experiment (Figure 7a, b), in comparison, indicated that length of survival in days was not statistically different for the two RNA interference lines and one of the overexpression lines (Table 1b). Only the 35S:ZAT10-14 line displayed a significantly altered survival in comparison to the wild-type, a difference which may be considered negligible (p-value = 0.049).

Discussion

Identification of a Photosynthetic Parameter Suitable for Assessment of Drought Progression

Here we have shown that F_v/F_m declines rapidly during late drought and can serve as an indicator of the latter phase of drought and subsequent loss of viability. Although it is possible that the other photosynthetic measurements obtained in this study could be employed as an indicator of viability, the F_v/F_m parameter is recommended for several reasons. First, as shown in Figure 1a, F_v/F_m values are typically very consistent between lines and individual plants; as such, any small decline is easily noticeable and signifies clearly that loss of viability is imminent. The consistency of the F_v/F_m parameter also increases the ease with which a threshold level can be defined. More importantly, unlike light-dependent parameters such as Φ_{PSII} and NPQ, F_v/F_m is obtained from specimens in the dark-adapted state, negating the need for an extended period of illumination prior to measurement. Thus, as measurement of F_v/F_m can be completed using a single saturating pulse, rapid screening of a large number of plants may be achieved.

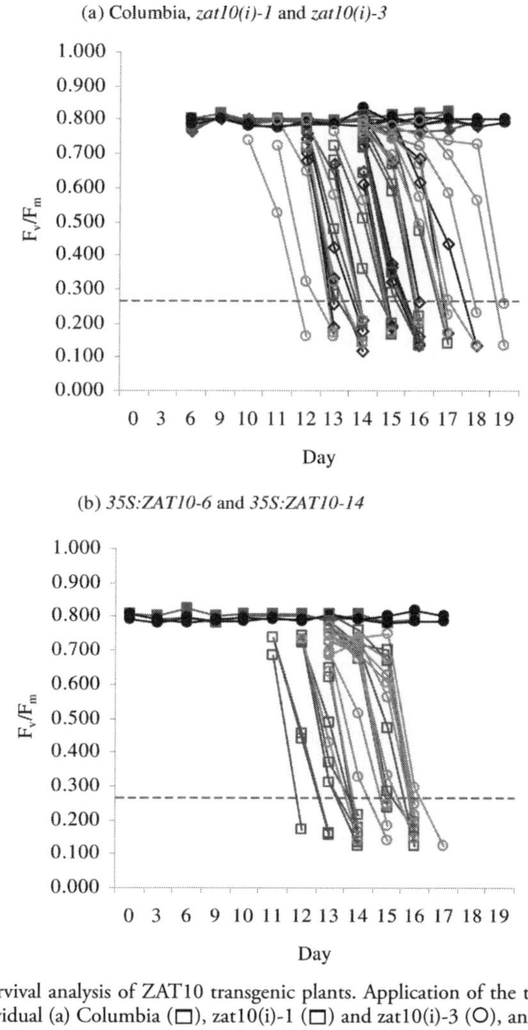

Figure 7. Drought survival analysis of ZAT10 transgenic plants. Application of the threshold test. The F_v/F_m measurements of individual (a) Columbia (□), zat10(i)-1 (□) and zat10(i)-3 (O), and (b) 35S:ZAT10-6 (□) and 35S:ZAT10-14 (O) specimens are shown. Filled symbols represent controls; empty symbols represent plants that failed to evidence signs of recovery within 3 days of rehydration. The 33% threshold for a typical average control F_v/F_m of 0.800 is shown as a dotted line. For control populations, n = 2 for each line; for drought-treated populations, n = 13 for each line.

Table 1. Drought survival analyses of ZAT10 transgenic plants using traditional and threshold test methods.

	Line	Col	*zat10(i)-1*	*zat10(i)-3*	*35S:ZAT10-6*	*35S:ZAT10-14*
(a)	# of viable plants post-rehydration	0/7	2/7	1/7	0/7	0/7
(b)	Survival time (days) ± s.d.	15.2 ± 1.5	15.3 ± 0.9	15.8 ± 2.5	15.5 ± 1.2	16.2 ± 1.1
	t-test	-	n.s.	n.s.	n.s.	~0.05

(a) Traditional rehydration survival test results. Plants were subjected to drought treatment for 20 days, after which watering was resumed; values indicate the number of plants that recovered within 3 days of rehydration. For each line, $n = 7$. (b) Threshold survival test results. For control populations, $n = 2$ for each line; for drought-treated populations, $n = 13$ for each line. s.d. = standard deviation. Pairwise *t*-tests were performed against survival times of wild-type Columbia plants, yielding *p*-values as shown; n.s. = not significant.

Quantification of Viability Using Chlorophyll Fluorescence Measurements

To employ the decline in F_v/F_m as a means of determining viability during drought, it was necessary to identify a threshold F_v/F_m level that would reflect a point at which recovery was no longer possible. As it is of course impossible to define an exact threshold level beyond which viability is lost, we identified a conservative threshold of 33% of control specimen measurements and showed that, in practice, decline of F_v/F_m below this level no plants were viable upon re-watering (Figure 3; Figure 4; Figure 5a, b; Figure 6a-d; Figure 7).

To validate the efficacy of the threshold test, the technique was employed to assess the drought performance of the alx8 and ost1-2 mutant lines previously identified as drought-resistant and drought-sensitive, respectively [42,43]. Using this method it was possible to monitor the viability of drought-affected plants and evaluate the survival times of individual plants in a precise and quantifiable manner (Figure 5). The robustness of the threshold test was further confirmed through its application in a drought survival analysis of three variegated lines of Arabidopsis. The variegated lines var1-1, var2-2 and alx13 are sensitive to photoinhibitory damage and consequently have impaired photosynthetic efficiencies. This impairment is manifest in reduced Fv/Fm levels in each of the three mutant lines, which in turn necessitated the application of their respective control Fv/Fm levels to calculate the 33% thresholds. The threshold test successfully ascertained loss of viability in specimens of all three mutants, demonstrating its utility even in situations where photodamage and differing photosynthetic capacities are present (Figure 6). Intriguingly, the test also indicated differences in drought survival between the mutants and wild-type, a discovery that is under further investigation.

As a demonstration of the advantages of the threshold test, the drought survival of ZAT10 transgenic lines were evaluated using both the threshold test technique and the traditional rehydration method. The limitations of the traditional rehydration test (Table 1a) are apparent: although three zat10(i) specimens remained viable at the end of the experiment, the extent of this increased survival is cannot be established as there is no indication of the time at which wild-type plants became inviable. Indeed, as this test does not yield survival data for individual specimens it is not possible to determine whether the surviving plants are outliers among their populations, nor can the variability in survival times within each population be estimated. It cannot be ascertained at all whether 35S:ZAT10 plants exhibit altered drought survival compared to the wild-type.

The threshold test, in contrast, provides a far more informative assessment of drought survival. From the data presented in Table 1b and Figure 7 it is immediately evident that the survival times of all of the lines in the threshold test experiment were very similar, with average survival times indicating that the loss of viability of all lines occurred within a 1-day period. Statistical assessment of the survival times of the transgenic lines indicated that 35S:ZAT10-14 plants may remain viable during drought for slightly longer than the wild-type, but also show that any increased viability is at most marginally significant. Note that the results shown in Table 1 are for the purposes of demonstrating differences in the interpretation of traditional and threshold survival test methods and do not represent a comprehensive analysis of the effect of altered ZAT10 expression on the drought response; such an investigation would require monitoring of additional parameters such as the extent of ZAT10 overexpression/suppression.

Applications and Suggestions for Using the Threshold Test for Measuring Viability

The threshold test offers a reliable, rapid and quantitative alternative to conventional studies of drought survival in Arabidopsis. As only minimal technical expertise and a basic understanding of chlorophyll fluorometry are required to obtain the necessary measurements, the threshold test may appeal to a broad spectrum of plant science laboratories. Further, this procedure does not require the use of expensive or esoteric equipment. Although the results presented in this analysis were produced using an IMAGING-PAM system (Walz; Effeltrich, Germany) and have also been validated using a Chlorophyll Fluorescence Imager (Technologica; Colchester, UK) (data not shown), a number of less costly devices are available. For example, the FluorPen (Photon Systems Instruments; Brno, Czech Republic) and Pocket PEA Chlorophyll Fluorimeter (Hansatech; Norfolk, UK) offer convenient means of monitoring F_v/F_m levels both in the laboratory and in the field at low cost. Instruments such as these are also amenable for determination of the OJIP fluorescence transient [29], and therefore offer the potential for assessment of plant performance during early and moderate phases of drought treatment [32] in addition to drought survivability. However, when employing a fluorescence probe it may be necessary to acquire several measurements in order to account for heterogeneities in photosynthetic efficiencies across the leaf surface of plants, particularly severely drought-stressed specimens.

While beyond the scope of this report, it is easily conceivable that the threshold test may be successfully applied to monitor the survival of plants under different forms of abiotic stress, particularly those that cause progressive deterioration of photosynthetic efficiencies. Prolonged cold or light stress, for example, can induce accumulative photoinhibitory damage to the photosynthetic machinery to the point at which a specimen is no longer able to sustain vital functions [49,50]. Likewise, it is quite likely that the threshold test could be adapted for use with other plant species. We have targeted Arabidopsis as this rapid test could be applied to mutant and genotype screens in advance of detailed analyses of water relations.

A Discussion of the Drought-Induced Changes in Chlorophyll Fluorescence Parameters in Arabidopsis

While an investigation of the physiological and photochemical bases of the observed drought-induced changes in chlorophyll fluorescence was not an objective of this report, they will be discussed briefly in this section. Measurements of the maximum and operating efficiencies of PSII, as represented by F_v/F_m and Φ_{PSII} respectively, indicated that there was no significant perturbation of PSII photochemistry or electron transport capacity within the photosystems despite the initial significant decreases in RWC (Figure 1a, b; Figure 2). Indeed, only when plant water reserves declined to critical levels (<20% RWC) were F_v/F_m measurements consistently reduced. These results are similar to observations in sunflower, where F_v/F_m was unchanged across a comparable range of water deficit stress [22], in pea, where only a slight decrease was noted despite RWC as low as 20% [51], and in triticale, where extended drought failed to alter F_v/F_m significantly [52]. Thus, although drought is known to cause gradual inhibition of assimilatory photochemistry through both stomatal [24] and metabolic [22] restriction of CO_2 availability, photosynthetic electron transport may be maintained throughout the course of drought treatment through dissipation of excess excitation energy by alternative electron sinks [40,53].

During prolonged water deficit, severe reduction of cellular water content results in enhanced leaf senescence, as reflected by elevated levels of reactive oxygen intermediates and chlorophyll degradation [54,55]. Thus, it is possible that the rapid decline in photosynthetic parameters observed during the latter stages of drought is attributable to senescence-induced chlorosis and disruption of the photosynthetic apparatus. The rapid photosynthetic decline during late drought may therefore be a consequence of the damage to PSII reaction centres or associated chlorophylls [56], although it has previously been suggested that drought-induced suppression of photosynthetic efficiencies may be due to the deterioration of an electron carrier at the donor side of PSII, rather than destruction of the PSII reaction centre or of chlorophyll molecules [57]. Chlorophyll fluorescence measurements may also be influenced by non-photosynthetic alterations in leaf physiology associated with prolonged drought, such as changes in leaf angle due to loss of turgor. Irrespective of the mechanisms responsible for the observed photosynthetic decline, though, the rapid change in the Fv/Fm parameter may nonetheless be employed via the threshold test as a means of estimating drought survival.

Conclusion

In this report, we describe a method of determining the survival of drought-treated Arabidopsis utilizing measurements of the F_v/F_m chlorophyll fluorescence parameter. Although photosynthetic parameters remained mostly unchanged during the first phase of drought treatment, a sudden deterioration in photosynthesis was observed to occur just prior to the terminal stages of drought and the loss of plant viability. By correlating the decline in the Fv/Fm parameter to this loss of viability, a procedure was developed to allow the monitoring of survival under water deficit conditions, namely defining a threshold of 33% of well-watered F_v/F_m values. The versatility of this technique was demonstrated through comparison of the drought performance of a number of Arabidopsis cultivars and to a variety of mutants with altered drought tolerance or photosynthetic capacity. As a rapid, non-invasive and inexpensive procedure, the threshold test for survival holds much value in screening for altered responses to drought in Arabidopsis germplasm. This procedure may complement existing methods of evaluating drought performance utilizing chlorophyll fluorescence [32], and increase the number of tools available for assessment of this and potentially other plant stresses.

Methods

Plant Growth Conditions and Drought Treatments

All Arabidopsis thaliana plants were cultivated under 100 ± 10 µmol photons m^{-2} s^{-1}, 8-hour photoperiod, 23°C/22°C day/night temperatures, 50%/70% day/night humidity. Seed were sown on a moistened, loosely-packed 3:1 mixture of soil:vermiculite, then vernalized at 4°C in darkness for 72 hours before transfer to growth conditions. Prior to initiating drought treatment plants were watered every second day, with every third watering supplemented with 0.5× Hoagland's

Fertilizer [58]. After 7 days' growth, seedlings were thinned to leave one plant per punnet. Drought treatments were initiated when plants were 28 days' old, at which time all specimens were at ~12-leaf stage, with the exception of assays involving variegated mutant lines and ZAT10 transgenic lines. In the experiments involving variegated mutants, in order to account for differences in developmental rates, Columbia populations began treatment at 30 days of age, alx13 at 33 days, and var1-1 and var2-2 at 37–40 days, at which times all plants were at ~14-leaf stage. For the traditional rehydration survival analysis of ZAT10 transgenic lines, drought treatment was initiated at 28 days' of age; after 20 days rehydration was performed as described below. For the threshold survival test of ZAT10 lines, drought treatment was initiated at 42 days' of age.

For drought treatments, all plants were first provided with a sufficiency of water. Punnets containing plants to be subjected to drought were then removed to water-free trays with spaces between specimens to allow air flow, and further watering withheld; all other environmental conditions were maintained as described above. Control plants remained under watered conditions for the duration of the experiment. Where rehydration was necessary, punnets were returned to watered trays for 72 hours. Plants that failed to exhibit any physical signs of recovery within this time were deemed to have lost viability.

Measurement of Photosynthetic Parameters

Chlorophyll fluorescence measurements were performed using an IMAGING-PAM chlorophyll fluorometer and ImagingWin software application (Walz; Effeltrich, Germany). For assessment of dark- and light-adapted parameters, a dark-light induction curve was performed. Dark-adapted plants were subjected to an initial saturating pulse of >1800 μmol photons m^{-2} s^{-1}, followed by a 40" delay in darkness and subsequently 10' of actinic illumination with saturating flashes at 20" intervals. An actinic irradiance of 100 ± 10 μmol photons m^{-2} s^{-1} was used to simulate growth conditions. The following parameters were derived from the final measurements obtained after the 10' light adaptation: Φ_{PSII}, Φ_{NO}, NPQ and ETR. F_v/F_m values were taken as the measurement of Φ_{PSII} at time zero. The four primary fluorescence signals – F_o, F_m, F_s' and F_m' – from which the above photosynthetic parameters were derived. For background information regarding photosynthetic parameters and theoretical aspects of chlorophyll fluorescence, refer to [25,26,39,41]. To account for variations in photosynthetic parameters across the surface of individual plants, the data presented are the average values obtained across individual rosettes. Note that, where false-colour images of the F_v/F_m parameter are shown, the ImagingWin software eliminates pixels in areas where Fm<0.040 in order to reduce background noise. For this reason, F_v/F_m images of certain severely drought-affected plants were unobtainable; in these instances the average F_v/F_m measurements alone are presented.

For experiments requiring only determination of Fv/Fm, measurements were obtained from application of a single saturating pulse to dark-adapted plants. All photosynthetic measurements were performed prior to dawn, after 12–16 hours' dark adaptation. For accurate measurement of Fv/Fm a dark adaptation of >15 minutes is typically sufficient.

Determination of Relative Water Content

For measurements of rosette RWC, the entire aerial parts of the plant were harvested using a single incision to the base of the stem, and the fresh weight (FW) of the rosette determined. The rosette was then floated on distilled water in darkness at 4°C for 24 hours before determination of turgid weight (TW). The rosette was then placed in a paper envelope and dried at 65°C for 24 hours, and the desiccated sample weighed once more to determine dry weight (DW). RWC was calculated from these measurements as follows:

$$RWC = ((FW-DW)/(TW-DW)) \times 100\%.$$

Competing Interests

The authors declare that they have no competing interests.

Authors' Contributions

NW conceived of the described procedure, performed all photosynthetic measurements and drought studies and prepared the manuscript. All authors participated in experimental design and data analysis. All authors read and approved the final manuscript.

Acknowledgements

We acknowledge the support of the Australian Research Council Centre of Excellence in Plant Energy Biology (CE0561495). We also thank Susanne von Caemmerer and Simon Dwyer for assistance with photosynthetic measurements, Pip Wilson and Gonzalo Estavillo for assistance with the alx8 and ost1-2 lines, and Peter Crisp and Tim Sloan-Gardner for assistance in the preparation of the manuscript.

References

1. Johnson N, Revenga C, Echeverria J: Managing water for people and nature. Science 2001, 292:1071–1072.
2. Somerville C, Briscoe J: Genetic engineering and water. Science 2001, 292:2217.
3. Zhang JZ, Creelman RA, Zhu J-K: From laboratory to field. Using information from Arabidopsis to engineer salt, cold, and drought tolerance in crops. Plant Physiol 2004, 135:615–621.
4. Saranga Y, Jiang CX, Wright RJ, Yakir D, Paterson AH: Genetic dissection of cotton physiological responses to arid conditions and their inter-relationships with productivity. Plant, Cell & Environment 2004, 27:263–277.
5. Vinocur B, Altman A: Recent advances in engineering plant tolerance to abiotic stress: achievements and limitations. Current Opinion in Biotechnology 2005, 16:123–132.
6. Umezawa T, Fujita M, Fujita Y, Yamaguchi-Shinozaki K, Shinozaki K: Engineering drought tolerance in plants: discovering and tailoring genes to unlock the future. Current Opinion in Biotechnology 2006, 17:113–122.
7. Catala R, Ouyang J, Abreu IA, Hu Y, Seo H, Zhang X, Chua N-H: The Arabidopsis E3 SUMO ligase SIZ1 regulates plant growth and drought responses. Plant Cell 2007, 19:2952–2966.
8. Fujita Y, Fujita M, Satoh R, Maruyama K, Parvez MM, Seki M, Hiratsu K, Ohme-Takagi M, Shinozaki K, Yamaguchi-Shinozaki K: AREB1 is a transcription activator of novel ABRE-Dependent ABA signaling that enhances drought stress tolerance in Arabidopsis. The Plant Cell 2005, 17:3470–3488.
9. Weele CM, Spollen WG, Sharp RE, Baskin TI: Growth of Arabidopsis thaliana seedlings under water deficit studied by control of water potential in nutrient-agar media. J Exp Bot 2000, 51:1555–1562.
10. Verslues PE, Agarwal M, Katiyar-Agarwal S, Zhu J, Zhu J-K: Methods and concepts in quantifying resistance to drought, salt and freezing, abiotic stresses that affect plant water status. The Plant Journal 2006, 45:523–539.
11. Bhatnagar-Mathur P, Vadez V, Sharma K: Transgenic approaches for abiotic stress tolerance in plants: retrospect and prospects. Plant Cell Rep 2008, 27(3):411–424.
12. Riga P, Vartanian N: Sequential expression of adaptive mechanisms is responsible for drought resistance in tobacco. Australian Journal of Plant Physiology 1999, 26:211–220.

13. Passioura JB: The perils of pot experiments. Functional Plant Biology 2006, 33:1075–1079.

14. Jones HG: Monitoring plant and soil water status: established and novel methods revisited and their relevance to studies of drought tolerance. J Exp Bot 2007, 58:119–130.

15. Poulson M, Boeger M, Donahue R: Response of photosynthesis to high light and drought for Arabidopsis thaliana grown under a UV-B enhanced light regime. Photosynthesis Research 2006, 90:79–90.

16. Bhatnagar-Mathur P, Devi M, Reddy D, Lavanya M, Vadez V, Serraj R, Yamaguchi-Shinozaki K, Sharma K: Stress-inducible expression of AtDREB1A in transgenic peanut (Arachis hypogaea L.) increases transpiration efficiency under water-limiting conditions. Plant Cell Reports 2007, 26:2071–2082.

17. Mane SP, Vasquez-Robinet C, Sioson AA, Heath LS, Grene R: Early PLDa-mediated events in response to progressive drought stress in Arabidopsis: a transcriptome analysis. J Exp Bot 2007, 58:241–252.

18. Karaba A, Dixit S, Greco R, Aharoni A, Trijatmiko KR, Marsch-Martinez N, Krishnan A, Nataraja KN, Udayakumar M, Pereira A: Improvement of water use efficiency in rice by expression of HARDY, an Arabidopsis drought and salt tolerance gene. Proceedings of the National Academy of Sciences 2007, 104:15270–15275.

19. Qin F, Kakimoto M, Sakuma Y, Maruyama K, Osakabe Y, Tran L-SP, Shinozaki K, Yamaguchi-Shinozaki K: Regulation and functional analysis of ZmDREB2A in response to drought and heat stresses in Zea mays L. The Plant Journal 2007, 50:54–69.

20. Chen M, Wang Q-Y, Cheng X-G, Xu Z-S, Li L-C, Ye X-G, Xia L-Q, Ma Y-Z: GmDREB2, a soybean DRE-binding transcription factor, conferred drought and high-salt tolerance in transgenic plants. Biochemical and Biophysical Research Communications 2007, 353:299–305.

21. Chaves MM: Effects of water deficits on carbon assimilation. [http://jxb.oxfordjournals.org/cgi/content/abstract/42/1/1] J Exp Bot 1991, 42:1–16.

22. Tezara W, Mitchell VJ, Driscoll SD, Lawlor DW: Water stress inhibits plant photosynthesis by decreasing coupling factor and ATP. Nature 1999, 401:914–917.

23. Flexas J, Medrano H: Drought-inhibition of photosynthesis in C3 plants: stomatal and non-stomatal limitations revisited. Ann Bot 2002, 89:183–189.

24. Cornic G: Drought stress inhibits photosynthesis by decreasing stomatal aperture – not by affecting ATP synthesis. Trends in Plant Science 2000, 5:187–188.
25. Oxborough K: Imaging of chlorophyll a fluorescence: theoretical and practical aspects of an emerging technique for the monitoring of photosynthetic performance. J Exp Bot 2004, 55:1195–1205.
26. Baker NR: Chlorophyll fluorescence: a probe of photosynthesis in vivo. Annual Review of Plant Biology 2008, 59:89–113.
27. Schreiber U, Vidaver W, Runeckles VC, Rosen P: Chlorophyll fluorescence assay for ozone injury in intact plants. Plant Physiol 1978, 61:80–84.
28. Conroy JP, Smillie RM, Kuppers M, Bevege DI, Barlow EW: Chlorophyll a fluorescence and photosynthetic and growth responses of Pinus radiata to phosphorus deficiency, drought stress, and high CO2. Plant Physiol 1986, 81:423–429.
29. Lazár D: The polyphasic chlorophyll a fluorescence rise measured under high intensity of exciting light. Functional Plant Biology 2006, 33:9–30.
30. Oukarroum A, Strasser RJ: Phenotyping of dark and light adapted barley plants by the fast chlorophyll a fluorescence rise OJIP. South African Journal of Botany 2004, 70:277–283.
31. Strauss AJ, Krüger GHJ, Strasser RJ, Heerden PDRV: Ranking of dark chilling tolerance in soybean genotypes probed by the chlorophyll a fluorescence transient O-J-I-P. Environmental and Experimental Botany 2006, 56:147–157.
32. Oukarroum A, Madidi SE, Schansker G, Strasser RJ: Probing the responses of barley cultivars (Hordeum vulgare L.) by chlorophyll a fluorescence OLKJIP under drought stress and re-watering. Environmental and Experimental Botany 2007, 60:438–446.
33. Omasa K, Shimazaki K-I, Aiga I, Larcher W, Onoe M: Image analysis of chlorophyll fluorescence transients for diagnosing the photosynthetic system of attached leaves. Plant Physiol 1987, 84:748–752.
34. Quilliam RS, Swarbrick PJ, Scholes JD, Rolfe SA: Imaging photosynthesis in wounded leaves of Arabidopsis thaliana. J Exp Bot 2006, 57:55–69.
35. Baker NR, Rosenqvist E: Applications of chlorophyll fluorescence can improve crop production strategies: an examination of future possibilities. J Exp Bot 2004, 55:1607–1621.
36. Savitch L, Barker-Åstrom J, Ivanov A, Hurry V, Öquist G, Huner N, Gardeström P: Cold acclimation of Arabidopsis thaliana results in incomplete recovery of

photosynthetic capacity, associated with an increased reduction of the chloroplast stroma. Planta 2001, 214:295–303.

37. Ehlert B, Hincha D: Chlorophyll fluorescence imaging accurately quantifies freezing damage and cold acclimation responses in Arabidopsis leaves. Plant Methods 2008, 4:12.

38. Muller-Moule P, Golan T, Niyogi KK: Ascorbate-deficient mutants of Arabidopsis grow in high light despite chronic photooxidative stress. Plant Physiol 2004, 134:1163–1172.

39. Maxwell K, Johnson GN: Chlorophyll fluorescence – a practical guide. J Exp Bot 2000, 51:659–668.

40. Lawlor DW, Cornic G: Photosynthetic carbon assimilation and associated metabolism in relation to water deficits in higher plants. Plant, Cell and Environment 2002, 25:275–294.

41. Kramer D, Johnson G, Kiirats O, Edwards G: New fluorescence parameters for the determination of Qa redox state and excitation energy fluxes. Photosynthesis Research 2004, 79:209–218.

42. Rossel JB, Walter PB, Hendrickson L, Chow WS, Poole A, Mullineaux PM, Pogson BJ: A mutation affecting ASCORBATE PEROXIDASE 2 gene expression reveals a link between responses to high light and drought tolerance. Plant, Cell and Environment 2006, 29:269–281.

43. Mustilli A-C, Merlot S, Vavasseur A, Fenzi F, Giraudat J: Arabidopsis OST1 protein kinase mediates the regulation of stomatal aperture by abscisic acid and acts upstream of reactive oxygen species production. Plant Cell 2002, 14:3089–3099.

44. Sakamoto W, Tamura T, Hanba-Tomita Y, Murata M: The VAR1 locus of Arabidopsis encodes a chloroplastic FtsH and is responsible for leaf variegation in the mutant alleles. Genes Cells 2002, 7:769–780.

45. Chen M, Jensen M, Rodermel S: The yellow variegated mutant of Arabidopsis is plastid autonomous and delayed in chloroplast biogenesis. J Hered 1999, 90:207–214.

46. Mittler R, Kim Y, Song L, Coutu J, Coutu A, Ciftci-Yilmaz S, Lee H, Stevenson B, Zhu J-K: Gain- and loss-of-function mutations in ZAT10 enhance the tolerance of plants to abiotic stress. FEBS letters 2006, 580:6537–6542.

47. Sakamoto H, Maruyama K, Sakuma Y, Meshi T, Iwabuchi M, Shinozaki K, Yamaguchi-Shinozaki K: Arabidopsis Cys2/His2-type zinc-finger proteins function as transcription repressors under drought, cold, and high-salinity stress conditions. Plant Physiol 2004, 136:2734–2746.

48. Rossel JB, Wilson PB, Hussain D, Woo NS, Gordon MJ, Mewett OP, Howell KA, Whelan J, Kazan K, Pogson BJ: Systemic and intracellular responses to photooxidative stress in Arabidopsis. Plant Cell 2007, 19:4091–4110.

49. Gray GR, Hope BJ, Qin X, Taylor BG, Whitehead CL: The characterization of photoinhibition and recovery during cold acclimation in Arabidopsis thaliana using chlorophyll fluorescence imaging. Physiologia Plantarum 2003, 119:365–375.

50. Rizza F, Pagani D, Stanca AM, Cattivelli L: Use of chlorophyll fluorescence to evaluate the cold acclimation and freezing tolerance of winter and spring oats. Plant Breeding 2001, 120:389–396.

51. Giardi MT, Cona A, Geiken B, Kuèera T, Masojídek J, Mattoo AK: Long-term drought stress induces structural and functional reorganization of photosystem II. Planta 1996, 199:118–125.

52. Hura T, Grzesiak S, Hura K, Thiemt E, Tokarz K, Wedzony M: Physiological and biochemical tools useful in drought-tolerance detection in genotypes of winter triticale: accumulation of ferulic acid correlates with drought tolerance. Annals of Botany 2007, 100:767–775.

53. Flexas J, Bota J, Escalona JM, Sampol B, Medrano H: Effects of drought on photosynthesis in grapevines under field conditions: an evaluation of stomatal and mesophyll limitations. Functional Plant Biology 2002, 29:461–471.

54. Flexas J, Escalona JM, Medrano H: Down-regulation of photosynthesis by drought under field conditions in grapevine leaves. Functional Plant Biology 1998, 25:893–900.

55. Rivero RM, Kojima M, Gepstein A, Sakakibara H, Mittler R, Gepstein S, Blumwald E: Delayed leaf senescence induces extreme drought tolerance in a flowering plant. Proceedings of the National Academy of Sciences 2007, 104:19631–19636.

56. Öquist G, Chow WS, Anderson JM: Photoinhibition of photosynthesis represents a mechanism for the long-term regulation of photosystem II. Planta 1992, 186:450–460.

57. Cornic G, Gouallec JL, Briantais JM, Hodges M: Effect of dehydration and high light on photosynthesis of two C3 plants (Phaseolus vulgaris L. and Elatostema repens (Lour.) Hall f.). Planta 1989, 177:84–90.

58. Hoagland DR, Arnon DA: The water-culture method of growing plants without soil. California Agricultural Experiment Station Circular 1938, 347:1–32.

CO_2 Assimilation, Ribulose-1,5-Bisphosphate Carboxylase/Oxygenase, Carbohydrates and Photosynthetic Electron Transport Probed by the JIP-Test, of Tea Leaves in Response to Phosphorus Supply

Zheng-He Lin, Li-Song Chen, Rong-Bing Chen, Fang-Zhou Zhang, Huan-Xin Jiang and Ning Tang

ABSTRACT

Background

Although the effects of P deficiency on tea (Camellia sinensis (L.) O. Kuntze) growth, P uptake and utilization as well as leaf gas exchange and Chl a

fluorescence have been investigated, very little is known about the effects of P deficiency on photosynthetic electron transport, photosynthetic enzymes and carbohydrates of tea leaves. In this study, own-rooted 10-month-old tea trees were supplied three times weekly for 17 weeks with 500 mL of nutrient solution at a P concentration of 0, 40, 80, 160, 400 or 1000 μM. This objective of this study was to determine how P deficiency affects CO_2 assimilation, Rubisco, carbohydrates and photosynthetic electron transport in tea leaves to understand the mechanism by which P deficiency leads to a decrease in CO_2 assimilation.

Results

Both root and shoot dry weight increased as P supply increased from 0 to 160 μM, then remained unchanged. P-deficient leaves from 0 to 80 μM P-treated trees showed decreased CO_2 assimilation and stomatal conductance, but increased intercellular CO_2 concentration. Both initial and total Rubisco activity, contents of Chl and total soluble protein in P-deficient leaves decreased to a lesser extent than CO_2 assimilation. Contents of sucrose and starch were decreased in P-deficient leaves, whereas contents of glucose and fructose did not change significantly except for a significant increase in the lowest P leaves. OJIP transients from P-deficient leaves displayed a rise at the O-step and a depression at the P-step, accompanied by two new steps at about 150 μs (L-step) and at about 300 μs (K-step). RC/CSo, TRo/ABS (or Fv/Fm), ETo/ABS, REo/ABS, maximum amplitude of IP phase, PIabs and PItot, abs were decreased in P-deficient leaves, while VJ, VI and dissipated energy were increased.

Conclusion

P deficiency decreased photosynthetic electron transport capacity by impairing the whole electron transport chain from the PSII donor side up to the PSI, thus decreasing ATP content which limits RuBP regeneration, and hence, the rate of CO_2 assimilation. Energy dissipation is enhanced to protect P-deficient leaves from photo-oxidative damage in high light.

Background

Phosphorus (P) is one of essential macronutrients required for the normal growth and development of higher plants. Plant roots acquire P as phosphate (Pi), primarily in the form of $H_2PO_4^-$, from the soil solution [1]. Although total Pi is abundant in many soils, the available Pi in the soil solution is commonly 1 – 2 μM due to its binding to soil mineral surfaces and fixation into organic forms [2]. Hence, P is one of the unavailable and inaccessible macronutrients in the soil [1] and is often the most limiting mineral nutrient in almost all soils [2]. Among the fertility constraints to crop production in China, low Pi availability is the primary limiting factor [3]. Pi availability is particularly limiting on the highly weathered acid soils of the tropics and subtropics, in which free iron and aluminum oxides bind native and applied Pi into forms unavailable to plants [2,3]. Therefore, Pi availability is often a major limiting factor for crop production in acid soils [2].

P deficiency affects photosynthesis in many plant species, including tea (Camellia sinensis (L.) O. Kuntze) [4], satsuma mandarin (Citrus unshiu Marc.) [5,6], pigeon pea (Cajanus cajan L. Millsp.) [7], soybean (Glycine max (L.) Merr.) [8], white clover (Trifolium repens L.) [9], sugar beet (Beta vulgaris L.) [10], tomato (Lycopersicon esculentum Mill.) [11], bean (Phaseolus vulgaris L.) [12], maize (Zea mays L.), sunflower (Helianthus annuus L.) [13]. In pigeon pea (cv. UPAS 120) [7] and tea [4], stomatal closure was at least partly responsible for the decreased photosynthetic rate under P deficiency, because the intercellular CO2 concentration was decreased. However, the lower CO2 assimilation in P-deficient leaves of soybean [14] and bean [12] was primarily caused by non-stomatal factors as the lower assimilation rate coincided with an increase of the intercellular CO2 concentration and the internal to ambient CO2 concentration ratio, respectively. Decreases in the activity and amount of Rubisco due to P deficiency have been reported for spinach (Spinacia oleracea L.) [15,16], sunflower [13], maize [17] and soybean [14,18]. However, experiments with sugar beet [10,19] and maize [13] showed that the effects of P deficiency on photosynthetic rate acted through RuBP regeneration rather than Rubisco activity. Jacob and Lawlor [20] concluded that the decreased CO2 assimilation in P-deficient sunflower and maize leaves was a consequence of a smaller ATP content and lower energy charge which limited the production of RuBP. A feedback inhibition of photosynthesis has been suggested as a cause of decreased CO2 assimilation at low P supply [21,22]. However, for tomato plants a decrease in starch accumulation and an increase in oxygen sensitivity of CO2 fixation with decreasing P supply suggest that feedback limitation is decreased under P deficiency [11,23]. P deficiency may also limit photosynthetic rate by altering leaf Chl and protein contents [24,25]. However, the decreased photosynthetic rate under P deficiency was not accompanied by decreased contents of Chl and protein per unit leaf area [10,15].

All oxygenic photosynthetic materials investigated so far using direct, time-resolved fluorescence measurement show the polyphasic rise with the basic steps of O-J-I-P [26-28]. The OJIP transient has been found to be a sensitive indicator of photosynthetic electron transport processes [29]. The kinetics of the OJIP are considered to be determined by changes in the redox state of QA [28,30], but at the same time, the OJIP transient reflects the reduction of the photosynthetic electron transport chain [31]. The OJ phase represents the reduction of the acceptor side of PSII [29,31]. The JI phase parallels the reduction of the PQ-pool [29,32] and the IP phase represents the fractional reduction of the acceptor side of PSI or the last step in the reduction of the acceptor side of PSII and the amplitude of the IP phase may be a rough indicator of PSI content [31,33]. Reports concerning the effects of P deficiency on photosynthetic electron transport activity are some conflicting. Abadia et al. [34] reported that low P had no major effect on the structure and function of the photosynthetic electron transport system or on photosynthetic quantum yield of sugar beet leaves. Jacob and Lawor [20] concluded that in vivo photosynthetic electron transport did not limit photosynthetic capacity in P-deficient sunflower and maize leaves. However, P-deficient citrus exhibited a 6% decrease in Fv/Fm and a 49.5% decrease in electron transport rate [5]. Recently, Ripley et al. [35] reported that P deficiency decreased TRo/ABS (Fv/Fm), ETo/ABS of sorghum (Sorghum bicolor (L.) Moench) leaves, but had no significant effect on electron transport flux per RC (ETo/RC). Thus, it is not well known how P deficiency affects photosynthetic electron transport in plants.

Tea is an evergreen shrub native to China and is cultivated in humid and sub-humid of tropical, sub-tropical, and temperate regions of the world mainly on acid soils [4]. P deficiency is frequently observed in tea plantations [36,37]. For this reason, P fertilizers are being used annually in tea plantations in order to raise tea productivity and improve tea quality [4]. Although Salehi and Hajiboland [4] investigated the effects of P deficiency on tea growth, P uptake and utilization as well as leaf gas exchange and Chl a fluorescence, very little is known about the effects of P deficiency on photosynthetic electron transport, photosynthetic enzymes and carbohydrates of tea leaves. The objective of this study was to determine how P deficiency affects CO2 assimilation, Rubisco, non-structural carbohydrates and photosynthetic electron transport in tea leaves to understand the mechanism by which P deficiency leads to a decrease in CO2 assimilation.

Results

Leaf P Content and Plant Growth Characteristics

As P supply decreased, leaf P content decreased curvilinearly (Fig. 1A). Both root and shoot dry weight increased as P supply increased from 0 to 160 μM, then remained unchanged (Fig. 1B and 1C). The ratio of root/shoot dry weight in the 0 to 80 μM P-treated trees was higher than in the 160 μM to 1000 μM P-treated ones (Fig. 1D).

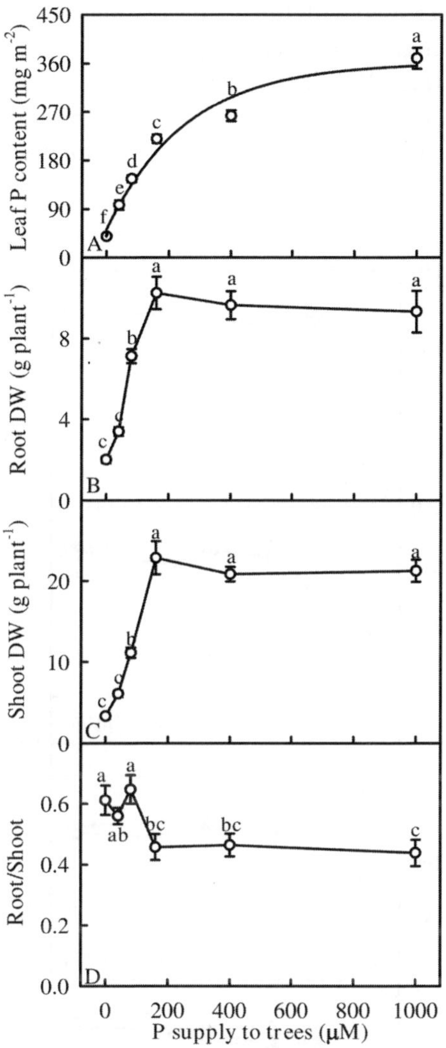

Figure 1. Effects of phosphorus (P) supply on leaf P content (A), root dry weight (B), shoot dry weight (C) and root/shoot dry weight ratio (D) of tea trees. Each point is mean ± standard error (n = 5 or 6). Regression equations: (A) y = 361.3948 – 308.8565 e-0.0039x (r2 = 0.9690, P = 0.0055). Different letters above or below standard error bars indicate significant difference at P < 0.05.

Specific Leaf Weight, Chl, Car, Total Soluble Protein and N

Specific leaf weight did not change significantly as leaf P content decreased from 369.3 mg m-2 to 97.5 mg m-2, then dropped significantly in the lowest P leaves (Fig. 2A). Leaf Chl (Fig. 2B), Car (Fig. 2C) and total soluble protein (Fig. 2D) contents did not change significantly as leaf P decreased from 369.3 mg m-2 to

146.0 mg m-2, then decreased with further decreasing leaf P content. Leaf N content remained little changed with decreasing leaf P content, except for a decrease in the lowest P leaves (Fig. 2D). The ratio of Chl a/b remained unchanged over the range of leaf P content examined (Fig. 2B). The ratio of Car/Chl remained relatively constant as leaf P content decreased, except for an increase in the lowest P leaves (Fig. 2C).

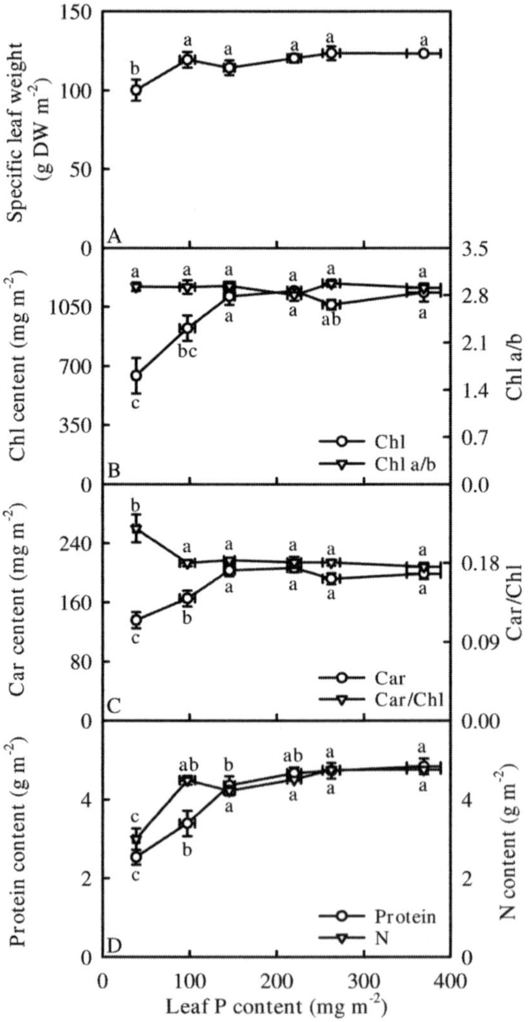

Figure 2. Specific leaf weight
(A), Chl content and Chl a/b ratio (B), carotenoid (Car) content and Car/Chl ratio (C), total soluble protein and N contents (D) in relation to P content in tea leaves. Each point is mean ± standard error for the leaf P content (horizontal, n = 6) and the dependent variable (vertical, n = 5 or 6). Different letters above or below standard error bars indicate significant difference at P < 0.05.

Leaf Gas Exchange and Rubisco

Both CO2 assimilation (Fig. 3A) and stomatal conductance (Fig. 3B) increased as leaf P content increased from 39.4 mg m-2 to 219.9 mg m-2, then remained relatively stable with further increasing leaf P content, whereas intercellular CO2 concentration decreased as leaf P content increased from 39.4 mg m-2 to 146.0 mg m-2, then did not change significantly with further increasing leaf P content (Fig. 3C).

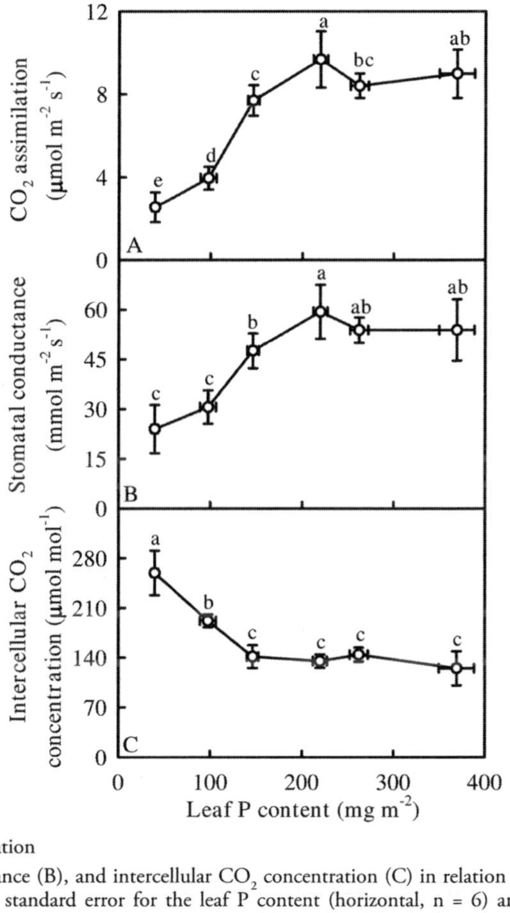

Figure 3. CO_2 assimilation (A), stomatal conductance (B), and intercellular CO_2 concentration (C) in relation to P content in tea leaves. Each point is mean ± standard error for the leaf P content (horizontal, n = 6) and the dependent variable (vertical, n = 5). Different letters above standard error bars indicate significant difference at $P < 0.05$.

On an area basis, both initial and total Rubisco activity kept relatively constant as leaf P content decreased from 369.3 mg m-2 to 219.9 mg m-2, then decreased with further decreasing leaf P content, whereas both initial and total activity expressed on a protein basis did not change significantly over the range of leaf P content examined, except for a slight decrease in initial activity in the lowest P leaves (Fig. 4A and 4B). Rubisco activation state remained unchanged as leaf P content decreased from 369.3 mg m-2 to 97.5 mg m-2, and then dropped in the lowest P leaves (Fig. 4C).

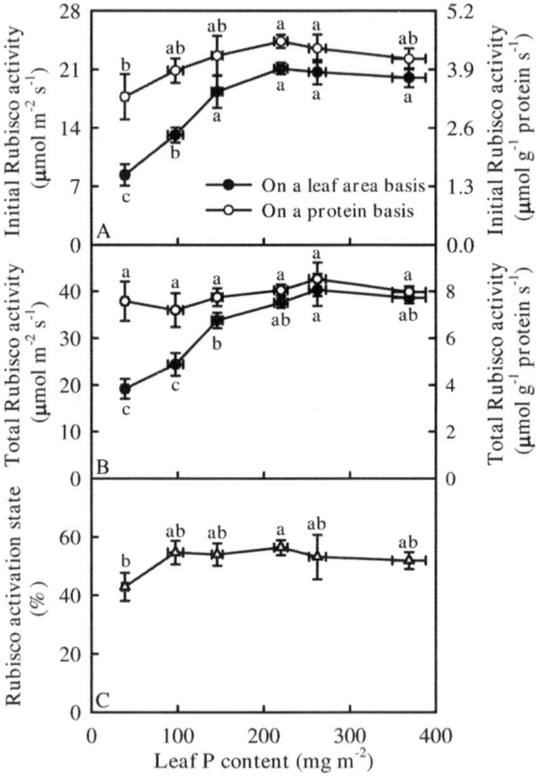

Figure 4. Initial ribulose-1,5-bisphosphate carboxylase/oxygenase (Rubisco) activity **(A), total** Rubisco activity (B), and Rubisco activation state (C) in relation to P content in tea leaves. Each point is mean ± standard error for the leaf P content (horizontal, n = 6) and the dependent variable (vertical, n = 5). Different letters above or below standard error bars indicate significant difference at $P < 0.05$.

Leaf Nonstructural Carbohydrates

On an area basis, contents of glucose and fructose did not change significantly over the range of leaf P content examined except for a significant increase in the lowest P leaves (Fig. 5A and 5B). Contents of sucrose and starch remained little changed as leaf P content decreased from 369.3 mg m-2 to 219.9 mg m-2, then decreased with further decreasing leaf P content (Fig. 5C and 5D). When expressed on a dry weight basis, sucrose content did not change significantly as leaf P content decreased from 369.3 mg m-2 to 146.0 mg m-2 except for a decrease in the 39.4 mg m-2 and 97.5 mg m-2 P leaves (Fig. 5G), whereas the other results expressed on a dry weight basis were similar to those expressed on an area basis (Fig. 5E, 5F and 5H).

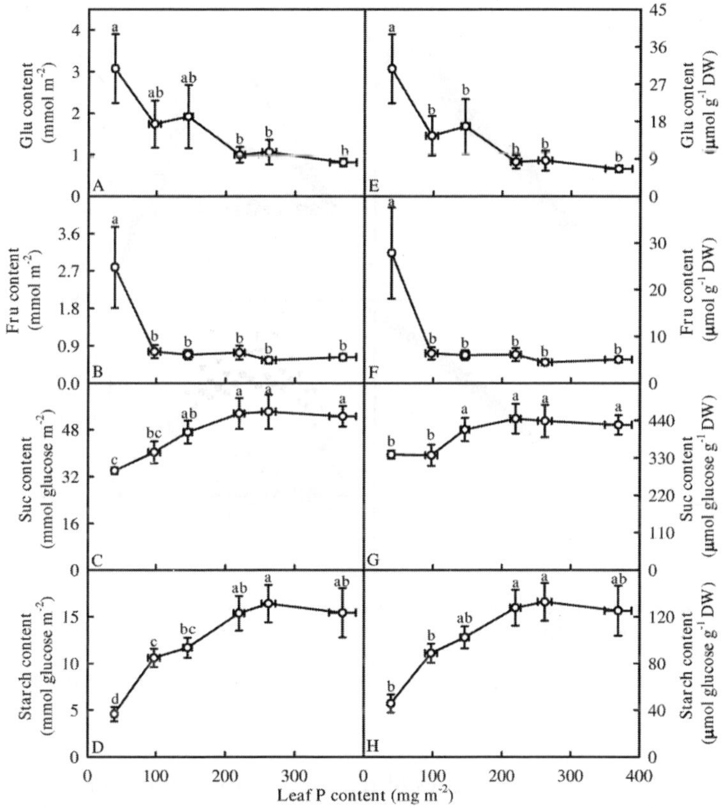

Figure 5. Glucose (Glu, A and E), fructose (Fru, B and F), sucrose (Suc, C and G), and starch (D and H) contents expressed on an area (A-E) or DW (F-J) basis in relation to P content in tea leaves. Each point is mean ± standard error for the leaf P content (horizontal, n = 6) and the dependent variable (vertical, n = 6). Different letters above standard error bars indicate significant difference at $P < 0.05$.

Leaf OJIP Transients and Related Parameters

All OJIP transients showed a typical polyphasic rise with the basic steps of O-J-I-P. OJIP transients of leaves from 0 and 40 μM P-treated trees showed a rise at the O-step and a large depression at the P-step (Fig. 6A).

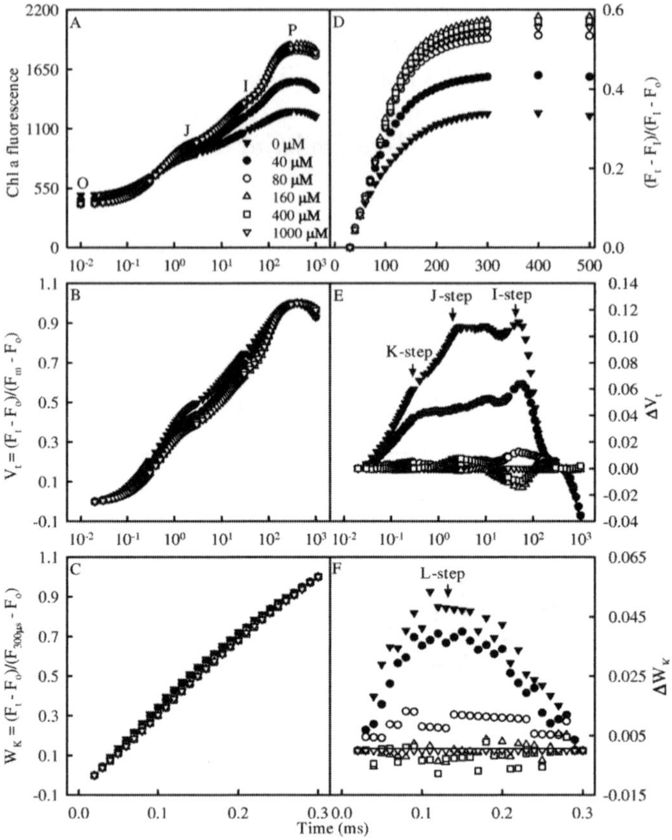

Figure 6. Effects of P supply on the average Chl a fluorescence (OJIP) transients (average of 7 – 15 samples, A) and the different expressions of relative variable fluorescence: (B) between F_o and F_m: $V_t = (F_t - F_o)/(F_m - F_o)$ and (E) the differences of the six samples to the reference sample treated with 1000 μM P (ΔVt), (C) between Fo and F300 μs: $W_K = (F_t - F_o)/(F_{300\mu s} - F_o)$ and (F) the differences of the six samples to the reference sample (ΔW_K), (D) IP phase: $(F_t - F_o)/(F_I - F_o) - 1 = (F_t - F_I)/(F_I - F_o)$ [71] in dark-adapted tea leaves.

Fig. 6B and 6E shows the kinetics of relative variable fluorescence at any time Vt = (Ft - Fo)/(Fm - Fo) and the differences of normalized P-treated transients minus 1000 μM P-treated transient (ΔVt). The differences revealed three obvious bands: increase in the K-step (300μs), in the 2 to 4 ms range J-step and in the 30 to 100 ms range I-step. The positive K-, J- and I-steps were very pronounced in the leaves from 0 and 40 μM P-treated trees. Fig. 6C and 6F depicts the relative variable fluorescence between Fo and F300μs (WK) and the differences of normalized P-treated transients minus 1000 μM P-treated transient (ΔWK). The differences showed a clear L-step. OJIP transients from 0 to 80 μM P-treated trees had decreased maximum amplitude of IP phase and rise time, and the end-levels were lowered by P deficiency (Fig. 6D).

Fig. 7 depicts the behavior patterns of 17 fluorescence parameters. For each parameter the values were normalized on that of the sample treated with 1000 μM P. Generally speaking, leaves from 0 to 80 μM P-treated plants had decreased ETo/TRo, REo/ETo, TRo/ABS, ETo/ABS, REo/ABS (Fig. 7A), TRo/CSo, RC/CSo, ETo/CSo, REo/CSo (Fig. 7B), REo/RC, ECo/RC, maximum amplitude of IP phase, PIabs and PItot, abs (Fig. 7C), but increased DIo/RC, DIo/CSo and DIo/ABS (ϕDo) (Fig. 7D).

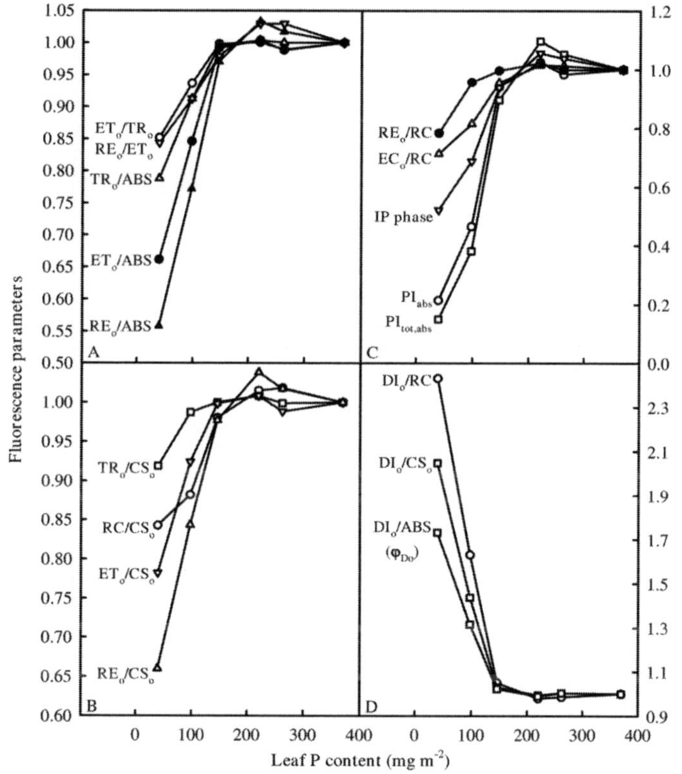

Figure 7. Seventeen fluorescence parameters derived by the JIP-test from the average OJIP transients of Fig. 6A in relation to P content in tea leaves.

All the values were expressed relative to the sample treated with 1000 μM P set as 1. Maximum amplitude of IP phase = $(F_m - F_o)/(F_I - F_o) - 1$ [71].

Leaf Maximum Amplitude of IP Phase, Pi_{abs} and Pi_{tot}, Abs in Relation to CO_2 Assimilation

Leaf CO2 increased linearly or curvilinearly with increasing maximum amplitude of IP phase (Fig. 8A), PI_{abs} (Fig. 8B) and PI_{tot}, abs (Fig. 8C), respectively.

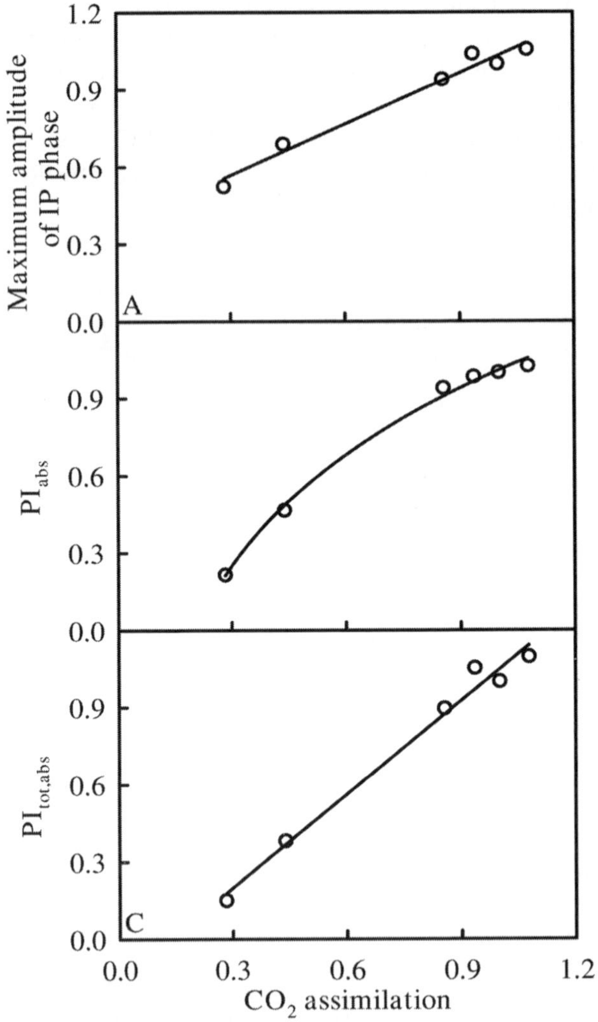

Figure 8. Maximum amplitude of IP phase (A), PI_{abs} (B) and PI_{tot}, abs (C) in relation to CO_2 assimilation in tea leaves.

All the values were expressed relative to the sample treated with 1000 µM P set as 1. Regression equations: (A) y = 0.5070 + 0.5208 × (r^2 = 0.9556, P = 0.0007); (B) y = -11.9070 + 12.9149 x0.0503 (y^2 = 0.9951, P = 0.0003); (C) y = -0.1650 + 1.2127 × (y^2 = 0.9839, P < 0.0001).

Discussion

Our results showed that 0, 40 and 80 µM P treatments decreased root and shoot dry weight (Fig. 1B and 1C), and foliar P content for the three treatments was lower than the sufficiency range of 1.9 to 2.5 mg g-1 DW [38]. In addition, nearly all physiological and biochemical activities reached their maximum in the leaves of about 220 mg m-2 from 160 μM P-treated trees (Figs. 2, 3, 4, 5, 6, 7). Based on these results, trees treated with 0, 40 or 80 μM P are considered P deficient. P deficiency resulted in an increase in the ratio of root/shoot dry weight (Fig. 1D), as previously observed in different plant species growing under different growth conditions [10,39-42]. The increase of the root/shoot dry weight ratio in response to P deficiency may be associated with stronger sink competition of the roots for P and photosynthates [7,40,43-45].

Despite decreased CO_2 assimilation, P deficiency causes increased starch content and decreased sucrose content in leaves of several plant species including soybean [44,46], tobacco (Nicotiana tabacum L.) [22], spinach, barley (Hordeum vulgare L.) [47] and Brachiaria hybrid [48]. Increased partitioning of photosynthetically fixed carbon into the starch at the expense of sucrose synthesis in leaves [22,44] and decreased demand from growth [22,46,49] have been shown to contribute to increased starch accumulation in P-deficient leaves. However, a simultaneous increase in starch and sucrose contents in the leaves of P-deficient soya (G. max (L.) Merr.) [47], bean [50] and sugar beet [51] plants has been observed while chloroplastic and leaf levels of sugar phosphates decreased markedly [19]. In our study, P-deficient leaves had decreased sucrose (Fig. 5C and 5G) and starch (Fig. 5D and 5H) contents, as previously found for trifoliate orange (Poncirus trifoliata (L.) Raf.), Swingle citrumelo (C. paradisi Macf. × P. trifoliata), Carrizo citrange (C. sinensis (L.) Osb. × P. trifoliata) [52] and rice (Oryza sativa L.) [48]. There appears to be considerable variation in the responses of leaf carbohydrate metabolism during P deficiency. Some of the variation may result from different degree of P deficiency, time of exposure to P deficiency, plant species, light intensities used in different studies [8,22,23,47,52]. It is noteworthy that specific leaf weight decreased in the lowest P leaves (Fig. 2A). This contrasts with previous data obtained for soybean [44] and sugar beet [10], whose leaves accumulated starch under P deficiency [10,44]. Regressive analysis showed that specific leaf weight decreased linearly with decreasing leaf starch content expressed on a leaf area basis ($P = 0.0053$, data not shown). Therefore, the decrease in specific leaf weight under P deficiency may be explained, at least in part, by the decrease in starch content.

The higher intercellular CO_2 concentration in P-deficient leaves indicates that the low CO_2 assimilation under P deficiency (Fig. 3A and 3C) is primarily caused by non-stomatal factors, as earlier reported for soybean [14] and bean [12]. However, Salehi and Hajiboland [4] proposed that lower stomatal conductance was the main cause for the decreased CO_2 assimilation rate in P-deficient tea leaves as the decrease in assimilation rate was accompanied by a decrease in the intercellular CO_2 concentration. Similar result has been obtained for pigeon pea (cv. UPAS 120) [7].

It has been suggested that low sink demand limits photosynthesis under P deficiency [21,22]. In our study, however, the decrease of assimilation CO2 rate under P deficiency was accompanied by a decrease in the starch accumulation (Fig. 3A, 5D and 5H), as previously reported for tomato grown in high light [23]. This indicates that the production, rather that the utilization of photosynthates, is limiting. Evidence shows that soluble sugars, specifically hexoses, may repress photosynthetic gene expression, particularly of the nuclear-encoded small subunit of Rubisco, thus decreasing Rubisco content and CO2 assimilation [53]. The lack of accumulation of sucrose and hexoses in the leaves from 40 and 80 μM P-treated trees (Fig. 5A–C and 5E–G) means that the feedback repression mechanism via accumulation of soluble sugars does not play a major role in determining the activity of Rubisco and the rate of CO2 assimilation in these leaves. However, this is not to deny that the decrease in CO2 assimilation in the lowest P leaves can be due to the accumulation of hexoses, because the levels of glucose + fructose observed was higher than the reported threshold level (4.5 mmol m-2) for hexose regulation of gene expression in tobacco [54]. The decrease in initial and total Rubisco activity expressed on an area basis in response to P deficiency was probably not the primary factor limiting CO2 assimilation, because there was a greater decrease in CO2 assimilation than in Rubisco activity (Fig. 3A, 4A and 4B). In our study, the observed lower initial and total Rubisco activity expressed on an area basis in P-deficient leaves could be associated with decreased total soluble protein content (Fig. 2D), because both initial and total activity expressed on a protein basis did not change significantly over the range of leaf P content examined, except for a slight decrease in the initial activity in the lowest P leaves (Fig. 4A and 4B). The decrease in CO2 assimilation in P-deficient leaves cannot be attributed to a decrease in Chl and protein contents, because the decrease in leaf Chl (Fig. 2B) and total soluble protein (Fig. 2D) contents was much less than CO2 assimilation (Fig. 3A). Similar results have been reported for spinach [15], sugar beet [10], and bean [12].

The presence of a positive L-step at ca. 150 μs in P-deficient leaves (Fig. 6F) means that the OJIP transients from P-deficient leaves are less sigmoidal than from P-sufficient ones and that the PSII units are less grouped or less energy is being exchanged between the independent PS II units. Because the grouped conformation is more stable than the ungrouped one, the decreased grouping implies that the PSII units of P-deficient leaves have lost stability and become more fragile. Similar results have been reported for N-deficient cowpea (Vigna unguiculata L.) [28] and Al-treated Citrus grandis (L.) Osbeck [55].

The decrease of Fv/Fm in P-deficient leaves was caused by both a decrease in Fm and an increase in Fo (Fig. 6A and 7A), as previously found for tea [4], satsuma mandarin [5] and sorghum [35]. The decrease in Fv/Fm under stress is considered to reflect the photoinhibitory damage to PSII complexes [56,57]. The higher Fo may be caused by both the damage of OEC and the inactivation of some of the PSII RCs [58,59], because P-deficient leaves had decreased RC/CSo (Fig. 7B) and increased damage to OEC, or it may be related to the accumulation of reduced QA [60], because the physiological fractional reduction of QA to QA -, as indicated by the increase in Mo (Fig. 6B and 6E), increased in P-deficient leaves. Quenching of Fm in P-deficient leaves may arise from the photoinhibitory quenching (qI), because an increase in Fo with a quenched Fm was observed in P-deficient leaves (Fig. 6A) [61] and from the xanthophyll cycle-dependent thermal energy dissipation, which was significantly higher in P-deficient satsuma mandarin leaves than in P-sufficient ones [6].

The J-step, I-step and IP phase of OJIP transients are correlated with the redox state of QA, the redox state of plastoquinone, and the redox state of end acceptors at PSI electron acceptor side, respectively [27,28,30,32]. The finding that P-deficient leaves had increased VJ and VI (Fig. 6B and 6E), but decreased maximum amplitude of IP phase (Fig. 6D) suggests that acceptor side of PSII became more reduced under P deficiency, but the acceptor side of PSI become more oxidized. P deficiency-induced photoinhibitory damage at PSII acceptor is also supported by the fact that Fv (Fv = Fm - Fo) was decreased in P-deficient leaves along with an increase in Fo (Fig. 6A), which is the characteristic of photoinhibitory damage at PSII acceptor side [62]. A positive K-step appeared at ca. 300 μs in the OJIP transients in P-deficient leaves. This means that the oxygen evolving complex (OEC) is damaged [63,64]. A positive K-step has also been found in N-deficient cowpea leaves [28].

Our result showed that P deficiency decreased the total electron carriers per RC (ECo/RC; Fig. 7C), the yields (TRo/ABS (Fv/Fm), ETo/TRo, REo/ETo, ETo/ABS, and REo/ABS; Fig. 7A), the fluxes (REo/RC and REo/CSo; Fig. 7B and 7C) and the fractional reduction of the PSI end electron acceptors, as indicated by the decreased maximum amplitude of IP phase (Fig. 6D), and damaged all of the photochemical and non-photochemical redox reactions, as indicated by the decreases in PIabs and PItot, abs (Fig. 7D). This means that leaves from P-deficient trees have a decreased capacity for electron transport, thus limiting ATP synthesis and RuBP regeneration. Lacking ATP has the consequence that Rubisco is not fully activated [65]. This might partly explain why P-deficient leaves had lower Rubisco activity and activation state (Fig. 4). Regressive analysis showed that CO_2 assimilation decreased linearly or curvilinearly with decreasing maximum amplitude of IP phase (Fig. 8A), PIabs (Fig. 8B) and PItot, abs (Fig. 8C), respectively. Therefore, we conclude that the decreased photosynthetic electron transport capacity, in conjunction with the lack of ATP which limit RuBP regeneration are probably the main factors contributing to decreased CO_2 assimilation under P deficiency.

Because P-deficient leaves only utilized a small fraction of the absorbed light energy in photosynthetic electron transport, as indicated by the decreases in ECo/RC, ETo/ABS and REo/ABS (Fig. 7A and 7C), compared with the P-sufficient ones, more excess excitation energy existed in P-deficient than in P-sufficient leaves in high light. Correspondingly, energy dissipation, as indicated by DIo/CSo, DIo/RC, and DIo/ABS (ϕDo), increased in P-deficient leaves (Fig. 7D). In addition to this, the excess absorbed light in turn can lead to the production of 1O2 and reduced active oxygen species, causing damage to photosynthetic apparatus and cell structure [35,66]. Indeed, photoinhibitory damage to both donor side and acceptor side has been demonstrated to increase the production of reactive oxygen species [61,67].

Conclusion

P deficiency decreased photosynthetic electron transport capacity by impairing the whole electron transport chain from the PSII donor side up to the PSI, thus decreasing ATP content which limits RuBP regeneration, and hence, the rate of CO2 assimilation. In addition to decrease light absorption by lowering Chl content, energy dissipation is enhanced to protect P-deficient leaves from photo-oxidative damage in high light.

Methods

Plant Culture and P Treatments

This study was conducted outdoors from March to November 2007 at Fujian Agriculture and Forestry University, Fuzhou. Own-rooted 10-mouth-old uniform tea (Camellia sinensis (L.) O. Kuntze cv. Huangguanyin) trees were transplanted into 6 L plastic pots containing sand. Each pot contained two trees, and was supplied twice weekly with 500 mL of 1/2 strength nutrient solution. Full-strength nutrient solution contained 1 mM $(NH_4)_2SO_4$, 0.8 mM K_2SO_4, 1 mM KNO_3, 2 mM $Ca(NO_3)_2$, 1 mM $NH_4H_2PO_4$, 0.05 mM $CaCl_2$, 0.6 mM $MgSO_4$, 46 μM H_3BO_3, 9 μM $MnSO_4$, 9 μM $ZnSO_4$, 2 μM $CuSO_4$, 2.6 μM Na_2MoO_4, and 30 μM Fe-EDTA. Six weeks after transplanting, the treatment was applied for 17 weeks: until the end of the experiment, each pot was supplied three times weekly with 500 mL of nutrient solution at a P concentration of 0, 40, 80, 160, 400 or 1000 μM from $NH_4H_2PO_4$ at pH of 5.5. N concentration was maintained at a constant by the addition of $(NH_4)_2SO_4$. At the end of the experiment, the fully-expanded (about seven weeks old) leaves from different replicates and treatments were used for all the measurements. Leaf discs (0.61 cm^2 in size) were collected at noon under full sun and immediately frozen in liquid N_2. Samples were stored at -80°C until they were used for the determination of Chl, carotenoids (Car), Rubisco, carbohydrates, and protein. Special care was taken to ensure that all samples were transferred directly from liquid N_2 to freezer of -80°C, at no time were any samples exposed to room temperature.

Measurements of Root and Shoot Dry Weight, and Specific Leaf Weight

At the end of the experiment, six trees per treatment from different pots were harvested. The trees were divided into roots and shoots. The plant materials were then dried at 80°C for 48 h and the dry weight measured. Specific leaf weight was calculated as the ratio of leaf dry weight to leaf area.

Determination of Leaf Chl, Car, Total Soluble Protein, and Total P

Chl, Chl a, Chl b and Car were assayed according to Lichtenthaler [68]. Total soluble protein was determined according to Bradford [69]. Total P was determined according to Fredeen et al. [44].

Leaf Gas Exchange Measurements

Measurements were made with a CI-301PS portable photosynthesis system (CID, WA, USA) at ambient CO_2 concentration with a natural photosynthetic photon flux density of 1500 ± 45 µmol m-2 s-1 between 10:30 and 12:00 on a clear day. During measurements, leaf temperature and ambient vapor pressure were 28.0 ± 1.0°C and 1.8 ± 0.1 kPa, respectively.

Measurements of Leaf OJIP Transients

OJIP transient was measured by a Handy Plant Efficiency Analyser (Handy PEA, Hansatech Instruments Limited, Norfolk, UK) according to Strasser et al. [26]. The transient was induced by red light of about 3,400 µmol m^{-2} s^{-1} provided by an array of three light-emitting diodes (peak 650 nm), which focused on the leaf surface to give homogenous illumination over the exposed area of the leaf. All the measurements were done with 3 h dark-adapted plants at room temperature.

JIP Test

OJIP transient was analyzed according to the JIP test. From OJIP transient, the extracted parameters (F_m, $F2_{0\ \mu s}$, $F_{50\ \mu s}$, $F_{100\ \mu s}$, $F_{300\ \mu s}$, F_J, F_I etc.) led to the calculation and derivation of a range of new parameters according to pervious authors [27,28,55,70,71] (see Table 1).

Table 1. Summary of parameters, formulae and their description using data extracted from chlorophyll a fluorescence (OJIP) transient.

Fluorescence parameters	Description
Fluorescence parameters	Description
F_t	Fluorescence intensity at time t after onset of actinic illumination
$F_{50\ \mu s}$ or $F_{20\ \mu s}$	Minimum reliable recorded fluorescence at 50 µs with the PEA- or 20 µs with Handy-PEA-fluorimeter
$F_{100\ \mu s}$ and $F_{300\ \mu s}$	Fluorescence intensity at 100 and 300 µs, respectively
F_J and F_I	Fluorescence intensity at the J-step (2 ms) and the I-step (30 ms), respectively
F_P (= F_m)	Maximum recorded (= maximum possible) fluorescence at P-step
Area	Total complementary area between fluorescence induction curve and $F = F_m$
Derived parameters	
Selected OJIP parameters	
$F_0 \cong F_{50\ \mu s}$ or $F_0 \cong F_{20\ \mu s}$	Minimum fluorescence, when all PSII RCs are open
$F_m = F_P$	Maximum fluorescence, when all PSII RCs are closed
$V_J = (F_{2\ ms} - F_o)/(F_m - F_o)$	Relative variable fluorescence at the J-step (2 ms)
$V_I = (F_{30\ ms} - F_o)/(F_m - F_o)$	Relative variable fluorescence at the I-step (30 ms)
$M_o = 4\ (F_{300\ \mu s} - F_o)/(F_m - F_o)$	Approximated initial slope (in ms^{-1}) of the fluorescence transient V = f(t)
$S_m = EC_o/RC = Area/(F_m - F_o)$	Normalized total complementary area above the OJIP (reflecting multiple-turnover Q_A reduction events) or total electron carriers per RC
Yields or flux ratios	
$\phi_{Po} = TR_o/ABS = 1-(F_o/F_m) = F_v/F_m$	Maximum quantum yield of primary photochemistry at t = 0
$\phi_{Eo} = ET_o/ABS = (F_v/F_m) \times (1 - V_J)$	Quantum yield for electron transport at t = 0
$\psi_{Eo} = ET_o/TR_o = 1-V_J$	Probability (at time 0) that a trapped exciton moves an electron into the electron transport chain beyond Q_A^-
$\phi_{Do} = DI_o/ABS = 1-\phi_{Po} = F_o/F_m$	Quantum yield at t = 0 for energy dissipation
$\delta_{Ro} = RE_o/ET_o = (1 - V_I)/(- V_J)$	Efficiency with which an electron can move from the reduced intersystem electron acceptors to the PSI end electron acceptors
$\phi_{Ro} = RE_o/ABS = \phi_{Po} \times \psi_{Eo} \times \delta_{Ro}\ \phi$	Quantum yield for the reduction of end acceptors of PSI per photon absorbed
Specific fluxes or activities expressed per reaction center (RC)	
$ET_o/RC = (M_o/V_J) \times \psi_{Eo} = (M_o/V_J) \times (1-V_J)$	Electron transport flux per RC at t = 0
$DI_o/RC = (ABS/RC) - (TR_o/RC)$	Dissipated energy flux per RC at t = 0
$RE_o/RC = (RE_o/ET_o) \times (ET_o/RC)$	Reduction of end acceptors at PSI electron acceptor side per RC at t = 0
$ET_o/CS_o = (ABS/CS_o) \times \phi_{Eo}$	Electron transport flux per CS at t = 0
$TR_o/CS_o = (ABS/CS_o) \times \phi_{Po}$	Trapped energy flux per CS at t = 0
$DI_o/CS_o = (ABS/CS_o) - (TR_o/CS_o)$	Dissipated energy flux per CS at t = 0
$RE_o/CS_o = (RE_o/ET_o) \times (ET_o/CS_o)$	Reduction of end acceptors at PSI electron acceptor side per CS at t = 0
Density of RCs	
$RC/CS_o = \phi_{Po} \times (ABS/CS_o) \times (V_J/M_o)$	Amount of active PSII RCs per CS at t = 0
Performance index	
$PI_{abs} = (RC/ABS) \times (\phi_{Po}/(1 - \phi_{Po})) \times (\psi_o/(1 - \psi_o))$	Performance index (PI) on absorption basis
$PI_{tot,\ abs} = (RC/ABS) \times (\phi_{Po}/(1-\phi_{Po})) \times (\psi_{Eo}/(1 - \psi_{Eo})) \times (\delta_{Ro}/(1 - \delta_{Ro}))$	Total PI, measuring the performance up to the PSI end electron acceptors

Leaf Rubisco Activity Measurements

Rubisco was extracted according to Chen et al. [72]. Rubisco activity was assayed according to Cheng and Fuchigami [73] with some modifications. For initial activity, 50 µL of sample extract was added to a cuvette containing 900 µL of an assay solution, immediately followed by adding 50 µL of 10 mM RuBP, then mixing well. The change of absorbance at 340 nm was monitored for 40 s. For total activity, 50 µL of 10 mM RuBP was added 15 min later, after 50 µL of sample extract was combined with 900 µL of an assay solution to fully activate all the Rubisco. The assay solution for both initial and total activity measurements, whose final volume was 1 mL, contained 100 mM HEPES-KOH (pH 8.0), 25 mM $KHCO_3$, 20 mM $MgCl_2$, 3.5 mM ATP, 5 mM phosphocretaine, 5 units NAD-glyceraldehyde-3-phosphate dehydrogenase (NAD-GAPDH, EC 1.2.1.12), 5 units 3-phosphoglyceric phospokinase (PCK, EC 2.7.2.3), 17.5 units creatine phosphokinase (EC 2.7.3.2), 0.25 mM NADH, 0.5 mM RuBP, and 50 µL sample extract. Rubisco activation state was calculated as the ratio of initial activity to total activity.

Measurements of Leaf Nonstructural Carbohydrates

Sucrose, fructose, glucose and starch were extracted 3 times with 80% (v/v) ethanol at 80°C and determined according to Jones et al. [74].

Experimental Design and Statistical Analysis

There were 20 pots trees per treatment in a completely randomized design. Experiments were performed with 5–15 replicates (one tree from different pots per replicate). Differences among treatments were separated by the least significant difference (LSD) test at $P < 0.05$ level.

Abbreviations

Chl: chlorophyll; CS: excited cross section; ET_o/ABS: quantum yield of electron transport at $t = 0$; N: nitrogen; OJIP: Chl a fluorescence; P: phosphorus; PI_{abs}: performance index; PI_{tot}, abs: total performance index; RC: reaction center; RC/CS_o: amount of active PSII RCs per CS at $t = 0$; RE_o/ABS: quantum yield of electron transport from QA- to the PSI end electron acceptors; Rubisco: ribulose-1,5-bisphosphate carboxylase/oxygenase; RuBP: ribulose-1,5-bisphosphate; TR_o/ABS or F_v/F_m: maximum quantum yield of primary photochemistry at $t = 0$; V_I: relative variable fluorescence at the I-step; V_J: relative variable fluorescence at the J-step.

Authors' Contributions

ZHL performed most of the experiments and wrote the manuscript. LSC designed and directed the study and revised the manuscript. RBC helped in designing the study. FZZ helped in making nutrient solution and cultivating trees. HXJ and NT helped in measuring CO_2 assimilation and Chl a fluorescence. All authors have read and approved the final manuscript.

References

1. Vance CP, Uhde-Stone C, Allan DL: Phosphorus acquisition and use: critical adaptations by plants for securing a nonrenewable resource. New Phytol 2003, 157:423–447.
2. Kochian LV, Hoekenga OA, Piñeros MA: How do crop plant tolerate acid soils? Mechanisms of aluminum tolerate and phosphorous efficiency. Annu Rev Plant Biol 2004, 55:459–493.
3. Yan X, Wu P, Ling H, Xu G, Xu F, Zhang Q: Plant nutriomics in China: an overview. Ann Bot 2006, 98:473–482.
4. Salehi SY, Hajiboland R: A high internal phosphorus use efficiency in tea (Camellia sinensis L.) plants. Asian J Plant Sci 2008, 7:30–36.
5. Guo Y-P, Chen P-Z, Zhang L-C, Zhang S-L: Effects of different phosphorus nutrition levels on photosynthesis in satsuma mandarin (Citrus unshiu Marc.) leaves. Plant Nutr Fert Sci 2002, 8:186–191.
6. Guo Y-P, Chen P-Z, Zhang L-C, Zhang S-L: Phosphorus deficiency stress aggravates photoinhibition of photosynthesis and function of xanthophyll cycle in citrus leaves. Plant Nutr Fert Sci 2003, 9:359–363.
7. Fujita K, Kai Y, Takayanagi M, El-Shemy H, Adu-Gyamfi JJ, Mohapatra PK: Genotypic variability of pigeonpea in distribution of photosynthetic carbon at low phosphorus level. Plant Sci 2004, 166:641–649.
8. Qiu J, Israel DW: Carbohydrate accumulation and utilization in soybean plants in response to altered phosphorus nutrition. Physiol Plant 1994, 90:722–728.
9. Hart AL, Greer DH: Photosynthesis and carbon export in white clover plants grown at various levels of phosphorus supply. Physiol Plant 1988, 73:46–51.
10. Rao IM, Terry N: Leaf phosphate status, photosynthesis, and carbon partitioning in sugar beet: I. Changes in growth, gas exchange, and Calvin cycle enzymes. Plant Physiol 1989, 90:814–819.
11. De Groot CC, Boogaard R, Marcelis LFM, Harbinson J, Lambers H: Contrasting effects of N and P deprivation on the regulation of photosynthesis in tomato plants in relation to feedback limitation. J Exp Bot 2003, 54:1957–1967.

12. Lima JD, Mosquim PR, Da Matta FM: Leaf gas exchange and chlorophyll fluorescence parameters in Phaseolus vulgaris as affected by nitrogen and phosphorus deficiency. Photosynthetica 1999, 37:113–121.

13. Jacob J, Lawlor DW: Dependence of photosynthesis of sunflower and maize leaves on phosphate supply, ribulose-1,5-biphosphate carboxylase/oxygenase activity and ribulose-1,5-bisphosphate pool size. Plant Physiol 1991, 98:801–807.

14. Lauer MJ, Pallardy SG, Blevins DG, Douglas D, Randall DD: Whole leaf carbon exchange characteristics of phosphate deficient soybeans (Glycine max L.). Plant Physiol 1989, 91:848–854.

15. Brooks A: Effect of phosphorus nutrition on ribulose-1,5-bisphosphate carboxylase activation, photosynthetic quantum yield and amounts of some Calvin cycle metabolites in spinach leaves. Aust J Plant Physiol 1986, 13:221–237.

16. Brooks A, Woo KC, Wong SC: Effects of phosphorus nutrition on the response of photosynthesis to CO2 and O2, activation of ribulose bisphosphate carboxylase and amounts of ribulose bisphosphate and 3-phosphoglycerate in spinach leaves. Photosynth Res 1988, 15:133–141.

17. Usuda H, Shimogawara K: Phosphate deficiency in maize. II. Enzyme activities. Plant Cell Physiol 1991, 32:1313–1317.

18. Sawada S, Usuda H, Tsukui T: Participation of inorganic orthophosphate in regulation of the ribulose-1,5-biphosphate carboxylase activity in response to changes in the photosynthetic source-sink balance. Plant Cell Physiol 1992, 33:943–949.

19. Rao IM, Arulanantham AR, Terry N: Leaf phosphate status, photosynthesis and carbon partitioning in sugar beet. II. Diurnal changes in sugar phosphates, adenylates and nicotinamide nucleotides. Plant Physiol 1989, 90:820–826.

20. Jacob J, Lawlor DW: In vivo photosynthetic electron transport does not limit photosynthetic capacity in phosphate-deficient sunflower and maize leaves. Plant Cell Environ 1993, 6:785–795.

21. Ciereszko I, Johansson H, Hurry V, Kleczkowski LA: Phosphate status affects the gene expression, protein content and enzymatic activity of UDP-glucose pyrophosphorylase in wild-type and pho mutants of Arabidopsis. Planta 2001, 212:598–605.

22. Pieters AJ, Paul MJ, Lawlor DW: Low sink demand limits photosynthesis under Pi deficiency. J Exp Bot 2001, 52:1083–1091.

23. De Groot CC, Marcelis LFM, Boogaard R, Lambers H: Growth and dry-mass partitioning in tomato as affected by phosphorus nutrition and light. Plant Cell Environ 2001, 24:1309–1317.

24. Plesničar K, Kastori R, Petrović N, Panković D: Photosynthesis and chlorophyll fluorescence in sunflower (Helianthus annuus L.) leaves as affected by phosphorus nutrition. J Exp Bot 1994, 45:919–924.

25. Usuda H: Phosphate deficiency in maize. V. Mobilization of nitrogen and phosphorus within shoots of young plants and its relationship to senescence. Plant Cell Physiol 1995, 36:1041–1049.

26. Strasser RJ, Srivastava A, Govindjee : Polyphasic chlorophyll a fluorescence transient in plants and cyanobacteria. Photochem Photobiol 1995, 61:32–42.

27. Strasser RJ, Srivastava A, Tsimilli-Michael M: The fluorescence transient as a tool to characterize and screen photosynthetic samples. In Probing Photosynthesis: Mechanisms, Regulation and Adaptation. Edited by: Yunus M, Pathre U, Mohanty P. London: Taylor and Francis; 2000:445–483.

28. Strasser RJ, Tsimilli-Micheal M, Srivastava A: Analysis of the chlorophyll a fluorescence transient. In Chlorophyll a Fluorescence: A Signature of Photosynthesis. Edited by: Papageorgiou GC, Govindjee. Dordrecht: Springer; 2004:321–362. [Govindjee (Series Editor): Advances in Photosynthesis and Respiration, vol. 19.] Return to text

29. Tóth SZ, Schansker G, Garab G, Strasser RJ: Photosynthetic electron transport activity in heat-treated barley leaves: The role of internal alternative electron donors to photosystem II. Biochim Biophys Acta 2007, 1767:295–305.

30. Lazár D: The polyphasic chlorophyll a fluorescence rise measured under high intensity of exciting light. Funct Plant Biol 2006, 33:9–30.

31. Schansker G, Tóth SZ, Strasser RZ: Methylviologen and dibromothymoquinone treatments of pea leaves reveal the role of photosystem I in the Chl a fluorescence rise OJIP. Biochim Biophys Acta 2005, 1706:250–261.

32. Schreiber U, Neubauer C, Klughammer C: Devices and methods for room-temperature fluorescence analysis. Phi Trans R Soc Lond B 1989, 323:241–251.

33. Schansker G, Tóth SZ, Strasser RJ: Dark-recovery of the Chl-a fluorescence transient (OJIP) after light adaptation: the qT-component of non-photochemical quenching is related to an activated photosystem I acceptor side. Biochim Biophys Acta 2006, 1757:787–797.

34. Abadia J, Rao IM, Terry N: Changes in leaf phosphate status have only small effects on the photochemical apparatus of sugar beet leaves. Plant Sci 1987, 50:49–55.

35. Ripley BS, Redfern SP, Dames J: Quantificatin of the photosynthetic performance of phosphorus-deficient Sorghum by means of chlorophyll a fluorescence kinetics. South Afr J Sci 2004, 100:615–618.

36. Tang J-F, Hu K-F, Yin J, Xiong J-W: Distribution of organic matter and available N-P-K in the tea garden soil of Xinyang. Henan Agri Sci 2007, (5):81–84.

37. Wei G, Zhang Q, Feng P: Soil fertility status of tea plantation in Guizhou. Guizhou Agri Sci 1996, (4):22–26.

38. Mills HA, Jones JB Jr: Plant Analysis Handbook II: A Practical Sampling, Preparation, Analysis, and Interpretation Guide. Georgia: Micromacro Publishing; 1996:186.

39. Halsted M, Lynch J: Phosphorus responses of C3 and C4 species. J Exp Bot 1996, 47:497–505.

40. Brahim MB, Loustau D, Gaudillère JP, Saur E: Effects of phosphate deficiency on photosynthesis and accumulation of starch and soluble sugars in 1-year-old seedlings of maritime pine (Pinus pinaster Ait). Ann Sci For 1996, 53:801–810.

41. Hammond JP, White PJ: Sucrose transport in the phloem: Integrating root responses to phosphorus starvation. J Exp Bot 2008, 59:93–109.

42. Hernández G, Ramírez M, Valdés-López O, Tesfaye M, Graham MA, Czechowski T, Schlereth A, Wandrey M, Erban A, Cheung F, Wu HC, Lara M, Town CD, Joachim Kopka J, Udvardi MK, Vance CP: Phosphorus stress in common bean: Root transcript and metabolic responses. Plant Physiol 2007, 144:752–767.

43. Ciereszko I, Gniazdowska A, Mikulska M, Rychter AM: Assmilatie translocation in bean plants (Phaseolus vulgaris L.) during phosphate deficiency. J Plant Physiol 1996, 149:343–348.

44. Fredeen AL, Rao IM, Terry N: Influence of phosphorus nutrition on growth and carbon partitioning in Glycine max. Plant Physiol 1989, 89:225–230.

45. Mimura T: Regulation of phosphate transport and homeostasis in plant cells. Int Rev Cytol Cell Biol 1999, 191:149–200.

46. Qiu J, Israel DJ: Diurnal starch accumulation and utilization in phosphorus-deficient soybean plants. Plant Physiol 1992, 98:316–323.

47. Foyer C, Spencer C: The relationship between phosphate status and photosynthesis in leaves. Planta 1986, 167:369–373.

48. Nanamori M, Shinano T, Wasaki J, Yamamura T, Rao IM, Osaki M: Low phosphorus tolerance mechanisms: Phosphorus recycling and photosynthate partitioning in the tropical forage grass, Brachiaria hybrid cultivar Mulato compared with rice. Plant Cell Physiol 2004, 45:460–469.

49. Usuda H, Shimogawara K: Phosphate deficiency in maize. III. Changes in enzyme activities during the course of phosphate deprivation. Plant Physiol 1992, 99:1680–1685.

50. Ciereszko I, Barbachowska A: Sucrose metabolism in leaves and roots of bean (Phaseolus vulgaris L.) during phosphate deficiency. J Plant Physiol 2000, 156:640–644.

51. Rao IM, Fredeen AL, Terry N: Leaf phosphate status, photosynthesis, and carbon partitioning in sugar Beet: III. Diurnal changes in carbon partitioning and carbon export. Plant Physiol 1990, 92:29–36.

52. Graham JH, Duncan LW, Eissenstat DM: Carbohydrate allocation patterns in citrus genotypes as affected by phosphorus nutrition, mycorrhizal colonization and mycorrhizal dependency. New Phytol 1997, 135:335–343.

53. Sheen J: Feedback control of gene expression. Photosynth Res 1994, 39:427–438.

54. Herbers K, Meuwly P, Frommer WB, Métraux J-P, Sonnewald U: Systemic acquired resistance mediated by the ectopic expression of invertase: possible hexose sensing in the secretory pathway. Plant Cell 1996, 8:793–803.

55. Jiang H-X, Chen L-S, Zheng J-G, Han S, Tang N, Smith BR: Aluminum-induced effects on photosystem II photochemistry in citrus leaves assessed by chlorophyll a fluorescence transient. Tree Physiol 2008, 28:1863–1871.

56. Maxwell K, Johnson GN: Chlorophyll fluorescence – a practical guide. J Exp Bot 2000, 51:659–668.

57. Baker NR, Eva Rosenqvist E: Applications of chlorophyll fluorescence can improve crop production strategies: An examination of future possibilities. J Exp Bot 2004, 55:1607–1621.

58. Chen L-S, Li P, Cheng L: Effects of high temperature coupled with high light on the balance between photooxidation and photoprotection in the sun-exposed peel of apple. Planta 2008, 228:745–756.

59. Yamane Y, Kashino Y, Koike H, Satoh K: Increases in the fluorescence Fo level and reversible inhibition of Photosystem II reaction center by high-temperature treatments in higher plants. Photosynth Res 1997, 52:57–64.

60. Bukhov NG, Sabat SC, Mohanty P: Analysis of chlorophyll a fluorescence changes in weak light in heat treated Amarenthus chloroplasts. Photosynth Res 1990, 23:81–87.

61. Gilmore AM, Hazlett TL, Debrunner PG, Govindjee : Comparative time-resolved photosystem II chlorophyll a fluorescence analyses reveal distinctive

differences between photoinhibitory reaction center damage and xanthophyll cycle-dependent energy dissipation. Photochem Photobiol 1996, 64:552–563.

62. Setlik I, Allakhverídiev SI, Nedbal L, Setlikova E, Klimov VV: Three types of Photosystem II photoinactivation. I. Damaging process on the acceptor side. Photosynth Res 1990, 23:39–48.

63. Srivastava A, Guisse B, Greppin H, Strasser RJ: Regulation of antenna structure and electron transport in Photosystem II of Pisum sativum under elevated temperature probed by the fast polyphasic chlorophyll a fluorescence transient: OKJIP. Biochim Biophys Acta 1997, 1320:95–106.

64. Hakala M, Tuominen I, Keränen M, Tyystjärvi T, Tyystjärvi E: Evidence for the role of the oxygen-evolving manganese complex in photoinhibition of Photosystem II. Biochim Biophys Acta 2005, 1706:68–80.

65. Streusand VJ, Portis AR Jr: Rubisco activase mediates ATP-dependent activation of ribulose bisphosphate carboxylase. Plant Physiol 1987, 85:152–154.

66. Chen L-S, Cheng L: Both xanthophyll cycle-dependent thermal dissipation and the antioxidant system are up-regulated in grape (Vitis labrusca BL. cv. Concord) leaves in response to N limitation. J Exp Bot 2003, 54:2165–2175.

67. Song YG, Liu B, Wang LF, Li MH, Liu Y: Damage to the oxygen-evolving complex by superoxide anion, hydrogen peroxide, and hydroxyl radical in photoinhibition of photosystem II. Photosynth Res 2006, 90:67–78.

68. Lichtenthaler HK: Chlorophylls and carotenoids: pigments of photosynthetic biomembranes. Methods Enzymol 1987, 148:350–382.

69. Bradford MM: A rapid and sensitive method for quantitation of microgram quantities of protein utilizing the principle of protein-dye binding. Anal Biochem 1976, 72:248–254.

70. Tsimilli-Michael M, Strasser RJ: In vivo assessment of stress impact on plant's vitality: applications in detecting and evaluating the beneficial role of mycorrhization on host plants. In Mycorrhiza: Genetics and Molecular Biology, Ecofunction, Biotechnology, Eco-physiology, and Structure and Systematics. Edited by: Varma A. Berlin: Springer; 2008:679–703.

71. Smit MF, Krüger GHJ, van Heerden PDR, Pienaar JJ, Weissflog L, Strasser RJ: Effect of trifluoroacetate, a persistent degradation product of fluorinated hydrocarbons, on C3 and C4 crop plants. In Photosynthesis. Energy from the Sun: 14th International Congress on Photosynthesis. Edited by: Allen JF, Gantt E, Golbeck JH, Osmond B. Dordrecht: Springer; 2008:1501–1504.

72. Chen L-S, Qi Y-P, Smith BR, Liu XH: Aluminum-induced decrease in CO2 assimilation in citrus seedlings is unaccompanied by decreased activities of key enzymes involved in CO2 assimilation. Tree Physiol 2005, 25:317–324.

73. Cheng L, Fuchigami LH: Rubisco activation state decreases with increasing nitrogen content in apple leaves. J Exp Bot 2000, 51:1687–1694.

74. Jones MGK, Outlaw WJ, Lowery OH: Enzymic assay of 10-7 to 10-14 moles of sucrose in plant tissues. Plant Physiol 1977, 60:379–383.

Copyrights

Index

B

E

F

G

H

I

J

K

L

M

N

O

P

Q

R

S

T

U

V

W